SPRINGER SERIES IN SURFACE SCIENCES 42

Springer
Berlin
Heidelberg
New York
Hong Kong
London
Milan
Paris
Tokyo

Physics and Astronomy ONLINE LIBRARY

http://www.springer.de

SPRINGER SERIES IN SURFACE SCIENCES

Series Editors: G. Ertl, H. Lüth and D.L. Mills

This series covers the whole spectrum of surface sciences, including structure and dynamics of clean and adsorbate-covered surfaces, thin films, basic surface effects, analytical methods and also the physics and chemistry of interfaces. Written by leading researchers in the field, the books are intended primarily for researchers in academia and industry and for graduate students.

38 **Progress in Transmission Electron Microscopy 1**
Concepts and Techniques
Editors: X.-F. Zhang, Z. Zhang

39 **Progress in Transmission Electron Microscopy 2**
Applications in Materials Science
Editors: X.-F. Zhang, Z. Zhang

40 **Giant Magneto-Resistance Devices**
By E. Hirota, H. Sakakima, and K. Inomata

41 **The Physics of Ultra-High-Density Magnetic Recording**
Editors: M.L. Plumer, J. van Ek, and D. Weller

42 **Islands, Mounds and Atoms**
Patterns and Processes in Crystal Growth Far from Equilibrium
By T. Michely and J. Krug

Volumes 1–37 are listed at the end of the book

T. Michely J. Krug

Islands, Mounds and Atoms

Patterns and Processes in Crystal Growth Far from Equilibrium

With 134 Figures
Including a Foldout Color Plate

Springer

apl. Professor Dr. Thomas Michely
I. Physikalisches Institut
RWTH Aachen
52056 Aachen, Germany
e-mail: michely@physik.rwth-aachen.de
http://ia.physik.rwth-aachen.de/michely/

Professor Dr. Joachim Krug
Fachbereich Physik
Universität Duisburg-Essen, Standort Essen
45117 Essen, Germany
e-mail: jkrug@theo-phys.uni-essen.de
http://www.theo-phys.uni-essen.de/tp/ags/krug dir/index.html

Series Editors:

Professor Dr. Gerhard Ertl
Fritz-Haber-Institute der Max-Planck-Gesellschaft, Faradayweg 4–6,
14195 Berlin, Germany

Professor Dr. Hans Lüth
Institut für Schicht- und Ionentechnik
Forschungszentrum Jülich GmbH,
52425 Jülich, Germany

Professor Douglas L. Mills, Ph.D.
Department of Physics, University of California,
Irvine, CA 92717, USA

Library of Congress Cataloging-in-Publication Data: Michely, T. (Thomas), 1961–. Islands, mounds and atoms: patterns and processes in crystal growth far from equilibrium/ T. Michely, J. Krug. p. cm. – (Springer series in surface sciences, ISSN 0931-5195; 42) Includes bibliographical references and index. ISBN 3-540-40728-6 (alk. paper) 1. Crystal growth. I. Krug, J. (Joachim), 1961– II. Title. III. Series. QD 921.M53 2003 548'.5–dc22 2003059073

ISSN 0931-5195
ISBN 3-540-40728-6 Springer-Verlag Berlin Heidelberg New York

Springer-Verlag Berlin Heidelberg New York
a member of BertelsmannSpringer Science+Business Media GmbH

http://www.springer.de

Printed in Germany

Typesetting: Frank Herweg, Leutershausen
Cover concept: eStudio Calamar Steinen
Cover production: *design & production* GmbH, Heidelberg

Printed on acid-free paper SPIN: 10672079 57/3141/ba - 5 4 3 2 1 0

Für Gisela und Barbara

Preface

Jede Wahrheit ist von Irrtum umgeben, wie die Erde von der Luft.[1]
Johann Wilhelm Ritter

Why did we decide to write this book? We asked ourselves this question while preparing the preface. Even at the end of our enterprise, it is hard to give a complete answer. It is the vision of understanding structure formation in nature on the atomic scale that is driving our research and which we want to communicate in this book. Perhaps it is also our fascination with the scientific process itself. Anything beyond this is difficult to pinpoint.

For one of us (T.M.) the roots of the book go back to end of the 1980s and the early 1990s, when he was a young scientist at the Institut für Grenzflächenforschung und Vakuumphysik (IGV) in Jülich and became involved in the observation of processes at surfaces in atomic detail with the scanning tunneling microscope. These early years were a time of great enthusiasm, when pictures and qualitative arguments seemed to suffice as a guide to the exploration of a new continent. During the later part of the last decade the field became more mature, turning to quantitative measurements, suitable for comparison with quantitative predictions. In this period of accumulation of knowledge it also became apparent how fallible the project of obtaining an atomic scale understanding of crystal growth was. The theorists' models occasionally missed the processes relevant in the real world and experimentalists occasionally investigated impurity effects instead of well behaved situations. Such setbacks are an unavoidable part of any progress in science.

The second author (J.K.) entered the field in 1986, when surface growth was discovered by the statistical physics community as a new arena in which to apply the concepts of scaling and universality that had proven so powerful in the investigation of equilibrium critical phenomena. The anticipation of universal behavior led many of us to expect that we could get away with very little attention to detail in devising our models. With hindsight, the physics of crystal growth turned out to be less simple but far richer. While the scaling description of surface roughness and morphology has permeated the field of thin film growth to a remarkable degree, we have had to learn time and again that quantitative modeling requires a thorough understanding of individual, often materials-specific atomistic processes.

Nevertheless, some central concepts have emerged and the broad outlines of a general picture can be discerned. In the last few years, the field of epitax-

[1] "Every truth is surrounded with error, like the earth is surrounded with air." From *Fragmente aus dem Nachlasse eines jungen Physikers* (Heidelberg, 1810).

ial growth kinetics has given rise to a coherent basis of reliable observations and adequate theoretical descriptions. Close cooperation between theorists and experimentalists was mandatory to reach this state of affairs. Once the results of different experimental techniques were found to agree with each other, a combination of ab initio calculations with growth simulations and statistical mechanics arguments was necessary to sort peculiarities from principles. We believe that the time is now ripe for the physics of growth kinetics to become fertile beyond mere model systems. It is our conviction that the concepts and examples discussed in this book will provide a basis for understanding kinetic effects in thin film and crystal growth in general, ranging from heteroepitaxial magnetic multilayers to organic thin films and crystals.

We expect that any physicist with an interest in crystal growth, growth kinetics and epitaxial or thin film growth will benefit from reading this book. It may serve as a state of the art introduction for students and scientists intending to perform research in the field. In addition, it is the hope of the authors that the visual appeal of the book might even attract a broader readership to our exciting field of research.

The book is the concerted effort of an experimentalist and a theorist. It reflects the tension that has always propelled advances in physics, the tension between paradigms and observations, between theory and experiment. Neither of us would have been able to write a book like this on his own. It represents the progress in the field of epitaxial growth kinetics from very personal, interacting viewpoints. We did not attempt to be objective in the selection of the topics and material, nor did we try to give an all-embracing review. The authors' own history in the field is visible throughout the book. Nevertheless, this monograph gives a fairly coherent picture, exemplifying how to link observations to models. We believe our approach will be useful to readers moving into the field or trying to apply the ideas formulated here to related areas.

Taking this approach necessarily means largely disregarding several important contributions to the field. These omissions should not be understood as a mark of disdain, but as a consequence of the authors' predilections and limitations. We welcome suggestions for improvements of the book to our e-mail addresses (see p. IV), which may enter into a future edition.

This book would not have been possible without the contributions of a large number of people.

For T.M., the person to thank here first is George Comsa. He enabled, stimulated and contributed to continuous research on the topics of this book at his institute, was an ever critical and analytically sharp dialog partner and always impressed with his imperturbable "science/not fashion" attitude. The stimulating environment at the Institut für Grenzflächenforschung und Vakuumphysik in Jülich until his retirement in 1997 provided the seed for this book. A few people were especially important in this productive environment: Karl Besocke, Hans Bonzel, Rudolf David, Jan Fremerey, Udo Linke,

Uffe Littmark, Bene Poelsema and Martin Teske. A great number of Diploma Students, Ph.D. students, guests and post-docs contributed in the groups of T.M. in Jülich and Aachen to the experimental work included in this book: Michael Bott, Carsten Busse, Rien Breeman, Stefanie Esch, Henri Hansen, Michael Hohage, Matthias Kalff, Winfried Langenkamp, Markus Morgenstern, Ansgar Petersen, Celia Polop, Marcel Rost, Christian Teichert, Zhenyu Zhang and Zhao Zhong. The theoretical collaborators of T.M. were of particular importance in developing a quantitative understanding of growth kinetics: Bob Averback, Peter Feibelman, Karl-Heinz Heinig, Miroslav Kotrla, Paolo Politi, Pavel Šmilauer, Matthias Strobel and Herbert Urbassek. Finally, the support of Deutsche Forschungsgemeinschaft through a Heisenberg Fellowship from 1997–2001 and the project *Atomare Prozesse beim homoepitaktischen Schichtwachstum* was indispensable for the creation of this book.

J.K. is grateful to Herbert Spohn for passing on his dedication to far from equilibrium physics, and to Paul Meakin for first introducing him to real thin film processes. The work on epitaxial crystal growth was begun in the group of Reinhard Lipowsky at the Institut für Festkörperforschung in Jülich, and continued at the University of Essen with the support of Deutsche Forschungsgemeinschaft within Sonderforschungsbereich 237 *Unordnung und grosse Fluktuationen*. Over the years, many students and collaborators have contributed to the project: Lothar Brendel, Claudio Castellano, Harvey Dobbs, Felix Hontinfinde, Harald Kallabis, Jouni Kallunki, Miroslav Kotrla, Philipp Kuhn, Sami Majaniemi, Michael Plischke, Paolo Politi, Martin Rost, Martin Schimschak, Martin Siegert, Pavel Šmilauer and Dietrich Wolf. During the early stages of writing this book, J.K. enjoyed the kind hospitality of the CAMP at the Danish Technical University and the Niels Bohr Institute in Copenhagen.

Both authors would like to acknowledge the colleagues, too numerous to be listed here, who in discussions influenced and shaped many of the ideas in this book. We thank all those, who agreed to reprint their figures in the book. Carsten Busse and Henri Hansen helped with several tables and figures, and Barbara Sacha checked the references. The book benefited from the critical reading of chapters in different stages of evolution by George Comsa, Peter Feibelman, Karsten Jacobsen, Eckhard Pehlke, Paolo Politi, and Dougald Sturges. Finally, special thanks are due to Barbara Maria Kloos for insisting on a non-technical title, and to the organizers of the workshop *Atomistic aspects of epitaxial growth* in June 2001 where, in the wee hours at Corfu airport, the present title was conceived.

The beauty of crystals has fascinated people since time immemorial. It is the hope of the authors that the present monograph will contribute to this fascination.

Menden and Köln,
July 2003

Thomas Michely
Joachim Krug

Contents

List of Frequently Used Symbols

a	Nearest neighbor distance between atoms
$a_\perp$	Projection of nearest neighbor distance normal to step direction, normal step displacement induced by a kink
A	Number of adsorption sites on an island
A-step	$\{001\}$-microfacetted step along a $\langle 1\bar{1}0\rangle$-direction
b^2	Step diffusivity, mean square step displacement per site
B-step	$\{111\}$-microfacetted step along a $\langle 1\bar{1}0\rangle$-direction
$c(\boldsymbol{x})$	Solute concentration in solution growth
C	Coordination number
$C(\boldsymbol{u})$	Two-point correlation function in three-dimensional space
$\mathcal{C}$	Lattice gas configuration in a KMC simulation
d	Layer thickness, atomic step height
d_s	Surface dimensionality
D	Surface diffusion coefficient
D_e	Diffusion coefficient of a step adatom along a close packed step
D_s	Diffusion coefficient of solute particles in solution
D_0	Diffusion prefactor
e	Elementary charge
E	Energy
E_a	Activation energy for a general thermally activated process
E_coh	Cohesive energy
E_D	Activation energy of adatom diffusion on a close-packed surface
E_s	Binding energy of a two-dimensional cluster of s atoms
E_S	Step edge barrier, activation energy for descent of an adatom from a step
F	Deposition rate (number of atoms deposited per unit time and surface area)
$\hat{F}$	Deposition rate (number of atoms deposited per unit time and lattice site)

Symbol	Meaning
$F_{1\mathrm{d}}$	Flux of adatoms to a step edge (number of adatoms per unit time and unit length)
g	Growth number [Eq. (4.6)]
$G(r)$	Height–height correlation function
$\tilde{G}(r)$	Height–difference correlation function
$\mathcal{G}, \tilde{\mathcal{G}}$	Height correlation scaling functions
h	Planck's constant
$h(x,t)$	Height function describing a one-dimensional surface profile
$h(\boldsymbol{r},t)$	Height function describing a two-dimensional surface profile
i^*	Size of critical nucleus (= largest unstable cluster)
$I(\boldsymbol{K})$	Scattering intensity
$j_{\mathrm{ES}}, \boldsymbol{j}_{\mathrm{ES}}$	Uphill adatom current caused by step edge barriers
j_{st}	Flux of adatoms along a step
k_{B}	Boltzmann's constant
$\boldsymbol{K}$	Scattering vector
l	Step spacing on a vicinal surface
l_D	Island spacing
l_{k}	Spacing between thermal kinks on a step in equilibrium
l_{ES}	Ehrlich–Schwoebel length [Eq. (4.33)]
$l_{1\mathrm{d}}$	Spacing between one-dimensional nuclei on a growing step [Eq. (4.94)]
L	Number of step edge sites of an island, edge length in nn-distances
$L_{1/2}$	Island edge length at which second layer nucleation has taken place with probability 1/2
L_c	Critical island edge length for the onset of a shape instability
$\boldsymbol{m}$	$= \nabla h$ surface slope vector
m	$= \lvert\boldsymbol{m}\rvert$ surface slope
$\boldsymbol{m}^*$, m^*	Selected mound slope
$n(\boldsymbol{r})$	Adatom density
$\boldsymbol{n}$	Two-dimensional unit vector normal to a step
$\boldsymbol{n}_{\mathrm{s}}$	Three-dimensional unit vector normal to a surface
n^{eq}	Equilibrium adatom concentration
n_s	Number density of clusters consisting of s atoms
$n_e(\boldsymbol{x})$	Electron density
nn	Nearest neighbor
N	Island density per unit area, average density of stable clusters
$\hat{N}$	Island density per adsorption site

$\mathcal{N}(\tau)$	Number of distinct adsorption sites visited by adatom in the time τ
p	Pressure
p_{enc}	Encounter probability for two adatoms on an island
p_{nuc}	Probability that an adatom participates in a nucleation event
q	Wavenumber (argument of spatial Fourier transform)
q_{F}	Fermi wavenumber
$\boldsymbol{r}$	Two-dimensional position vector
R	Radius of circular island
R_c	Critical island radius for second layer nucleation
R^*	Critical island radius for the onset of a shape instability
R_{top}	Radius of the top terrace of a mound
t	Time
T	Temperature
T_{m}	Melting temperature
T_{onset}	Onset temperature for a thermally activated process
v	Advancement speed of a step
V	Advancement speed (growth rate) of a crystal or thin film surface
W	Surface width in units of the atomic layer thickness [Eq. (4.4)]
$\boldsymbol{x}$	Three-dimensional position vector
z	Dynamic exponent
$1/z$	Coarsening exponent
Z	Canonical partition function
α	Static roughness exponent
β	Dynamic roughening exponent
γ	Specific surface free energy (surface free energy per unit area)
Γ	Rate of a general atomistic process
δ	Specific step free energy (step free energy per unit length)
$\tilde{\delta}$	Step stiffness [Eq. (3.8)]
δ_0	Specific step free energy at $T = 0\,\mathrm{K}$
ΔE_{S}	$= E_{\mathrm{S}} - E_{\mathrm{D}}$ additional step edge barrier
Δt	Time between subsequent arrivals of atoms on an island
Δt_{e}	Time between subsequent arrivals of adatoms at an island edge
ϵ	Kink energy

ϵ_{F}	Fermi energy
ζ	Island shape parameter [Eq. (4.45)]
$\zeta(y,t)$	Configuration of a step oriented along the y–axis
θ_n	Layer coverage of layer n
Θ	Coverage, total deposited amount
κ	Local surface curvature
κ_{st}	Local step curvature
λ	Mound separation, characteristic length scale of mound morphology
μ	Surface chemical potential
μ_{st}	Step chemical potential
ν	Jump rate of an adatom on a close-packed surface
ν'	Jump rate for the descent of an adatom over a step
ν_0	Attempt frequency for a jump on a close-packed surface
ν_0'	Attempt frequency for the descent of an adatom over a step
$\xi(t)$	Correlation length in kinetic roughening
σ_s	Capture number of an s-cluster
σ_{st}	Adatom mobility at a kinked step
τ	Lifetime or residence time of an adatom on an island
τ_{1d}	Residence time of a step adatom on a straight edge
τ_D	Diffusion time for an adatom to reach a step or island boundary
τ_{e}	Time for a step adatom to reach a kink
τ^*	Onset time of mound instability
φ_n	Exposed coverage of layer n
χ	Scaling exponent of the island number density [Eq. (2.19)]
ω	Nucleation rate on an island (number of nucleation events per unit time)
$\omega(q)$	Growth rate of surface or step pertubation with wavenumber q
$\omega(\phi,q)$	Growth rate of step train perturbation with phase shift ϕ and wavenumber q
Ω	Area of an adsorption site

1. Introduction

To him the patterns were highly remarkable and of the utmost importance, for the chip of glass in his eye made him see them that way.
H.C. Andersen, The Snow Queen

While the focus of this monograph is on developments of the past decade, our subject has a long and rather eventful history. It therefore seems appropriate to begin with a few historical remarks, which emphasize the important role that visual evidence has played in advancing the atomistic understanding of crystal growth. In subsequent parts of this introductory chapter we place our subject into the broader contexts of crystal growth, thin film deposition and epitaxy, and we provide a brief outline of the book.

1.1 Atoms, Crystals and Visualization

According to Richard Feynman, the single most significant insight of modern science is that *"all things are made of atoms"* [1]. The atomic hypothesis has been around, in one form or the other, since the beginning of scientific thought, but it was generally accepted in the scientific community only a century ago. The main obstacle to persuading the tenacious skeptics of the existence of atoms was, of course, the fact that atoms are very small; the direct imaging of atomic structure had to wait for the invention of the techniques of field ion microscopy (FIM) and scanning tunneling microscopy (STM) in the second part of the last century.

But also among the accumulating indirect evidence in favor of the atomic hypothesis, those pieces turned out to be particularly effective which had a distinct visual component. The final breakthrough came with Jean Perrin's experiments on Brownian motion [2]. They were particularly compelling not only because they provided, through the theory developed by Einstein and Smoluchowski, an estimate of Avogadro's number, but also because the incessant quiver of the Brownian particles, observed through a light microscope, visualizes the agitated motion of the invisible molecules.

A hundred years earlier, another crucial piece of evidence had been presented with the discovery of crystal isomorphism and polymorphism [3]. The fact that different materials may condense into the same lattice structure (isomorphism), while on the other hand a single material may crystallize into different forms (polymorphism), disproved the then governing dogma that the shape of a crystal is somehow encoded in its chemical constituents. Instead, it could be said that the geometric, macroscopic shape of a crystal visualizes, as it were, the spatial arrangement of its atomic building blocks.

An early formulation of this idea is due to Johannes Kepler in his treatise on the shape of snowflakes [4]. Walking across the Charles bridge in Prague in the winter of 1609, Kepler noticed that the snowflakes falling on his coat were all hexagonal. He first came up with a model that was based on the cubic close packing of spheres in three dimensions. He conjectured that the six points of the snowflake are actually the corners of an octahedron, spanned by three crossing needles. A few days later, another flurry of snowfall forced him to refute this model, because he could clearly see that the snowflakes were in fact flat platelets. Kepler therefore proposed that their shape could be associated with the hexagonal close packing of spheres in the plane[1].

Kepler's approach summarizes in a nutshell the research strategy that will be adopted in this book. Indeed, it could be argued that our understanding of crystal growth has always progressed in this way: through the closely intertwined development of models and ever more powerful imaging techniques.

If the regular atomic arrangement of a crystal lattice is a source of fascination, it is perhaps even more remarkable that such perfection can arise from the blind, random processes of molecular diffusion, attachment and detachment. The scope of this question could be appreciated, and its resolution attempted only once the atomic theory of matter was firmly in place. The first important steps were taken by W. Kossel, I. N. Stranski and others in the twenties and thirties of the last century [6]. They developed an atomistic theory which provided a kinetic underpinning of the thermodynamic view of nucleation and growth pioneered by J. W. Gibbs. The synthesis of the kinetic and thermodynamic approaches was completed by Max Volmer in 1939 [7]. In its broad outlines, the resulting picture of the growth of a high symmetry crystal surface through two-dimensional nucleation differs little from that presented in this book.

A few years later, however, it became clear that the accepted theory fails spectacularly in predicting the growth rates of many real crystals; for the example of iodine crystals growing from the vapor, the discrepancy between theory and experiment amounted to a factor of about 10^{1000}! The paradox was resolved by the ingenious suggestion of F. C. Frank that two-dimensional nucleation can be avoided in the presence of screw dislocations [8]. A crystal with a screw dislocation does not consist of independent layers, each requiring nucleation before it can grow, but just of one layer, which overlaps itself helicoidally. Such a crystal can grow simply by turning the screw, through the advancement of the spiral step emanating from the dislocation.

[1] The history of the study of snowflakes has been described in [5].

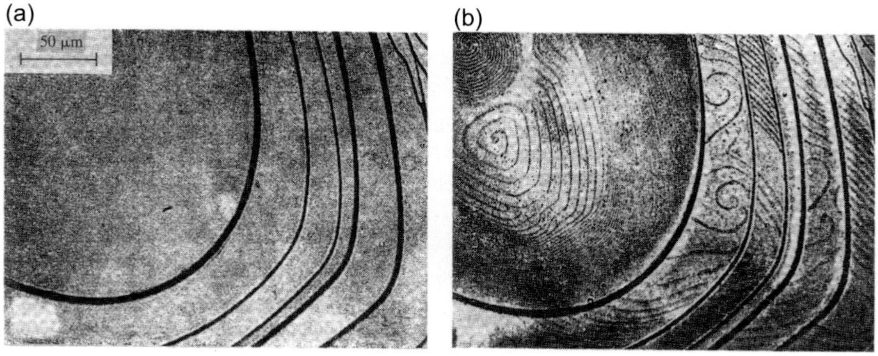

Fig. 1.1. The (0001) face of a SiC crystal, (**a**) prior to, and (**b**) after decoration with NH_4Cl crystallites precipitated from the vapor phase. The lines visible in (**a**) are macrosteps with a height of tenths of a micron, while the thin lines in (**b**) are atomic height steps forming growth spirals (Fig. 1 of [9], originally published in [10])

R. A. Kaischew has described how Frank's hypothesis burst like a bombshell at the meeting where it was first announced in 1949 [6]. It was met with skepticism on the side of the champions of the classical theory of two-dimensional nucleation, foremost I. N. Stranski, because spiral steps and the associated growth hillocks had not been seen experimentally[2]. But thanks to new visualization tools, such as phase contrast microscopy and dew decoration, the issue was soon settled in favor of the dislocation theory [3]. The dew decoration technique, discovered by George L. Lemmlein when he exhaled onto his microscope slide in the cold war winter of 1944 in Moscow [9], allows to visualize atomic height steps on crystal surfaces, because the droplets condense preferentially at the steps (Fig. 1.1). In this way it was demonstrated that growth spirals, rather than pathological exceptions, are the rule for many kinds of crystals.

Imaging methods, most notably electron microscopy, have played an equally central role in the early development of thin film science [12, 13]. While basic concepts of nucleation and film growth were formulated already in the 1920's, the detailed atomistic theory of these processes (which will be outlined in Sect. 2.3.1) was worked out only in the sixties and seventies, after electron microscopy studies had provided pictorial evidence for the postulated scenario and posed a host of new questions.

The first technique capable of imaging single atoms on a crystal surface, field ion microscopy[3] (FIM), was introduced by E. W. Müller [14, 15] around the time of Frank's spiral growth hypothesis. Perhaps not surprisingly, al-

[2] In fact growth spirals in paraffin crystals were reported by Charles M. Heck as early as 1937 [11], but their importance was not understood at the time.

[3] For an explanation of the principle of FIM see Sect. 6.2.2. The methodology of surface diffusion studies with FIM is described in Sect. 2.2.3.

ready one of the first FIM studies of surface diffusion lead to a discovery of fundamental importance for our understanding of the growth of crystals and thin films. Observing the diffusion of tungsten atoms on a tungsten surface, G. Ehrlich and F. Hudda found that the atoms tended to be confined within the boundaries of the crystal plane onto which they had been deposited [16]. They postulated that the atoms experience an additional energy barrier – the *step edge barrier* – when attempting to cross a step. The existence of such a barrier is plausible in a hard sphere model with pair interactions: When diffusing across a cliff edge, the atom loses the coordination to its neighbors, which increases its potential energy. This model is hardly more sophisticated than that used by Kepler. Nevertheless it took the direct, visual observation by FIM to put the researchers on the trail of this important atomistic process.

A few weeks after the publication of the paper by Ehrlich and Hudda, R. L. Schwoebel and E. J. Shipsey submitted an article where some of the consequences of the step edge barrier for the growth of stepped surfaces were spelled out [17]. But the full extent of the role that step edge barriers play in crystal growth has been appreciated only during the last decade, and the subject is still evolving rapidly[4]; we will devote a large part of Chap. 4 to this issue. Meanwhile, the notion that the crossing of a boundary is encumbered by an additional energy barrier has been generalized to one-dimensional diffusion along step edges, where the atoms can be reluctant to cross kinks or corners (see Sect. 3.7.1), and to the diffusion between different facets of a three-dimensional crystal [18].

The most recent chapter in the history of visual exploration in the atomic world was opened with the invention of the scanning tunneling microscope (STM) in 1981 [19, 20]. There can be no doubt about the fact that the STM has revolutionized surface science. Together with the budding technological applications of nanostructures, it has sparked off a mighty revival of interest in the atomistic aspects of crystal growth, which continues unabashed to the present day, and which forms the raison d'être for this book. The STM appears to be uniquely suited for investigations into the relationship between atomistic processes and growth morphology, because it combines atomic resolution (previously available only to the FIM) with the capability of imaging the surface morphology over large areas (as was traditionally done using electron microscopy). As a consequence, STM studies have further emphasized the importance of atomic scale defects in crystal growth – both of "good" defects such as steps and kinks, which provide the growth sites for the continuation of the crystal structure, and of "bad" defects such as impurities. Both aspects are going to recur many times in the following chapters.

[4] It is interesting to note that the papers of Ehrlich and Hudda [16] and of Schwoebel and Shipsey [17] have each accumulated about 50 citations per year since 1996, while the citation level in earlier years was considerably lower.

1.2 Crystal Growth Far from Equilibrium

In this section we will position our topic in relation to the two established fields of crystal growth and thin film deposition. As stated in the subtitle of the book, we consider crystal growth far from equilibrium. An alternative way of phrasing this subject is to say that we deal with the growth of a thin film on its own crystalline substrate. Our subject thus connects the realms of crystal growth and thin film deposition in a diagonal way: It contains one key aspect from each field, the far from equilibrium situation of thin film growth and the chemical homogeneity and crystallinity of the materials of crystal growth; and at the same time it is distinguished from both fields, because crystal growth usually proceeds close to equilibrium, while thin film deposition is performed mostly on a chemically different substrate, which often does not even posses a crystalline structure (the same may be true for the deposited film).

The thermodynamic meaning of growth far from equilibrium is illustrated by the "Gedankenexperiment" sketched in Fig. 1.2. Two single crystal sheets of identical material are mounted parallel to each other separated by a small gap. They are placed in a vacuum so that the mean free path for atoms in the gas phase becomes considerably larger than the gap between the plates. The temperatures of the plates are selected such as to transport material from one of the plates to the other through the vapor phase, i.e. to perform crystal growth or thin film deposition on one of the plates[5].

As an example for estimating numbers, we consider platinum as material for the substrate and deposit plates. Typical thin film deposition conditions are established by keeping the Pt substrate at room temperature and exposing it to a Pt vapor with a pressure of the order of 10^{-5} mbar, resulting in a deposition rate of the order of one atomic layer per second. Such a vapor pressure $p(T)$ is established by heating the deposit Pt plate to a temperature $T = T_1$ of about 1900 K. The vapor pressure of Pt at room temperature can be estimated from the standard Gibbs free energy for evaporation and is of the order of $p(T_2 = 300\,\mathrm{K}) \approx 10^{-90}$ mbar. The *supersaturation* σ defined as

$$\sigma = \ln(p(T_1/p(T_2)) \tag{1.1}$$

is the relevant dimensionless quantity used in crystal growth to characterize the driving force for the phase transition. In the present situation $\sigma \approx 200$. The resulting difference in chemical potential between the Pt vapor phase and the Pt substrate $\Delta\mu = \sigma k_{\mathrm{B}} T$ is about 5 eV per atom. This is an enormous driving force, similar in magnitude to the total binding energy of an atom in a Pt crystal.

In typical crystal growth experiments carried out close to equilibrium, σ is limited to a value of the order of unity (often it is only a few percent [22]), i.e

[5] This Gedankenexperiment was actually realized for studies of the current driven step motion on Si(001) in equilibrium with its vapor [21].

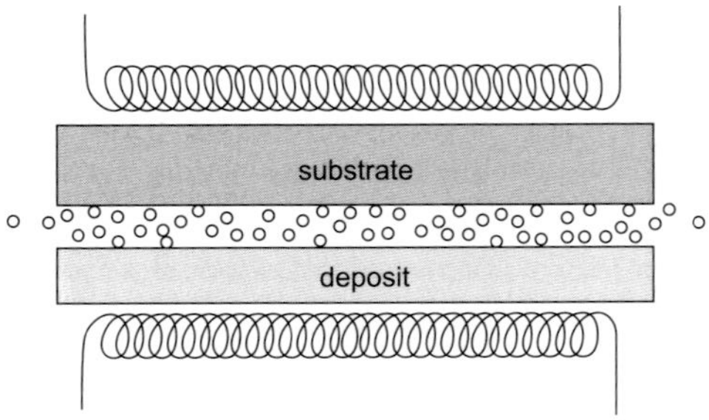

Fig. 1.2. Gedankenexperiment to illustrate the transition from crystal growth to thin film deposition by changing the supersaturation (see text). Two thin crystal plates are separated by a small gap and can be heated independently. Growth of the substrate plate proceeds by transfer of atoms from the deposit plate through the vapor phase in the separating gap

the vapor pressure ratio is limited to factors of about 3. In order to produce near equilibrium crystal growth in the situation sketched in Fig. 1.2, high temperatures are necessary for both the substrate and the deposit plates, because otherwise no particles would be exchanged at all. To achieve crystal growth rates of the order of one atomic layer per second, the plates must be heated to temperatures around 1900 K. In order to keep the supersaturation within the crystal growth regime, the temperature difference between the two plates has to be limited to less than 50 K. The resulting difference in chemical potential driving the particles over the phase boundary[6] is then of the order of 10 meV. As under crystal growth conditions the substrate itself has a non-negligible vapor pressure, it is clear that particles may cross the phase boundary several times, while this is not the case in thin film deposition.

The large difference in supersaturation between thin film deposition and crystal growth implies a conceptual difference in the description of these two situations. For atoms passing the phase boundary from the vapor to the solid phase under crystal growth conditions only the optimum binding sites of the crystal are stable binding sites. Metastable binding sites with higher chemical potential can only be accessed temporarily, as otherwise the energy balance between the phases would be invalidated. Atoms in metastable binding sites either recross the phase boundary or rapidly switch to stable binding sites. Thermodynamics is therefore sufficient to give an adequate description of crystal growth.

[6] In practice crystal growth from the melt rather than from the vapor phase is the method of choice, and the difference in chemical potential $\Delta\mu$ between the growing crystal and the undercooled melt is given by $\Delta\mu = \Delta T \Delta H_{\mathrm{m}}/T_{\mathrm{m}}$, where T_{m} is the melting temperature, $\Delta T = T_{\mathrm{m}} - T$ is the undercooling, and ΔH_{m} is the enthalpy of melting. Typically, ΔT is limited to values of the order of 1 K and thus $\Delta\mu$ is again of the order of 10 meV.

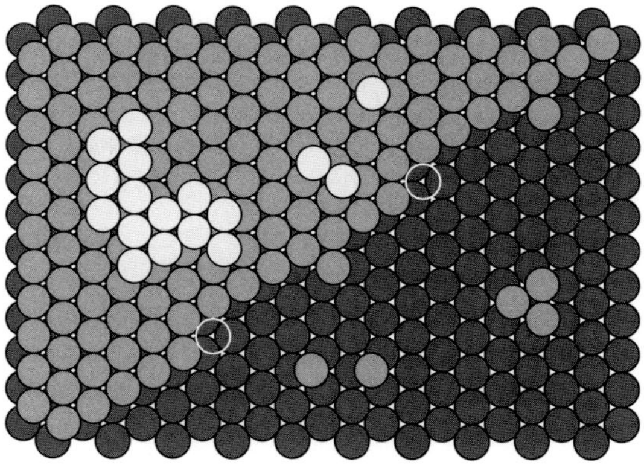

Fig. 1.3. Ball model of a growing crystal surface. The open white circles indicate kink positions, the only sites of stable atom incorporation during crystal growth close to thermal equilibrium. Adatoms, adatom clusters and step adatoms shown in the ball model are unstable for typical crystal growth conditions, but may be stable for typical thin film deposition conditions far from equilibrium (see text)

In contrast, in thin film deposition the large difference in chemical potential between the two phases, 5 eV/atom in the example of Pt, allows to populate a manifold of metastable sites on the substrate, which are higher in chemical potential than for an atom in the bulk, but still significantly lower than the gas phase chemical potential. Due to the temperature dependence of the kinetic rates connecting different metastable sites, and due the interaction of atoms in metastable sites, complex, temperature dependent growth morphologies will arise. In their manifold patterns, these metastable morphologies store some of the energy which was available for the phase transition. A thermodynamic description – although setting the boundary conditions of the process – is thus unable to describe thin film deposition. Instead, a kinetic description based on the rates of atomistic processes is the appropriate tool for describing crystal growth far from equilibrium.

The ball model of a crystal surface shown in Fig. 1.3 illustrates the difference between crystal growth and thin film deposition on the atomistic level. In crystal growth near equilibrium, all single adsorbed atoms (adatoms), smaller or larger clusters of atoms, as well as atoms adsorbed to the straight step edge (step adatoms) are unstable and will quickly be incorporated into *kink* positions, the positions indicated by a white circle. The kink is the repetitive position in crystal growth: Incorporation of an atom does not remove the kink and causes binding of the atom with the average bond strength of the crystal. The kink position therefore defines the chemical potential of the crystal phase. The concept of the kink or half-crystal position (in German: *Halbkristalllage*) introduced by Kossel in 1927 [23] was a major step towards an atomistic understanding of crystal growth [6].

Under thin film deposition conditions, the situation is completely different: All the adatoms, clusters and step adatoms shown in Fig. 1.3 are stable at sufficiently low temperatures. Depending on temperature, adatoms may be mobile and aggregate to form clusters, clusters or step adatoms may move along the step edge and all these processes occur with specific rates leading to complex growth morphologies.

1.3 Epitaxy

The thin film deposition situation considered here is a special case of epitaxial growth. The term *epitaxy*, introduced by L. Royer in 1928 [24], is used for the growth of a crystalline layer upon (*epi*) a crystalline substrate, where the crystalline orientation of the substrate imposes an order (*taxis*) on the orientation of the deposit layer[7]. In practice, an epitaxial growth system is characterized by the two parallel contact planes of the two crystal structures and by parallel crystallographic directions within these planes. The deposit orientation is selected by the constraint to minimize that part of the interface energy which arises from the posssible mismatch between the substrate and deposit crystal structures.

If the two materials are different, the term *heteroepitaxy* is used, while for the situation under consideration here – growth of a crystalline layer on a chemically identical, crystalline substrate – the term *homoepitaxy* is commonly employed. Evidently homoepitaxy extends the crystal structure of the substrate, such that the epitaxial relation is given by two identical planes and two identical directions within these planes. Homoepitaxy of Si on Si(001) is of technical importance in device fabrication [25]. Even for deposition of a crystalline thin film which is thick enough to screen the influence of the foreign substrate, further growth on this film is effectively homoepitaxial[8].

In 1958, Ernst Bauer [26] introduced a classification of epitaxial growth modes based on thermodynamic considerations, which therefore apply to growth near thermodynamic equilibrium. Different growth modes may be distinguished according to the balance between the surface free energy γ_{D} of the deposit, the surface free energy γ_{S} of the substrate, and the interface free energy γ_{int}[9]. For the case

$$\gamma_{\mathrm{S}} < \gamma_{\mathrm{D}} + \gamma_{\mathrm{int}} \tag{1.2}$$

[7] See [12] for a historical review.

[8] In general, however, such a thin film contains a certain number of defects like dislocations or grain boundaries. Their density defines an effective length scale below which growth may be considered as undisturbed homoepitaxial growth

[9] The interface energy is defined by the Dupré–relation: $\gamma_{\mathrm{int}} = \gamma_{\mathrm{S}} + \gamma_{\mathrm{D}} - \beta_{\mathrm{SD}}$, where β_{SD} is the work required to separate substrate and deposit crystals in their actual orientation and with the actual atomic distances [27].

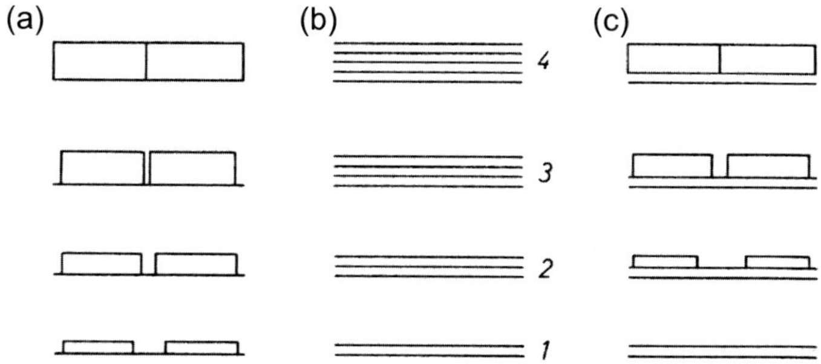

Fig. 1.4. The three epitaxial growth mechanisms close to thermodynamic equilibrium: (**a**) Volmer–Weber growth mode, (**b**) Frank–van-der-Merwe growth mode and (**c**) Stranski–Krastanov growth mode (Fig. 3 of [26])

the energy balance requires to minimize the area covered by the deposit material and the deposit will grow in the form of three-dimensional islands on the substrate. This mode is named *Volmer–Weber growth* (see Fig. 1.4a). If, on the other hand,

$$\gamma_S > \gamma_D + \gamma_{int} \tag{1.3}$$

the energy balance requires to maximize the area covered by the deposited material and the deposit will initially grow smoothly, one atomic layer at a time. In general, in this initial state of growth the deposit material will be strained and store elastic energy. After deposition of a certain number of layers the condition (1.3) must break down. Considering the n-th layer as the new substrate, the chemical influence of the original substrate will be largely screened and the two surface free energies γ_S and γ_D will be nearly identical. However, the elastic influence of the original substrate, imposing a strained n-th layer, will give rise to a positive, nonzero interface energy γ_{int} between layer n and $n+1$, which invalidates condition (1.3) and leads back to the case (1.2). Thus beginning with the deposition of layer $n+1$, three dimensional islands will form on top of the n smoothly grown layers (see Fig. 1.4c). This scenario defines the *Stranski–Krastanov growth* mode. Although in reality a number of subtle details complicate the picture[10], the faster decay of the substrate chemical influence compared to the substrate elastic influence at the growth front must lead to a breakdown of layer–by–layer growth and eventually to the formation of three dimensional islands for any epitaxial system that initially satisfies (1.3).

The limiting condition

$$\gamma_S = \gamma_D + \gamma_{int} \tag{1.4}$$

can only be fulfilled for homoepitaxial growth systems, where $\gamma_{int} = 0$ by definition. The corresponding *Frank–van-der-Merwe* growth mode (see Fig. 1.4b)

[10] There are several recent reviews on strain-induced self organisation and strained hetereoepitaxy of semiconductor materials [28–30].

is characterized by layer-by-layer growth of unlimited duration. Any deviation from perfect layerwise growth would increase the deposit surface area, thus causing an unnecessary increase of the total free energy, which according to the thermodynamic requirements would rapidly be driven back.

Bauer's fundamental classification scheme of near-equilibrium growth modes is frequently misconceived in the literature on epitaxial growth. Observed growth morphologies created by heteroepitaxy *far* from equilibrium are often simply associated with the Volmer–Weber, Stranski–Kranstanov or Frank–van der Merwe scenarios. However, a heteroepitaxial growth morphology may well have a predominantly kinetic origin, and should not be categorized in terms of near-equilibrium growth modes without further consideration. In heteroepitaxy energetic effects (e.g., due to the lattice mismatch and differences in surface free energies) and kinetic effects due to the far from equilibrium growth conditions are usually intertwined, making it impossible to clearly distinguish the two contributions. The descriptive use of the terms Volmer–Weber, Stranski–Kranstanov or Frank–van der Merwe growth in order to characterize a morphology of three-dimensional islands, a morphology of islands on top of a smooth layer or a layer-by-layer growth morphology is thus a misuse, which creates needless confusion on the roles of energetics and kinetics in growth.

For homoepitaxy close to equilibrium the thermodynamic prediction is Frank–van-der-Merwe growth. Thus any deviation from perfect layer-by-layer growth must have a kinetic origin. For this reason homoepitaxial growth systems are the optimal choice for studying the kinetics of growth in thin film deposition. Understanding growth kinetics in homoepitaxial systems is a key to separate the entangled contributions of kinetics and energetics in heteroepitaxial growth situations. In this way the study of homoepitaxy provides also a kinetic basis for understanding and manipulating heteroepitaxial thin film growth.

1.4 About this Book

The organization of the monograph largely follows the natural sequence through which a thin film grows. In the next chapter, the arrival of atoms at the surface, their diffusion and aggregation into two-dimensional islands is treated. Chapter 3 is focused on the shapes of these islands, essentially a two-dimensional problem. In the subsequent Chap. 4 the growth continues into the third dimension. Here the crucially important processes of interlayer transport are addressed, and it is explained why mounds often form in far from equilibrium growth. Chapter 5 describes ways in which mound formation can be avoided, and layer-by-layer growth (which, as was explained above, is the natural homoepitaxial growth mode near equilibrium) can be enforced by manipulating the growth conditions. Finally, Chap. 6 compiles some of the experimental and theoretical methods that are employed in the study of homoepitaxial growth; it should be usable as a reference source while reading the other chapters.

Chapters are subdivided into sections and subsections. While sections within a chapter and, even more so, subsections within a section obviously refer to each other, we have attempted to write the different portions of the book in a reasonably self-contained way. This should allow readers with some background knowledge to select those sections that are of particular interest to them. Chapters 2–4 conclude with a section on somewhat specialized topics, which supplement the discussion in the preceding, main sections, and which can be left out during a first reading. We should concede, however, that the distinction between "main" and "supplementary" topics is neither particularly systematic nor objective; the reader should simply pick and choose according to his/her own preferences.

The book emphasizes the kinetic and atomistic description of growth. Occasionally a coarse grained phenomenological continuum description is appropriate, e.g. for the description of mound coarsening in Chap. 4. Neglecting the exchange of adatoms with the gas phase, it is also sometimes useful and possible to define local thermodynamic equilibria, e.g., in the analysis of two-dimensional equilibrium island shapes in Chap. 3.

Kinetic effects in growth are pronounced at low substrate temperatures and high deposition rates (large supersaturations), which result in a small lateral length scale for the film morphology. This is one of the reasons why the availability of variable temperature scanning tunneling microscopes, which allow for high resolution morphological imaging down to cryogenic temperatures, gave a great boost to the field. The present monograph consistently uses STM topographs to illustrate and develop the concepts of growth kinetics.

A further goal of the authors has been to treat theoretical and experimental results on an equal footing. Whenever possible, theoretical ideas are exemplified for homoepitaxial growth on Pt(111). The restriction to one specific growth system establishes a coherent picture and keeps the discussion of experimental details to a minimum. The Pt(111) system demonstrates not only the present level of understanding of crystal growth kinetics, but also illustrates the origin of theoretical and experimental limitations in the description of growth kinetics in a transparent way.

The clear focus on growth kinetics, the integrated experimental and theoretical treatment of the topic as well as the use of a paradigmatic model system to illustrate the concepts are key elements which distinguish the present monograph from a number of recent books on thin film growth phenomena [31–37]. These books discuss many other aspects which are not covered here.

2. Condensation, Diffusion and Nucleation

Every move you make
Every bond you break
Every step you take
I'll be watching you
Gordon M. Sumner

The early stage of thin film growth involves the arrival and accommodation of atoms on the surface (*condensation*), their migration along the surface (*diffusion*) and their aggregation into two-dimensional adatom islands (*nucleation*). The following three sections treat basic aspects of these processes, while the last section is devoted to a selection of advanced topics of current interest.

2.1 Arriving at the Surface: Sticking and Transient Mobility

The condensation of an atom at a surface proceeds through a sequence of steps in which kinetic energy is transferred to the substrate and dissipated into lattice vibrations. The kinetic energy E_{kin} of an atom created by evaporation is typically of the order of 0.1 eV [1]. As the atom approaches the surface, it is accelerated in the attractive potential of the substrate. The associated gain in kinetic energy is of the order of the cohesive energy[2] E_{coh}, which much exceeds E_{kin}.

In order to stick at the surface, rather than being reflected into the atmosphere, the atom has to dissipate a sufficient fraction of its total energy during the collision with the substrate to place it below the adsorption en-

[1] The energy scale used throughout the book for atomic processes is electron volts. 1 eV is the energy needed to move an elementary charge against a potential of 1 V. It is useful to remember that the temperature corresponding to an energy of $k_{\mathrm{B}}T = 1\,\mathrm{eV}$ is $T = 11606\,\mathrm{K}$.

[2] The cohesive energy is the energy required to disperse a solid body into its single atoms. For a crystal it is the energy needed to move an atom from the kink position to infinity at a temperature of 0 K. It amounts to 5.86 eV per atom for Pt [38].

ergy[3] $E_{\rm ad}$, which is some fraction of $E_{\rm coh}$ (see Sect. 2.3.4 for a quantitative estimate). The probability for this to occur defines the *sticking coefficient* of the atom. For a metal atom impinging on a cold substrate of the same material the sticking coefficient is very close to unity, indicating that the energy transfer is highly efficient. In an oversimplified picture this can be attributed to the well-known property of classical hard spheres of equal mass to completely interchange their energy and momentum in a collision [39]. Model calculations for an atom impacting on a one-dimensional harmonic chain show trapping even when the initial kinetic energy amounts to more than 20 times the adsorption energy [40].

Once the atom has been adsorbed, becoming an *adatom*, the next relevant energy scale is the activation barrier $E_{\rm D}$ for surface diffusion. As $E_{\rm D}$ is typically small – only a few tenths of an eV – compared to the adsorption energy, it is conceivable that a freshly adsorbed atom retains enough of its energy to keep moving ballistically over the surface. This idea has created a famous ghost in epitaxial growth, the concept of *transient mobility*. Motivated by observations of low temperature layer-by-layer growth, Egelhoff and Jacob speculated [41] that the kinetic energy gained in condensation allows the atom to skip several lattice sites away from its point of first encounter prior to immobilization, giving rise to significant (but transient) adatom mobility far below the onset of thermally activated diffusion. Such nonthermal motion is well documented for rare gas atoms on metal surfaces, for example Xe on Pt(111) [42].

The importance of transient mobility in epitaxial metal growth has remained somewhat controversial, perhaps because definite statements are hard to come by both theoretically and experimentally. Theoretically, one has to rely on molecular dynamics simulations of the condensation process. For a variety of fcc(001) metal systems, Sanders and De Pristo [43] found that 99.9 % of the deposited atoms remained localized in the first unit cell they encountered, while for fcc(111) substrates this number dropped to 67 %, and 1 % of the atoms traversed three units cells before coming to rest [44]. Inspection of the particle trajectories shows that atoms which collide head-on with a substrate atom are those that tend to move far away from the impact point [45]. These results are fairly insensitive to the choice of the interaction potential, but they depend crucially on the use of a sufficient number of atomic layers in the simulation. If the simulated solid is too small, the impact energy is refocused on the incident atom rather than being dissipated.

Experimentally, information about transient mobility can be inferred from *post mortem* observations of the spatial arrangement of the deposited atoms [46]. In a very clear experiment Wang and Ehrlich [47] analyzed the distribution of adatoms (Ir,Re,W,Pd) deposited onto the Ir(111) surface at low temperatures over the two inequivalent adsorption sites on a fcc(111)

[3] The adsorption energy is the energy needed to remove an adsorbed atom from the surface to infinity.

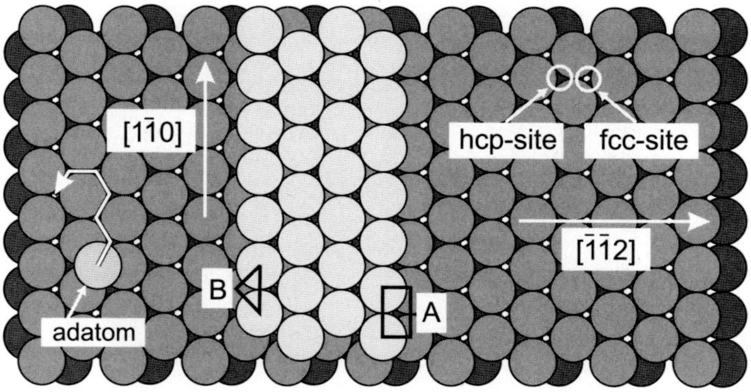

Fig. 2.1. Ball model of an fcc(111) surface. Condensation or adsorption of an adatom on an fcc-site reproduces the correct stacking sequence for a face centered cubic crystal, while adsorption on an hcp-site corresponds to the correct stacking sequence for a hexagonal close packed crystal and introduces a stacking fault into the fcc crystal. Adatom adsorption to fcc-sites is energetically preferred and adatom diffusion proceeds via intermediate hcp-sites as indicated. Dense packed steps are oriented along the $\langle 1\bar{1}0 \rangle$ directions, the $[1\bar{1}0]$ direction is indicated. The two types of different dense packed steps on (111) are named B-step (a microfacet on a $\{111\}$-plane) and A-step (a microfacet on a $\{001\}$-plane)

surface: the fcc-site continuing the bulk stacking to the surface layer and the hcp-site which would be the correct adsorption site for a hexagonal close packed (hcp) crystal (see Fig. 2.1). They found, for temperatures below the onset of thermal hopping, a random distribution of the adatoms between the two types of sites, although the associated binding energies differ considerably. This finding contradicts the idea of transient mobility, which would have implied a more frequent occupation of the more favorable binding site.

Figure 2.2 shows a scanning tunneling microscope image after deposition of a small quantity of Pt–adatoms onto Pt(111) at 23 K. According to the calibration of the evaporator the number of white dots corresponds within the statistical uncertainties to the number expected from a random condensation of the deposited adatoms without transient mobility. Similar STM–experiments for the condensation of Ag–adatoms on Pt(111) [48] and Au–adatoms on Au(110) [49] yield the same result: The adatom distributions are random, inconsistent with the assumption of significant transient mobility. On Pt(111) the regular fcc-site is energetically the more favorable site for Pt adatom adsorption: At 45 K adatoms initially adsorbed at hcp-sites hop to fcc-sites [50].

Taken together, evidence from experiments and simulations suggests that transient mobility on metal surfaces is a small effect, which is unlikely to play a major role in low temperature homoepitaxy. The efficient energy dissipation mechanism responsible for the high sticking probability is capable also of removing most of the residual kinetic energy before it can induce significant

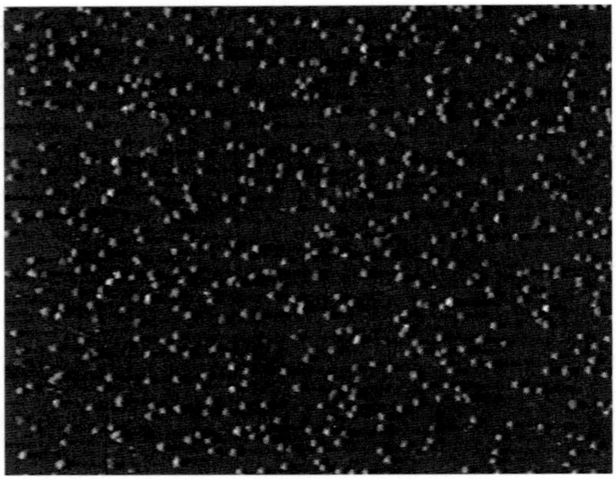

Fig. 2.2. STM topograph obtained after deposition of 4.2×10^{-3} monolayers of Pt-adatoms at 23 K. One monolayer (ML) corresponds to the number of atoms needed to form an atomic layer and is 1.504×10^{19} atoms/m^2 for Pt(111). The image size is 920 Å × 700 Å

adatom motion. An alternative explanation for the layer-by-layer growth observed by Egelhoff and Jacob [41] is described below in Sect. 2.4.1. For the purpose of the present discussion, we conclude that condensation on a flat substrate results in a random distribution of adatoms, which serves as the starting point for all subsequent evolution. Deviations from the random distribution caused by steps and other morphological features on the substrate will also be addressed in Sect. 2.4.1.

2.2 Moving on: Surface Diffusion

The diffusion of adatoms across the surface is a decisive factor in all stages of epitaxial growth, because it provides the mechanism by which atoms find their way to the incorporation sites required to reproduce the correct crystal structure. The sensitivity of growth morphology to the surface temperature derives largely from the fact that surface diffusion is a thermally activated process. Accordingly, the microscopic physics of surface diffusion and the determination of surface diffusion parameters has been the focus of much experimental and theoretical effort [51–53].

2.2.1 The Surface Diffusion Coefficient

The most simple view of surface diffusion (which will essentially suffice for the purposes of this book) regards the migration of an adatom as a two-dimensional random walk on the lattice of preferential adsorption sites. For such a random walk $\boldsymbol{r}(t)$ the mean square displacement grows linearly in time according to

$$\langle (\boldsymbol{r}(t) - \boldsymbol{r}(0))^2 \rangle = \nu \ell^2 t = 4Dt \tag{2.1}$$

where ℓ^2 denotes the mean square distance covered in a single jump, ν is the jump rate and D denotes the *surface diffusion coefficient*. The factor 1/4 in the relationship

$$D = \frac{1}{4}\ell^2 \nu \tag{2.2}$$

is a convention reflecting the two-dimensional nature of surface diffusion[4]. It ensures that in the low coverage limit, D coincides with the *collective surface diffusion coefficient* D_c entering the diffusion equation

$$\frac{\partial n}{\partial t} = D_\mathrm{c} \nabla^2 n. \tag{2.3}$$

While (2.1) applies to the motion of a single, marked or "tagged" particle[5], (2.3) describes the time evolution of a density profile $n(\boldsymbol{r}, t)$ of adsorbed atoms. In the presence of adatom–adatom interactions at finite coverage, D and D_c are quite distinct quantities, both numerically and conceptually [51,53]. Here our attention will be restricted to the low coverage regime where diffusing adatoms may be treated as independent (see however Sect. 2.4.5). Note that in this limit the diffusion equation (2.3) can also be taken to describe the time evolution of the positional probability distribution of a single adatom.

To obtain an expression for D, we need to focus on the individual move of an adatom between two adsorption sites. We assume throughout that the adatom motion can be described classically[6]. Apart from jumps between adsorption sites, classical surface diffusion can also proceed through the exchange of the adatom with a substrate atom (see Sect. 2.4.2). Most of the concepts discussed in this subsection apply equally to jump and exchange diffusion, but we will use the language of jump diffusion for simplicity. The transition path connecting the initial and final configurations of the adatom passes over a saddle point in configuration space, which is separated from the initial configuration by the energy barrier E_D. The jump rate is then expected to follow the universal Arrhenius law for thermally activated processes [55],

$$\nu = \nu_0 \mathrm{e}^{-E_\mathrm{D}/k_\mathrm{B}T}, \tag{2.4}$$

where the preexponential factor ν_0 is often referred to as the *attempt frequency*, and k_B denotes Boltzmann's constant. The attempt frequency is a measure for the time scale of the dynamics within an adsorption well. A necessary condition for the applicability of (2.4) – and, indeed, of the whole

[4] For diffusion in d_D dimensions, the factor is $1/2d_D$.

[5] This is sometimes emphasized by referring to D as the *tracer surface diffusion coefficient* [51]. The term *tracer* is derived from the use of radioactive tracers for low concentrations of tagged particles in bulk diffusion studies.

[6] Quantum tunneling contributes significantly to the diffusion of light adatoms, such as hydrogen, at low temperatures ($T < 100$ K) [53,54].

random walk picture – is that jumps are rare events, in the sense the $\nu \ll \nu_0$; only then does the adatom spend sufficient time between jumps to equilibrate within a well and lose the memory of where it came from. At room temperature, this implies that the surface diffusion barrier should be at least 0.1 eV.

2.2.2 Transition State Theory

The derivation of (2.4) from microscopic, classical or stochastic dynamics is the goal of a long and still ongoing research effort in theoretical physics and physical chemistry [56]. One of the earliest and most widespread approaches to the problem is the *transition state theory* (TST), which is based on the assumption of thermal equilibrium between the adatom populations in the adsorption well and at the saddle point[7] (the transition state) [55]. The rate is therefore proportional to the ratio of the partition functions of the two states, which yields the expression [57]

$$\nu^{\mathrm{TST}} = \frac{k_{\mathrm{B}}T}{h} \frac{Z_{\mathrm{s}}}{Z_{\mathrm{a}}} \mathrm{e}^{-E_{\mathrm{D}}/k_{\mathrm{B}}T}. \tag{2.5}$$

While the potential energy difference E_{D} between the two states appears in the exponential factor, $Z_{\mathrm{s,a}}$ denote the *vibrational* partition functions of the saddle point and the adsorption site, respectively. Evaluating the vibrational partition functions within a harmonic approximation, one obtains the preexponential factor

$$\nu_0^{\mathrm{TST}} = \frac{\prod_{i=0}^{N_{\mathrm{mode}}} \omega_i^{(\mathrm{a})}}{\prod_{i=1}^{N_{\mathrm{mode}}} \omega_i^{(\mathrm{s})}} \tag{2.6}$$

in terms of the normal mode frequencies $\omega_i^{(\mathrm{a,s})}$ of the stable vibrations at the adsorption site and at the saddle [58].

The expression (2.6) raises the question about the number of degrees of freedom effectively involved in the jump of a single atom. The most naive picture of an atom moving in a completely rigid substrate potential is certainly oversimplified, since nearby substrate atoms can relax considerably in response to the migration of the adatom. In a density functional theory study of the diffusion of Ag on Ag(111), Ratsch and Scheffler found that at least 5 atoms had to be included in order to get reasonable convergence of the attempt frequency [59]. Liu et al. [60] computed preexponential factors for a range of fcc metal surfaces using two versions of the embedded atom method, and compared the results obtained from (2.6) with the vibration frequency $\omega_0^{(\mathrm{a})}$ along the transition path at the adsorption site. The difference between

[7] A second, somewhat technical assumption is that every trajectory reaching the saddle point actually makes the transition to the other well. Because the recrossing probability is in fact nonzero, the TST rate always overestimates the true transition rate [56].

the two estimates was typically less than a factor of two, with the full TST formula (2.6) giving larger values.

As an alternative to (2.6), the partition functions in (2.5) can be written in terms of the vibrational free energies $F_{\mathrm{s,a}} = U_{\mathrm{s,a}} - TS_{\mathrm{s,a}}$. Since there is always one more stable vibrational mode at the adsorption site than at the saddle point, the classical equipartition theorem gives $U_{\mathrm{a}} - U_{\mathrm{s}} = k_{\mathrm{B}}T$ [61], and the preexponential factor

$$\nu_0^{\mathrm{TST}} = \frac{k_{\mathrm{B}}T}{h} \mathrm{e}^{1+\Delta S/k_{\mathrm{B}}} \tag{2.7}$$

is seen to reflect the entropy difference $\Delta S = S_{\mathrm{s}} - S_{\mathrm{a}}$ between the two states.

The beauty of transition state theory is that it allows one to compute a kinetic rate from the knowledge of the potential energy surface alone. On the other hand, it is difficult to assess its limits of applicability from within the theory. The thermal equilibrium assumption of TST requires not only that the adatom spend sufficient time in the well between subsequent jumps, but also that its interaction with the substrate be strong enough to ensure fast equilibration both in the well and at the saddle. The simplest approach that explicitly includes the effect of the friction between adatom and substrate is the Kramers model for a single degree of freedom coupled to a heat bath, which is represented by stochastic Langevin forces [56, 62]. In the Kramers model, a rough criterion for the applicability of the strong friction regime (where TST holds) can be formulated as [56]

$$\omega_{\mathrm{b}}\tau_{\mathrm{s}} < E_{\mathrm{D}}/k_{\mathrm{B}}T, \tag{2.8}$$

where τ_{s} is the adatom–substrate relaxation time and ω_{b} denotes the absolute value of the (imaginary) unstable vibration frequency at the saddle point, proportional to the curvature of the potential along the unstable direction. When (2.8) is violated, the rate is reduced by a factor $E_{\mathrm{D}}/(k_{\mathrm{B}}T\omega_{\mathrm{b}}\tau_{\mathrm{s}})$ compared to the TST prediction. The accuracy of TST has also been tested against molecular dynamics simulations. Apart from the expected deviations when the diffusion barrier becomes comparable to $k_{\mathrm{B}}T$ [63], good agreement has been found [64].

The physical content of the attempt frequency in (2.4) is a matter of some debate in the literature. Often it is simply identified with a "typical" surface phonon frequency of the substrate, a point of view that is supported by the TST expression (2.6). Nevertheless the suggestion that ν_0 contains nontrivial physics beyond the TST has repeatedly appeared. A venerable example is the Meyer–Nedel compensation law, which postulates a systematic increase of the attempt frequency with increasing energy barrier for classes of related activated processes[8]. Evidence for such an effect, based on molecular dynamics simulations of diffusion on various fcc(111) and fcc(100) surfaces,

[8] The notion of related processes is not sharply defined. For example, it is not clear if exchange and hopping processes should be considered related in this sense, as assumed in [65].

was presented by Boisvert et al. [65], but neither recent density functional calculations [59], nor earlier studies using the embedded atom method in conjunction with the TST [60] seem to support this trend. In the context of diffusion of different metal atoms on the W(211) surface, Wang and Ehrlich [66] have shown experimentally that there is little variation in preexponential factors, and that earlier reports on anomalous prefactors can be attributed to insufficient data. Other possible mechanisms contributing to observations of (apparently) anomalous preexponential factors will be discussed below in Sects. 2.4.5 and 2.4.6.

2.2.3 Direct Observation of Adatom Diffusion

The classical method for the experimental determination of D is field ion microscopy (FIM) [52]. If applicable, this method appears still to be the most accurate one. The displacement of an adatom adsorbed to the apex plane of a field emitter tip is recorded after a migration interval of duration t, during which the tip is heated to the temperature T of interest. The observation occurs at a lower temperature $T' < T$, where diffusion is entirely frozen. Repeated imaging and heating cycles yield an estimate of the mean square displacement (2.1), from which the diffusion coefficient is extracted. By performing measurements at different temperatures T, both the diffusion barrier E_D and the preexponential factor can be obtained. The slope of the linear regression of the data in a plot of $\ln D$ versus $1/T$ equals E_D, while the intercept yields the diffusion prefactor $D_0 = \ell^2 \nu_0/4$ [cf. (2.2)].

A careful investigation [67] of Pt–adatom diffusion on Pt(111) resulted in an activation energy of $E_\mathrm{D} = 0.260 \pm 0.003\,\mathrm{eV}$ and a prefactor of $D_0 = 2.0(\times 1.4^{\pm 1}) \times 10^{-7}\,\mathrm{m^2 s^{-1}}$. For Pt–adatoms on Pt(111) the diffusion jumps are to a good approximation limited to hops between nearest neighbor fcc-sites at distance $\ell = a = 2.775$ Å [50]. The minimum energy path between adjacent fcc-sites is via twofold coordinated bridge positions and the intermediate hcp-site, as indicated in Fig. 2.1. Thus the value of $D_0 = (1/4)a^2\nu_0$ corresponds to an attempt frequency $\nu_0 = 1.0 \times 10^{13}\,\mathrm{s^{-1}}$. An earlier FIM analysis [68], based on a measurement of the mean square displacement at a single temperature and assuming a universal prefactor [52] $D_0 = 10^{-7}\,\mathrm{m^2 s^{-1}}$ produced the estimate $E_\mathrm{D} = 0.25 \pm 0.02\,\mathrm{eV}$, in good agreement with the full analysis.

An analogous method, in which migration at temperature T is separated from the observation stage at temperature $T' < T$, was used by Mo [69] for the determination of surface diffusion parameters by scanning tunneling microscopy (STM). However, while the measurement of absolute adatom displacements on the apex plane of a field emitter *after* a heating interval is straightforward, it is quite tedious to do the same by STM: Changes in sample temperature are fast for the small mass of a field emitter tip, but they are slow for the large sample in an STM apparatus, which needs to be rigidly mounted. The measurement times and the danger of adsorbate contamination

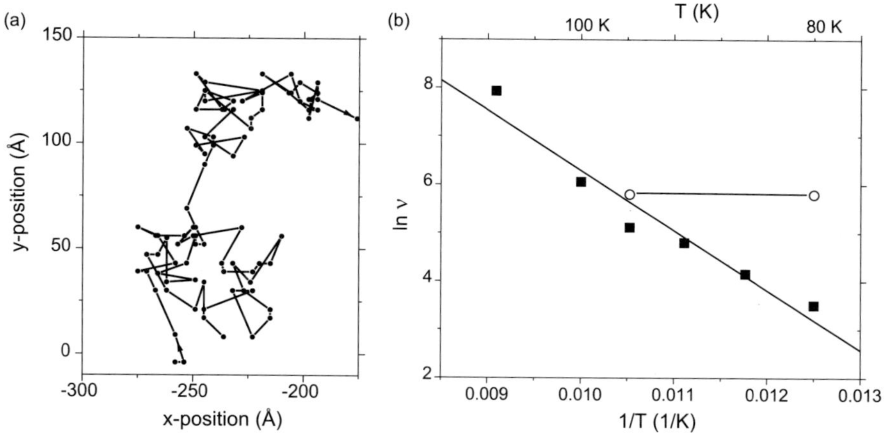

Fig. 2.3. (**a**) Map of consecutive adatom positions on STM topographs taken at 85 K in time intervals of 30 s. (**b**) Temperature dependence of the adatom hopping frequency ν determined by maps as in (**a**). *Full squares*: Data taken with a tunneling resistance of $R_\mathrm{t} = 150\,\mathrm{M\Omega}$ ($U_\mathrm{t} = 0.03\,\mathrm{V}$, $I_\mathrm{t} = 0.2\,\mathrm{nA}$). *Full line*: Linear regression to the data. *Open circles*: Data with $R_\mathrm{t} = 15\,\mathrm{M\Omega}$ ($U_\mathrm{t} = 0.03\,\mathrm{V}$, $I_\mathrm{t} = 2\,\mathrm{nA}$). Line to guide the eye

of the experiment are thus much greater. In addition, retrieving a particular sample location after heating is a nontrivial task.

For the STM, the acquisition of consecutive images *at* the migration temperature appears to be much easier and was expected to allow an accurate determination of the mean square displacement. For Pt(111) this was attempted by taking STM–movies with a fixed time interval $t = 30\,\mathrm{s}$ between the images at temperatures $80\,\mathrm{K} < T < 110\,\mathrm{K}$ [70]. Figure 2.3a shows the map of adatom positions obtained from such a movie. As expected for a random walk, the displacements along the two orthogonal coordinate axes follow a binomial distribution. From the temperature dependence of the mean square displacement, the activation energy $E_\mathrm{D} = 0.11\,\mathrm{eV}$ and the unusually low attempt frequency $\nu_0 = 1 \times 10^8\,\mathrm{s}^{-1}$ were obtained (Fig. 2.3b). These values are in clear contradiction to the field ion microscopy results.

Even more important, a change in the tunneling resistance significantly affects the diffusion behavior of the adatoms, as can be seen in Fig. 2.3. The tunneling tip thus interacts with the adsorbed adatom. Similar effects of the tunneling tip on the mobility of diffusing species have been found in other contexts [71–73]. Standard procedures for reducing the tip–adatom interaction are an increase in the tunneling resistance (the tip is then further away from the adatom) or a decrease in the interaction time between tip and adatom (short scanning times followed by long waiting times) [74]. The nature of the tip–adatom interaction is generally complex. Possible origins are the interaction of the STM tip field with static or dynamic dipole moments of the adatom, direct electronic interactions of the tip–adatom orbitals, or the population of adatom or tip atom states by tunneling electrons leading to temporal bond weakening or breaking [71, 75, 76].

Despite the difficulties encountered in the direct observation of adatom motion on Pt(111), there are a number of systems for which the diffusion of adatoms, impurity atoms or dimers has been followed directly by STM with negligible tip influence. Examples include the determination of the diffusion parameters for a Pt adatom on Pt(110) [77], for the Si dimer on Si(100) [78] and for Pd or In atoms in the Cu(100) surface layer [79, 80]. All these cases are distinguished from the situation of Pt adatom diffusion on Pt(111) by an activation energy that is larger by at least a factor of three, making the position of the atomic scale object less susceptible to the tip influence. In addition, typically a smaller tunneling current was used during observation, which reduces the current density and field strength at the location of the object.

A remarkable alternative to these experiments are measurements with the atom tracking STM developed by Swartzentruber [78], where the tip is not scanning but locked to the position of the object under interest by a lateral feedback mechanism. If conditions of negligible interaction of the tip and the object under concern are established, with this method a detailed characterization of the motion of atomic scale objects, including site occupation maps and residence time as well as jump width distributions, is possible with a time resolution on the order of milliseconds. This allows to identify complex diffusion mechanisms and details of the potential energy landscape [80, 81].

2.2.4 The Onset Method

A new method for the determination of adatom diffusion parameters using the STM, which avoids the disadvantages of the previously described approaches, was proposed and successfully applied to the Pt/Pt(111) system by Bott et al. [82]. It requires only one short diffusion interval at the migration temperature T. Imaging is carried out at $T' < T$, where diffusion is entirely frozen, thereby ruling out contamination as well as tip effects influencing the results.

The key experiment of the *onset method* is the deposition of a small quantity of adatoms at a temperature below or above the onset of adatom diffusion on the time scale of the deposition interval. At the end of deposition the sample is quenched and subsequently imaged by STM. The small number of deposited adatoms allows an accurate determination of the object number density by STM. The full squares in Fig. 2.4 show the object number density as a function of temperature for Pt-deposition on Pt(111). Below approximately 110 K, the object density is constant, but sharply bends down to lower densities at higher temperature. Obviously above 110 K adatom diffusion becomes efficient enough to allow for the formation of adatom clusters.

In principle, the quantitative relationship between the temperature dependence of the object density and the diffusion parameters of interest could be worked out using the concepts of nucleation theory to be introduced in

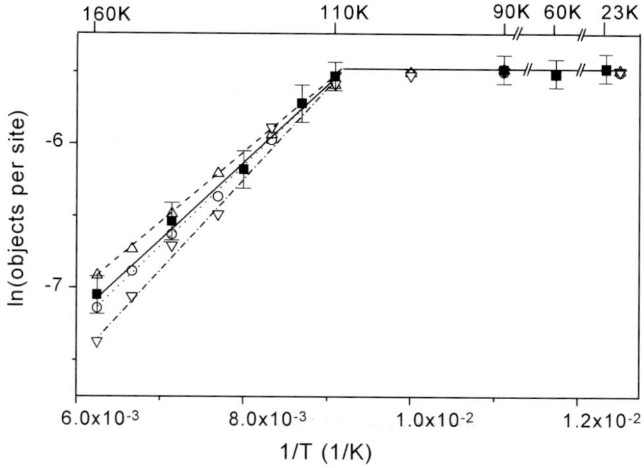

Fig. 2.4. Object densities determined by STM after deposition of $(4.2 \pm 0.4) \times 10^{-3}$ ML Pt on Pt(111) with a rate of $(6.6 \pm 0.7) \times 10^{-4}$ MLs^{-1} at the indicated temperatures and after subsequent quenching to 20 K (*full squares and lines*). Best fit to the experimental data results from kinetic MC simulations with $E_\mathrm{D} = 0.26\,\mathrm{eV}$ and $\nu_0 = 5 \times 10^{12}\mathrm{s}^{-1}$ (*open circles and dotted line*). The sensitivity of the simulation to the fitting parameters is indicated by the results for the parameter sets $E_\mathrm{D} = 0.24\,\mathrm{eV}$ and $\nu_0 = 6 \times 10^{11}\mathrm{s}^{-1}$ (*open up-triangles and dashed line*) and $E_\mathrm{D} = 0.28\,\mathrm{eV}$ and $\nu_0 = 4 \times 10^{13}\mathrm{s}^{-1}$ (*open down-triangles and dash-dotted line*). Lines to guide the eye

the next section. However, in order to avoid any bias associated with the simplifying assumptions of nucleation theory, a kinetic Monte Carlo (KMC) simulation of the deposition experiment is employed instead. Realistic simulations pose no major problem because of the low temperatures and low coverages of the experiment. The simulation takes place on a triangular lattice. Atoms are deposited in accordance with the experimental values of the flux and the deposition time, and adatoms reaching the nearest neighbor site of an occupied site are assumed to become immobile. The adatom hopping rate ν is taken to be of the Arrhenius form (2.4). This leaves the attempt frequency ν_0 and the diffusion barrier E_D as the only free parameters which are then determined iteratively by fitting the object number densities resulting from the Monte Carlo simulation to the experimental results. In Fig. 2.4 it can be seen that small differences in the simulation parameters give rise to significant differences in the simulated object densities. The sensitivity of the procedure allowed for Pt(111) the determination of the adatom diffusion parameters within small error margins: $E_\mathrm{D} = 0.26 \pm 0.01\,\mathrm{eV}$ and $\nu_0 = 5 \times 10^{12 \pm 0.5}\,\mathrm{s}^{-1}$. The agreement with the FIM values mentioned above is remarkable in view of the fact that the applied methods are completely different.

To summarize the physical assumptions underlying the onset method, as embodied in the accompanying KMC simulation: (i) Adatoms bind to each

Table 2.1. Pt adatom diffusion parameters from experiment (EXP) and calculated by density functional theory (DFT) or by semi-empirical methods (SEM).

EXP	1994 [68]	1996 [82]	1998 [67]	2000 [84]
E_D (eV)	0.25 ± 0.02	0.26 ± 0.01	0.260 ± 0.003	0.5^{a}
ν_0 (s^{-1})	-	$5 \times 10^{12\pm0.5}$	$1.0(\times 1.4^{\pm1}) \times 10^{13}$	-

DFT	1994 [68]	1996 [85]	1998 [86]	1998 [87]
E_D (eV)	0.38	0.36	0.33	0.29

SEM	1978 [88]	1991 [60]	1994 [89]	1994 [90]
E_D (eV)	0.07	0.078	0.038	0.08
ν_0 (s^{-1})	-	2×10^{12}	-	-

SEM	1994 [91]	1994 [92]	1996 [93]	1996 [94]
E_D (eV)	0.13	0.16	0.08	0.22

SEM	1999 [95]
E_D (eV)	0.17

[a] This result of electron diffraction spot profile analysis is likely to be due to a misconception in data analysis.

other. (ii) The mobility of adatoms exceeds that of all adatom clusters. In particular, it is larger than that of dimers. (iii) The adatoms do not interact with each other via long-range, repulsive or attractive pair interactions. While assumption (i) is self evident, assumption (ii) is in agreement with the present experimental knowledge on the diffusion of adatoms and adatom clusters [52, 83] and was explicitly verified for Pt(111) [67]. Only assumption (iii) is critical, and we will return to it below in Sect. 2.4.5.

2.2.5 Theoretical Estimates of Diffusion Parameters on Pt(111)

The three independent microscopic experiments measuring the Pt-adatom diffusion parameters on Pt(111), two using FIM and one using the STM onset method, agree with each other within the error bars. This is the first and, so far, only example for the quantitative determination of an atomic process by different high precision experiments with full agreement of the results. This fact will not only serve, in the next section, as a stringent test of nucleation theory, but it is also a cornerstone for a judgement on the reliability of theoretical methods used to calculate atomic rate parameters.

Table 2.1 compares the experimental values (first row) to the results of density functional calculations (second row) and the results of semi-empirical calculations of different types (third and fourth row) for the diffusion parameters of Pt adatom diffusion on Pt(111). Apparently density functional theory (DFT) calculations tend to yield too large values for E_{D}, while methods using semi-empirical potentials (SEM) underestimate the barrier. The only available calculation of the attempt frequency employs transition state theory for an embedded atom potential [60]. The DFT values exhibit a consistent improvement with time, and the latest result [87] deviates only by 0.03 eV from the experimental value. Also the SEM values have a tendency to approach the experimental result. However, even the latest value still deviates considerably by 0.09 eV from the experimentally determined barrier. Whereas DFT calculations appear to be developing into a tool of predictive power, it seems questionable whether semi-empirical calculations will be able to serve as a reliable guide to the true energetics in situations involving low coordinated atoms.

2.3 Getting Together: Two-Dimensional Nucleation

The formation of an atomic layer on a high symmetry substrate without steps or defects has to proceed through the aggregation of mobile adatoms into stable clusters, which subsequently grow by accretion. The process is analogous to the nucleation of a condensed phase out of a supersaturated gas, as described by thermodynamic nucleation theory [7]. The central object of this theory is the *critical nucleus*, which defines the free energy barrier that has to be surmounted to reach the stable phase. The size of the critical nucleus is inversely proportional to the supersaturation.

In far from equilibrium growth the critical nucleus size may reach atomic dimensions, thus precluding a straightforward application of thermodynamic concepts and necessitating the development of an *atomistic* theory of nucleation kinetics [96]. This was achieved in the 1960's and 70's by Zinsmeister, Stowell, Venables and others; extensive reviews are available [13,97–99]. Precise experimental tests of the theory have become possible only recently[9], and they constitute the core content of this section. In addition, large scale computer simulations play an increasingly important rôle in establishing the validity and limitations of nucleation theory, as they allow for separate scrutiny of the various assumptions that go into the theory.

2.3.1 Atomistic Nucleation Theory

The classical approach to nucleation kinetics starts from balance or rate equations for the areal concentrations n_s of clusters consisting of s atoms; n_1 is

[9] Recent reviews are [48,100].

the adatom density, n_2 the density of dimers, and so on. To be precise, we define n_s as the number of clusters per surface area, averaged over a region containing a large number of clusters. If the adatoms are the only mobile species (the mobility of larger clusters is negligibly small), then clusters grow solely by aggregation of single adatoms[10]. Defining Γ_s to be the net rate at which $s+1$-clusters form from s-clusters, we have for $s \geq 2$

$$\frac{\mathrm{d}n_s}{\mathrm{d}t} = \Gamma_{s-1} - \Gamma_s \quad (s \geq 2). \tag{2.9}$$

The net formation rates Γ_s can be written as

$$\Gamma_s = \sigma_s D n_1 n_s - \gamma_{s+1} n_{s+1}, \tag{2.10}$$

where γ_s is the rate at which adatoms detach from an s-cluster and the dimensionless *capture number* σ_s accounts for the propensity of an s-cluster to absorb an adatom (see below). The chain (2.9) of aggregation equations is fed by the adatom density n_1. If desorption from the surface can be neglected (the *complete condensation* limit [97, 99]), adatoms are lost only through dimer formation and capture at larger clusters, and the adatom rate equation reads

$$\frac{\mathrm{d}n_1}{\mathrm{d}t} = F - 2\Gamma_1 - \sum_{s \geq 2} \Gamma_s. \tag{2.11}$$

The deposition rate F is defined as the number of atoms deposited per unit time and surface area.

In principle, (2.9-2.11) provide a complete description of the nucleation process; in practice, they contain far too many (generally unknown[11]) kinetic parameters to be useful. This difficulty is commonly circumvented by introducing a distinction between stable and unstable clusters, and postulating a separation of time scales between the kinetics of the two kinds of clusters. Stable clusters of sizes $s \geq i^* + 1$ are assumed not to decay, i.e. $\gamma_s = 0$ for $s > i^*$, while the detachment of adatoms from unstable clusters with $s \leq i^*$ occurs sufficiently rapidly to establish thermodynamic equilibrium between the different cluster size populations. It is important to note that, in contrast to thermodynamic nucleation theory, the critical cluster size i^* introduced here contains a *kinetic* element, as it refers to stability and equilibration only on the time scale relevant to the deposition experiment. This implies that i^* depends on flux as well as on temperature.

Next the total density N of stable clusters, also referred to as *islands* in the following, is introduced through

$$N = \sum_{s=i^*+1}^{\infty} n_s. \tag{2.12}$$

[10] For a discussion of the effects of cluster mobility, see Sect. 2.4.4.

[11] In fact, for Pt(111) cluster diffusion coefficients and dissociation rates up to $s = 8$ have been determined [101], see Sect. 2.3.4.

Summing (2.9) from $s = i^* + 1$ to infinity, this is seen to evolve according to

$$\frac{\mathrm{d}N}{\mathrm{d}t} = \sigma_{i^*} D n_1 n_{i^*}. \tag{2.13}$$

The assumption of thermal equilibrium among unstable clusters implies that the net formation rates Γ_s vanish for $1 \leq s \leq i^* - 1$, and induces the detailed balance relations (also known as the *Walton relation*) [96]

$$\Omega n_s \approx (\Omega n_1)^s \mathrm{e}^{E_s/k_\mathrm{B}T} \qquad (2 \leq s \leq i^*) \tag{2.14}$$

between the unstable cluster concentrations and the adatom concentration. Here Ω denotes the area of an adsorption site[12] . The left hand side of (2.14) is the probability that an adsorption site is occupied by an s-cluster. This is equated to the probability $(\Omega n_1)^s$ of finding s independent adatoms at adjacent sites, enhanced by the Boltzmann factor containing the cluster binding energy E_s which is gained in forming the s-cluster. Equivalently, E_s is the energy needed to disperse the cluster into single adatoms; note that $E_1 = 0$.

Using (2.14) the nucleation rate on the right hand side of (2.13) can be expressed in terms of the adatom density. To complete the description, the rate equation for n_1 is simplified by introducing the average capture number for stable islands

$$\bar{\sigma} = N^{-1} \sum_{s=i^*+1}^{\infty} n_s \sigma_s. \tag{2.15}$$

Equation (2.11) can then be written in the form

$$\frac{\mathrm{d}n_1}{\mathrm{d}t} = F - \sigma_{i^*} D n_1 n_{i^*} - \bar{\sigma} D n_1 N. \tag{2.16}$$

Together (2.13), (2.14) and (2.16) form a closed set of equations from which the island and adatom densities can be computed.

Before turning to the solution of these equations, some remarks concerning the capture numbers σ_s are in order. In the early literature on atomistic nucleation theory, a geometric view of the capture numbers as effective cross sections for atoms colliding with clusters prevailed [102]. More appropriately, they are defined through the requirement that the flux of adatoms to a cluster of size s should be, on average, equal to $\sigma_s D n_1$. The capture number of a cluster evidently depends on its size, but in general it depends on the sizes and locations of surrounding clusters as well, because these affect the adatom concentration field (see Sect. 3.6). The calculation of σ_s involves the solution of a diffusion equation for the adatom concentration, with appropriate boundary conditions representing both the capture of adatoms at the cluster of interest, and the presence of other clusters far away[13]. In this sense

[12] For a square lattice, e.g. a fcc(001) surface, $\Omega = a^2$, while for a triangular lattice, such as the fcc(111) surface, $\Omega = (\sqrt{3}/2)a^2$.

the capture numbers retain some information about the spatial arrangement of the clusters, which is otherwise not accounted for in the rate equations[14].

Two approximate schemes for computing the σ_s appear in the literature [97]. In the *uniform depletion approximation* the surrounding clusters are included through a spatially uniform adatom loss term, while the *lattice approximation* assumes a regular, periodic arrangement of identical clusters [103]. The former approximation maximizes the competition between clusters whereas the latter minimizes it; consequently the two approaches yield lower and upper bounds on the capture numbers, respectively. A fully self-consistent implementation of the uniform depletion approximation, in which the capture numbers are evolved numerically along with the cluster concentrations, produces island and adatom densities in excellent agreement with kinetic Monte Carlo simulations [104, 105].

Due to the scale invariance of the stationary diffusion equation, one expects the capture number σ_s to depend primarily on the ratio of the island radius to the island distance, i.e. on the combination sN [104]. For the large islands which dominate the average capture number (2.15), island radii are comparable to island distances in the late stages of nucleation, and therefore $\bar{\sigma}$ should show little dependence on N, whereas the capture numbers of small islands, in particular σ_{i^*}, will generally vary with the island density [see (2.31)]. In practice, both σ_{i^*} and $\bar{\sigma}$ turn out to be slowly varying functions which can be replaced by constants for many (though not all) purposes. We adopt this simplification for the present discussion.

The solutions of the coupled equations (2.13) and (2.16) display two temporal regimes. In the early time, *transient nucleation regime*, the loss terms on the right hand side of (2.16) are negligible. The adatom concentration increases proportional to the total coverage $\Theta = \Omega F t$, and the island density grows rapidly as $N \sim \Theta^{i^*+2}$. This regime ends when capture of adatoms at stable islands becomes appreciable, at a coverage $\Theta_{\mathrm{tr}} \sim (F/D)^{2/(i^*+3)} \exp[-E_{i^*}/(i^*+3)k_{\mathrm{B}}T]$, which can be estimated by comparing the first and last terms on the right hand side of (2.16). In the subsequent *steady state nucleation regime* these two terms balance completely, and the adatom density is determined by the island density through

$$n_1 \approx \frac{F}{D\bar{\sigma}N}. \tag{2.17}$$

Inserting this into (2.13, 2.14) and integrating by separation of variables yields the central result of nucleation theory,

$$\hat{N} \equiv \Omega N \approx \eta(\Theta, i^*) \left(\frac{\Omega^2 F}{D}\right)^{\frac{i^*}{i^*+2}} \mathrm{e}^{E_{i^*}/(i^*+2)k_{\mathrm{B}}T}, \tag{2.18}$$

[13] Similar calculations are carried out in Sect. 4.3.

[14] In the jargon of statistical mechanics, the rate equation description constitutes a *mean field* approximation. See Sects. 2.4.3 and 4.9.4 for a discussion of effects which require a treatment beyond the mean field approximation.

with $\eta = [(i^*+2)\sigma_{i^*}\bar{\sigma}^{-(i^*+1)}\Theta]^{1/(i^*+2)}$, and $\hat{N}$ denoting the number of islands per adsorption site. We see that the island number density continues to slowly grow in the steady state regime, as a power of Θ with a small exponent $1/(i^*+2)$ which decreases with increasing i^*. The coverage dependence of η is further diminished by a more accurate treatment of the capture numbers, which shows in particular an increase of $\bar{\sigma}$ with increasing Θ [99].

The most important feature of (2.18) is that it takes the form of a *scaling relation*

$$N \sim \left(\frac{F}{D}\right)^{\chi} \tag{2.19}$$

between the island number density and the ratio of the two basic kinetic rates D and F of the deposition process, with the *scaling exponent* χ taking the value $\chi = i^*/(i^*+2)$ (for a generalized expression see (2.32)).

The rate equations formulated above are restricted to the low coverage regime, since finite coverage effects such as direct impingement and cluster–cluster coalescence have been neglected. In the absence of cluster mobility, coalescence provides the only mechanism by which the island density can decrease. This produces a maximum in $N(\Theta)$ which is often used as a convenient reference point for the experimental determination of cluster densities. Adding an appropriate coalescence term to the right hand side of (2.13), expressions for the maximal island density $N_{\max}$ and the corresponding coverage $\Theta_{\max}$ can be derived [99]. Typically $\Theta_{\max} \approx 0.05 - 0.2$, with little dependence on island density or deposition parameters. Therefore (2.18) can be used also for the maximum or saturated island number density $N_{\max}$, with a prefactor $\eta_{\max}$ of order unity. It is worth pointing out that $N_{\max} \gg N(\Theta_{\mathrm{tr}})$, which means that the vast majority of the islands nucleate in the steady state regime.

2.3.2 The Island Size Distribution

Whereas classical nucleation theory has been primarily concerned with the average island number density N, the recent availability of real space images with atomistic resolution has led researchers to ask more detailed questions about the spatial and size distributions of the islands. An important conceptual innovation is the *scaling hypothesis* for the island size distribution, which was first formulated, in this context[15], by Bartelt and Evans [107]. The scaling hypothesis states that the island statistics in the steady state nucleation regime, prior to coalescence, depends on the coverage Θ and the deposition parameters D and F only through the mean island size

$$\bar{s} = N^{-1} \sum_{s=i^*+1}^{\infty} s n_s, \tag{2.20}$$

[15] The application of scaling ideas to growth and aggregation phenomena has become a vast field, see e.g. [32, 36, 106] for reviews.

provided that $\bar{s} \gg 1$. In particular, the island size distribution n_s should then be a function only of the ratio $s/\bar{s}$, which is usually written in the form[16]

$$n_s = (\Theta/\bar{s}^2) f(s/\bar{s}). \tag{2.21}$$

Together with the normalization conditions[17]

$$\int_0^\infty \mathrm{d}x\ f(x) = \int_0^\infty \mathrm{d}x\ x f(x) = 1 \tag{2.22}$$

this ensures that the relation (2.12) for the total density of stable clusters, and the relation $\Theta = N\bar{s}$ are fulfilled; the latter expresses the fact that the stable clusters essentially exhaust the entire coverage in the late stage of nucleation.

The scaling function $f(x)$ in (2.21) can be computed from the rate equations (2.9), if the capture numbers σ_s are specified [97,98]. Assuming constant capture numbers $\sigma_s \equiv \bar{\sigma}$ and a critical nucleus size i^*, one obtains [107,108]

$$f_{i^*}(x) = \begin{cases} \frac{1}{i^*+2}\left(1 - \frac{i^*+1}{i^*+2}x\right)^{-i^*/(i^*+1)} & : \quad 0 \le x < \frac{i^*+2}{i^*+1} \\ 0 & : \quad x \ge \frac{i^*+2}{i^*+1}. \end{cases} \tag{2.23}$$

This proves the existence of a scaling function for a simple case, and shows that in general $f(x)$ should be expected to depend on the critical nucleus size i^*. However the form (2.23) is physically unsatisfactory because of the sharp cutoff and the associated divergence at $x = (i^* + 2)/(i^* + 1)$, and the island size distribution approaches the scaling form only for extremely large values of $D/\Omega^2 F$; moreover the scaling functions (2.23) do not agree with KMC simulations. While the discrepancy is not surprising in view of the assumption of constant capture numbers, even a fully self-consistent calculation within the uniform depletion approximation fails to produce a size distribution that matches simulations [104]. In contrast to the island number density, an accurate description of the island size distribution appears to be out of reach for conventional rate equation approaches.

A possible remedy consists in an additional scaling function for the capture numbers, which takes the form [109]

$$\sigma_s/\bar{\sigma} = \mathcal{C}_\sigma(s/\bar{s}). \tag{2.24}$$

Evidence for such scaling comes from KMC simulations and STM observations [110]. The scaling function $\mathcal{C}_\sigma$ is roughly constant for $x = s/\bar{s} < 1$ and increases linearly for $x > 1$. The latter implies a strong correlation between size and capture number for islands that are larger than the average. This correlation is in fact rather obvious if one identifies the capture numbers with the

[16] The reader should be cautioned that in some articles the distribution $p_s = sn_s/\Theta$ is considered instead, which describes the probability that a randomly chosen deposit atom resides in an island of size s.

[17] Since $\bar{s} \gg 1$, the scaled island size $x = s/\bar{s}$ can be treated as a continuous variable.

(relative) sizes of the *capture zones*, from which the growing islands recruit their adatoms: Islands with large capture zones are larger simply because they have grown faster in the past [111]. Using (2.24) in (2.9), an expression for the island size scaling function $f(x)$ in terms of the scaling function $\mathcal{C}_\sigma(x)$ can be derived, which (under plausible assumptions on $\mathcal{C}_\sigma$) does not display the unphysical singularity of (2.23) and agrees well with direct KMC simulations. A fully self-consistent treatment of the problem can be formulated by coupling the rate equations for the island densities to additional evolution equations for the capture zones, and deriving the joint probability distribution of island sizes and capture areas [112–114].

An empirical formula for the scaling functions $f_{i^*}(x)$ has been proposed by Amar and Family [108] on the basis of their KMC simulations. It reads

$$f_{i^*}(x) = C_{i^*} x^{i^*} \exp[-i^* a_{i^*} x^{1/a_{i^*}}], \tag{2.25}$$

where the coefficients a_{i^*} and C_{i^*} are fixed through the normalization conditions (2.22). Although this expression has been used repeatedly in the analysis of experimental and simulational data, the fact that it vanishes at $x = 0$ presents a problem in view of the theoretical approach based on (2.24), which predicts that generally $f_{i^*}(0) > 0$ [109].

2.3.3 Experimental Test of Nucleation Theory

Much of the recent interest in nucleation theory derives from the hope to reliably extract atomistic energy parameters, such as the barrier and attempt frequency for surface diffusion, from experimental observations of island densities on mesoscopic or microscopic scales. For this purpose it is most suitable to determine the island density at low temperature, where the encounter of two adatoms results irreversibly in the formation of an immobile dimer. If dimers are stable and immobile, i. e. $i^* = 1$, the expression (2.18) simplifies to

$$\hat{N} = \eta \left(\frac{\Omega^2 F}{D} \right)^{1/3} = \eta \left(\frac{4\Omega^2 F}{a^2 \nu_0} \right)^{1/3} \mathrm{e}^{E_\mathrm{D}/3k_\mathrm{B}T}. \tag{2.26}$$

An Arrhenius plot of $\log N$ versus $1/T$ thus yields the activation energy E_D from the slope and the prefactor D_0 (or attempt frequency ν_0) from the intercept of the straight line fit. For the case $i^* = 1$, the maximum value η_max of the dimensionless prefactor $\eta(\Theta, i^*)$ is estimated to be $\eta_\mathrm{max} = 0.25$ [99].

Experimentally the optimal coverage for the determination of island number densities is obtained from a sequence of deposition experiments at a fixed temperature and with successively higher coverages Θ. STM topographs from such a series [82] are shown in Fig. 2.5. The corresponding island number densities are shown in Fig. 2.6. These figures illustrate the transient and steady state nucleation regimes described above in Sect. 2.3.1. Initially, the stable clusters are widely spaced, such that continued deposition allows new nucle-

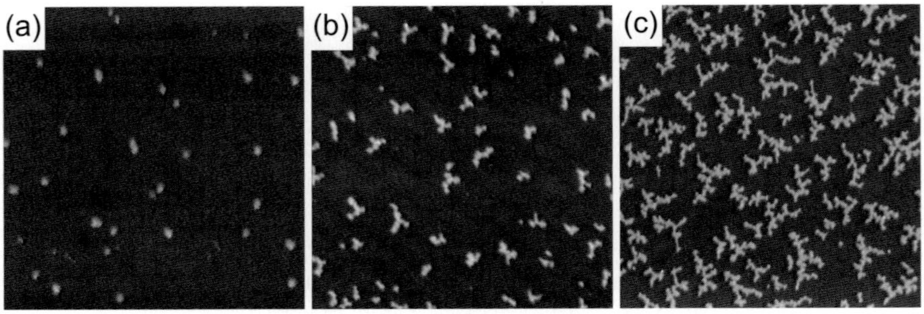

Fig. 2.5. STM topographs after deposition of (**a**) 0.002 ML, (**b**) 0.01 ML, and (**c**) 0.1 ML of Pt on Pt(111) at 180 K with a rate of $(6.6 \pm 0.7) \times 10^{-4}$ MLs^{-1}. Topograph size is 960 Å× 960 Å

ation in the spaces between existing clusters. As the island number density increases, capture at existing island edges becomes the most probable fate of freshly deposited adatoms, and the nucleation rate dwindles. Finally at even higher coverage, the islands start to coalesce and the observable island number density decreases. In Fig. 2.6 a broad maximum of the island density at about $\Theta_{\mathrm{max}} \approx 0.1$ ML can be seen. Also shown are the island number densities as obtained from KMC simulations with the adatom diffusion parameters determined by the onset method. Simulations and experiments apparently agree nicely.

Having identified a suitable reference coverage $\Theta_{\mathrm{max}} \approx 0.1$ ML, the temperature dependence of the saturated island density is determined experimentally. Figure 2.7 exhibits STM topographs of the adatom islands obtained at various temperatures. The most striking feature of these images is the irregular branched island shape of threefold symmetry, which will be analyzed in

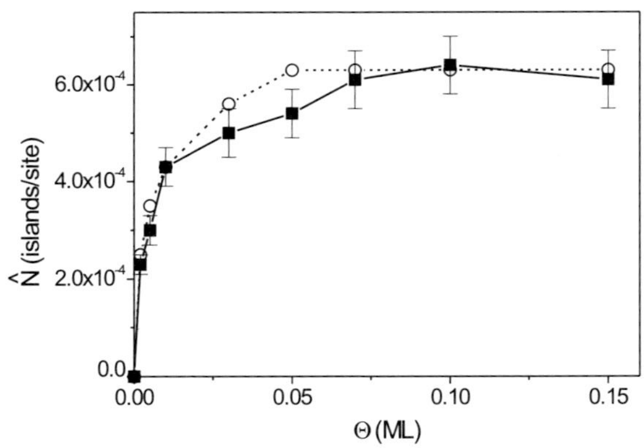

Fig. 2.6. *Full squares*: Experimental island number densities after deposition at 180 K with $F = 6.6 \pm 0.7 \times 10^{-4}$ MLs^{-1}. *Open circles*: KMC simulations of island number densities using the deposition parameters of Fig. 2.5 and diffusion parameters $E_{\mathrm{D}} = 0.26$ eV and $\nu_0 = 5 \times 10^{12}\,\mathrm{s}^{-1}$

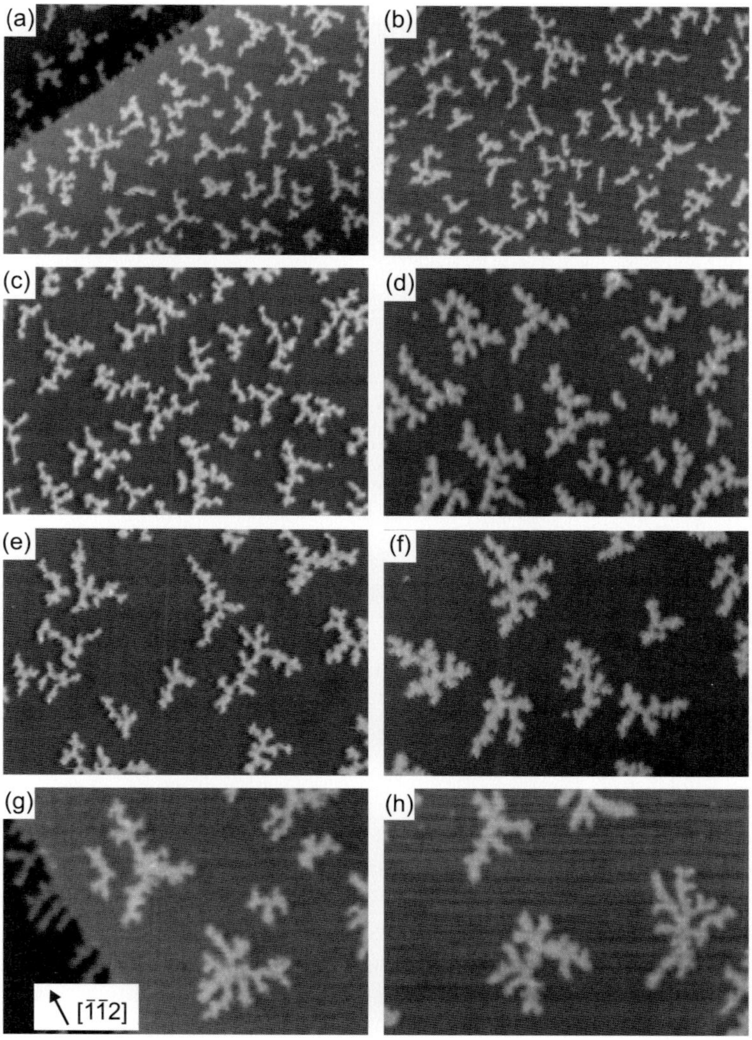

Fig. 2.7. STM topographs of adatom islands on Pt(111) after deposition of 0.1 ML Pt at (**a**) 150 K, (**b**) 165 K, (**c**) 180 K, (**d**) 195 K, (**e**) 210 K, (**f**) 225 K, (**g**) 245 K, (**h**) 265 K. The deposition rate is $F = 6.6 \times 10^{-4}\,\mathrm{MLs}^{-1}$, and the topograph size is 800 Å × 500 Å (Fig. 9 of [115])

detail in the next chapter. The decrease of N with increasing T is apparent as well. Figures 2.7a and 2.7g show branches growing away from preexistent step edges, which act as preferential nucleation sites. Since the presence of such step edges influences the island number density in their vicinity, for the quantitative determination of N only areas without preexistent steps were chosen. The results are shown in Fig. 2.8. Linear regression to the experimental data

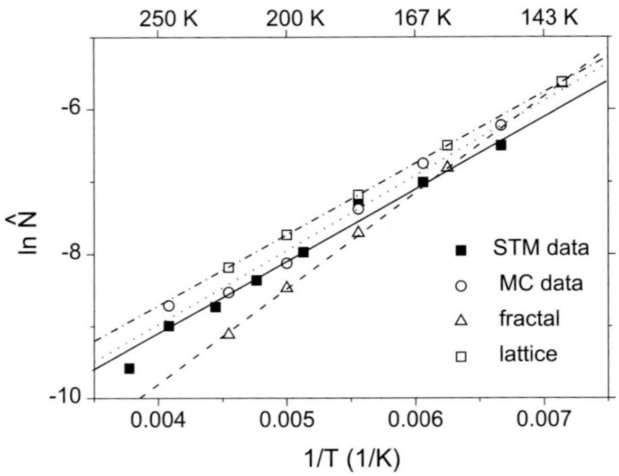

Fig. 2.8. *Full squares*: Island number densities $\hat{N}$ (in units of islands per lattice site) in the temperature range from 150 K to 265 K, obtained in the experiments visualized in Fig. 2.7. *Full line*: Linear regression to the data between 150 K and 245 K. *Open circles and dotted line*: KMC simulations for $\hat{N}$ with the parameters determined by the onset method and deposition conditions as in the experiment, together with a linear regression to the simulation data. *Open squares and dash dotted line*: Data and linear regression resulting from the numerical solution of the rate equations (2.13), (2.16) using the lattice approximation for the average capture number $\bar{\sigma}$. *Open triangles and dashed line*: Data and linear regression resulting from the numerical solution of the rate equations (2.13), (2.16) using an average capture number $\bar{\sigma} \sim (\bar{s})^{1/1.7}$

and application of (2.26) results in $E_D = 0.26\,\mathrm{eV}$ and $\nu_0 = 6 \times 10^{12}\,\mathrm{s}^{-1}$ in excellent agreement with the values obtained by the onset method (Sect. 2.2.4). This confirms the validity of the nucleation theory approach to the determination of adatom diffusion parameters.

For comparison Fig. 2.8 presents island number densities obtained from various theoretical approaches. In all calculations the diffusion parameters determined by the onset method were employed. The displayed KMC data suffer somewhat from poor statistics, but the slope and intercept of the linear regression coincide almost perfectly with the experimental data. Results from numerical solutions of the rate equations (2.13), (2.16) with two different choices of capture numbers are shown. In the first set of calculations the dimer formation rate was set to $\sigma_1 = 3$ and the expression $\bar{\sigma}$ derived from the lattice approximation [103] was used, while the second set employed a geometrically motivated[18] capture number $\bar{\sigma} \sim (\bar{s})^{1/d_\mathrm{f}}$, where $d_\mathrm{f} = 1.7$ is the fractal dimension of diffusion-limited aggregates [48,100] (see Sect. 3.1). The first choice fits the experimental data very well, while the second one does not; this underlines that a geometric interpretation of the capture numbers is not

[18] The spatial extent of a fractal island of size s is proportional to s^{1/d_f}.

quantitatively tenable. Raising the attempt frequency used in the solution of the 'lattice' rate equations to about $\nu_0 = 1 \times 10^{13}\,\mathrm{s}^{-1}$ would lead to a perfect match between experiment and theory. The discrepancies of the various fits for ν_0 might appear as a point of concern. However, they are within the error margin and one may consider an agreement in ν_0 to within a factor of two or three as perfect in view of the experimental uncertainties.

In conclusion, determination of the temperature dependence of the saturated island number density at low temperatures, where $i^* = 1$, is a straightforward way to determine adatom diffusion parameters. This conclusion is based on the excellent agreement between the results from nucleation theory, onset method and FIM measurements. Fits of the numerical solutions of the rate equations (using the lattice approximation for the capture numbers) to the data or direct use of the approximate scaling law (2.26) yield virtually identical results. The nucleation theory method is applicable as long as its underlying assumptions are fulfilled. The most critical ones are the absence of adatom–adatom interactions, the immobility of the dimers and the absence of contamination effects. Examples where the nucleation theory method appears to fail are discussed below in Sects. 2.4.5 and 2.4.6.

2.3.4 Binding Energies from Island Densities

Besides the determination of the adatom diffusion parameters, it is of crucial importance to understand the energetics of bonding between surface atoms, as it fixes the energy and temperature scale of a host of activated surface processes during growth. One possible experimental route for obtaining information on the binding energetics is supplied by nucleation theory. Recall that, for the critical cluster size $i^* > 1$, the fundamental result (2.18) of nucleation theory provides a relationship between the binding energy E_{i^*} of the critical cluster and the saturated island number density N. The slope in an Arrhenius plot of N versus $1/T$ is predicted to be $(i^* E_\mathrm{D} + E_{i^*})/(i^* + 2)$, which allows the determination of E_{i^*}, once E_D and i^* are known.

In a plot of $\ln N$ versus $1/T$ one thus expects distinct slope changes whenever a new scaling regime for the island number density is entered due to a change in i^*. With increasing temperature the scaling regime for $i^* = 1$ is left as soon as the single bond formed between the dimer atoms no longer guarantees their stability. A new scaling regime for N is entered when each cluster atom forms two bonds within the cluster. On a surface with a triangular lattice, such as fcc(111), this is realized for a trimer ($i^* = 2$), while on a surface of square symmetry, e.g. fcc(100), only a tetramer ($i^* = 3$) configuration allows two bonds per cluster atom (in a trimer on a square lattice two of the three atoms possess only one nearest neighbor bond).

A number of results for the binding energies in small clusters on their own substrate have been reported, which are mostly based on the analysis of slope changes in island density curves. For Fe(001) a value of $0.55 \pm 0.45\,\mathrm{eV}$ for the nearest neighbor bond strength $E_\mathrm{b} = E_3/2$ in an Fe trimer was derived in an STM–study [116]. In agreement with this estimate, KMC simulations reproducing the experimental island densities had to assume a single bond energy of $0.6 \pm 0.1\,\mathrm{eV}$ [108]. These results were criticized on the basis of consistency arguments derived from FIM data for the W dimer binding energy on W(110) [117]. Nevertheless, recent growth studies employing ion scattering essentially confirmed the STM–result for the bond strength in an Fe trimer on Fe(001): a value of $0.50 \pm 0.12\,\mathrm{eV}$ was obtained [118]. Likewise, for trimers on Cu(001) 0.06 eV [119], and for Au(001) 0.5–0.7 eV [120] were given as estimates of the bond strength. For the bond strength in a Rh dimer on Rh(111) an STM study resulted in $E_2 = 0.6 \pm 0.4\,\mathrm{eV}$ [121]. Comparison of measured island densities to KMC simulations allowed to estimate $E_\mathrm{b} = 0.3\,\mathrm{eV}$ in an Ag trimer on Ag(001) [122]. For Pt(110), direct STM observations yield $E_2 = 0.07\,\mathrm{eV}$ [123] while DFT calculations find $E_2 = 0.21\,\mathrm{eV}$ [124].

The quoted values exhibit an unphysical scatter in magnitude – the bond strength on Cu(001) is found to be ten times smaller than that on Au(001)! – and in some cases they have extremely large error margins. These features reflect the following, fundamental difficulties in the application of nucleation theory for the determination of bond strengths:

• *Cluster mobility.* It can plausibly be argued that the activation energy for dimer diffusion should typically be smaller than that for dimer dissociation [125]. This implies that upon increasing temperature, a regime of mobile dimers with a distinct island density scaling (derived in Sect. 2.4.4) is encountered before the critical nucleus size becomes $i^* = 2$ (surface of triangular symmetry) or $i^* = 3$ (surface of square symmetry). Failure to identify this intermediate regime will lead to erroneous estimates for the binding energy of the dimer or trimer, respectively. Moreover, even if the scaling regime for $i^* = 2$ or $i^* = 3$ is safely established, the measured slope in the $\ln N$ versus $1/T$ plot will be increased if dimers and larger cluster diffuse sufficiently rapidly. As independent information on cluster diffusion is generally not available, the entanglement of cluster diffusion with the cluster binding problem makes the determination of binding energies a formidable task already for $i^* = 2$ or $i^* = 3$, and practically impossible for larger critical cluster sizes.

• *Small binding energy differences.* If the differences between the cluster binding energies are small, the critical nucleus size changes so rapidly with temperature that no extended temperature range exists in which the scaling law (2.18) can be applied. A rate equation analysis of the transition from $i^* = 1$ to $i^* = 3$ in fcc(100) epitaxy shows that a well-defined scaling regime with $i^* = 3$ occurs only if the ratio of the bond energy E_b to the diffusion barrier is not small compared to unity [126]. For the example of Pt(111) described in detail below, the nucleation theory estimates imply $E_\mathrm{b}/E_\mathrm{D} \approx 2.5$, so the assumption of an extended scaling regime is at least self-consistent. In contrast, for fcc(100) surfaces in many cases no well defined scaling regime with $i^* = 3$ exists, because the stronger corrugation implies that E_D is comparable to, or even larger than E_b [52, 116, 118, 119, 122, 126, 127]. For these surfaces it is then not possible to use 2.18) to determine E_b from the slope of $\ln N$ versus $1/T$. Some information may nevertheless be gained from the temperature at which the scaling regime for $i^* = 1$ terminates [126].

• *Nucleation without a critical nucleus.* It must always be kept in mind that the notion of a critical nucleus size separating (absolutely) stable from (absolutely) unstable clusters is to some extent fictitious, since atoms can detach from clusters of all sizes. A powerful demonstration of this point was provided by KMC simulations with simple bond-counting energetics [128, 129], in which both the exponent χ in the scaling law (2.19), and the island size scaling function $f(x)$ in (2.21) were found to vary continuously with temperature, in a manner that is difficult to reproduce within rate equation theory.

• *Heterogeneous nucleation.* As the island number density decreases towards higher temperatures, even a very low concentration of defects may become relevant as centers for heterogeneous nucleation. Defects may be chemically different substitutional surface atoms, localized adsorbates or strain fields resulting from subsurface defects such as vacancies and impurity atoms (see Sect. 2.4.6).

Keeping these problems in mind, in the following an extended set of island number density data for Pt/Pt(111) ranging from 150 K up to 640 K shown in Fig. 2.9 will be analyzed in order to obtain estimates for the dimer binding energy [130]. The data is from two independent measurements. The full circles represent the highest quality data [131, 132], for which extremely clean deposition conditions were achieved, such that heterogeneous nucleation can be largely ruled out. The data set consisting of the full squares [82] is identical to that of Fig. 2.8, but renormalized to the deposition rate used in the first set; for $i^* = 1$ one has $N' = N(F'/F)^{1/3}$. The two data sets match perfectly, and are therefore a sound basis for the application of nucleation theory. The experimental island number densities up to 400 K lie on an almost straight line, with a slight downward curvature becoming apparent at temperatures above 250 K. Between 400 K and 440 K the slope of the curve changes significantly, suggesting a change in the size of the critical nucleus. The data between 440 K and 600 K is again positioned on a nearly straight line. In the last section $E_\mathrm{D} = 0.26\,\mathrm{eV}$ was derived by application of (2.26) to the data below 245 K. The gradual slope change above 250 K is due to dimer diffusion becoming relevant to island nucleation. As shown in Sect. 2.4.4, the asymptotic rate equation expression for the saturated island density N in the presence of adatom and dimer diffusion, but in the absence of dimer dissociation, reads [133]

$$N \sim F^{2/5} \mathrm{e}^{(E_\mathrm{D}+E_{\mathrm{D},2})/5k_\mathrm{B}T} . \tag{2.27}$$

Application of (2.27) to the two data points at 300 K and 400 K yields the estimate $E_{\mathrm{D},2} = 0.34 \pm 0.04\,\mathrm{eV}$ for the activation energy for dimer diffusion, in good agreement with direct FIM measurements which give $E_{\mathrm{D},2} = 0.37 \pm 0.02\,\mathrm{eV}$ [83]. For both estimates of $E_{\mathrm{D},2}$ the condition formulated in [133] for

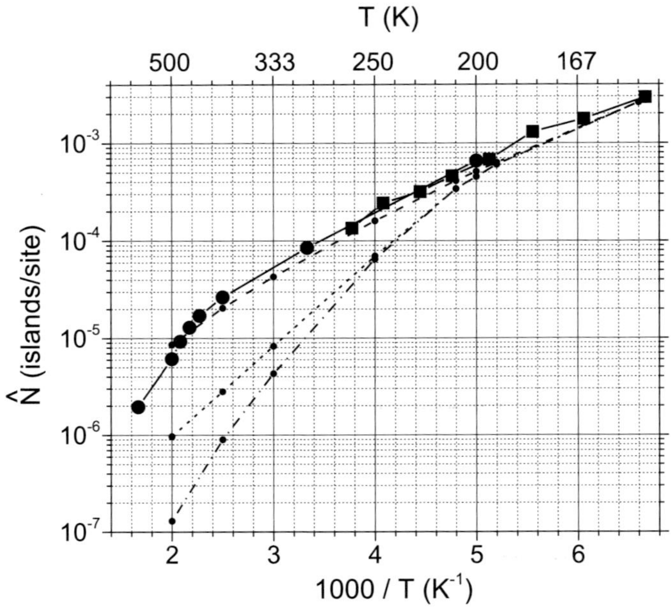

Fig. 2.9. Platinum island number densities on Pt(111) based on STM measurements and numerical solutions of rate equations with FIM data as input. *Full circles*: STM data [131]; $\Theta = 0.17 \pm 0.01\,\mathrm{ML}$, $F = 5.0 \pm 0.3 \times 10^{-3}\,\mathrm{ML\,s^{-1}}$. *Full squares*: STM data [82] as shown in Fig. 2.8, but renormalized to $F = 5.0 \times 10^{-3}\,\mathrm{ML\,s^{-1}}$ assuming $i^* = 1$; $\Theta = 0.10 \pm 0.01\,\mathrm{ML}$. *Dashed line*: Solution of rate equations assuming cluster mobility as measured by FIM, but dimer stability in the entire temperature range [101]. *Dotted line*: Solution of rate equations assuming dimer dissociation in accord with FIM data, other clusters stable and no cluster diffusion [101]. *Dash-dotted line*: Solution of rate equations assuming cluster dissociation in accord with FIM measurements, but no cluster diffusion [101]

the existence of the intermediate, dimer diffusion induced scaling regime is indeed fulfilled above 270 K, justifying a posteriori the application of (2.27). Lacking an analytic rate equation expression for $i^* = 2$ that includes cluster diffusion, the island densities for $T \geq 440\,\mathrm{K}$ are fitted to (2.18) with $i^* = 2$, which results in a dimer binding energy $E_2 = 0.74 \pm 0.10\,\mathrm{eV}$. This value is likely to be too large, since cluster diffusion increases the slope of the $\ln N$ versus $1/T$ curve due to cluster coalescence.

As simple analytic formulae for the island density in the presence of cluster diffusion *and* dimer dissociation are not available, it is worthwhile to look for alternative methods to determine E_2 from island density data. We sketch here a simple approach based on a comparison of the relevant time scales [130]. The key observation is that the information about the change in i^* is contained already in the distinct slope change of $\ln N$ versus $1/T$. It may be characterized by a transition temperature T_t, delimiting the regimes of stable and unstable dimer nuclei, which should depend only marginally on the extent of dimer and cluster diffusion. A good estimate of T_t in the present case is 430 K, the temperature where the linear fits to the low and high temperature data intersect. At the transition temperature T_t the dimer lifetime in the absence of growth, $\tau_{\rm diss}$, should become comparable to the average time interval Δt_2 between dimer formation and the arrival of the next atom stabilizing it to a trimer. The time interval Δt_2 depends on the time Δt between subsequent deposition events in the capture zone of the dimer, which is of order $\Delta t \approx N/F$ [compare to (3.47)], and the diffusion time $\tau_D \approx 1/(DN)$ required for the freshly deposited atom to reach the dimer. It follows from the scaling law (2.19) evaluated at $i^* = 1$ that $\tau_D \ll \Delta t$, hence deposition is the limiting step and $\Delta t_2 \approx \Delta t \approx N/F$. Setting $1/\Delta t \approx 1/\tau_{\rm diss} = \nu_0 {\rm e}^{E_{\rm diss,2}/k_{\rm B}T}$ at $T = T_t$, we find that the activation energy for dimer dissociation $E_{\rm diss,2}$ can be estimated as

$$E_{\rm diss,2} \approx -k_{\rm B}T_t \ln \frac{F}{\nu_0 N}. \tag{2.28}$$

Assuming the frequency factor $\nu_0 = 5 \times 10^{12}\,{\rm s}^{-1}$ derived for monomer diffusion [82] to hold also for dimer dissociation one obtains $E_{\rm diss,2} = 0.88\,{\rm eV}$. To first approximation $E_{\rm diss,2} = E_2 + E_{\rm D}$ [83, 134] and thus $E_2 = 0.62\,{\rm eV}$. The transition temperature method was tested against the numerical solution of rate equations [130] and KMC simulations and yielded excellent agreement with the input values.

A second method that requires as input only the transition temperature T_t, without any information about the slope in the $\ln N$ versus $1/T$ plot, follows from the fact that at T_t, (2.18) must be valid both for $i^* = 1$ and $i^* = 2$. Equating the two expressions allows to determine E_2 from the knowledge of the adatom diffusion parameters and yields $E_2 = 0.65\,{\rm eV}$, in good agreement with the time scale approach.

To assess the reliability of our estimate for E_2, we refer to the possible pitfalls listed above. The issues of cluster diffusion and heterogeneous nucleation were addressed already. For the present system the difference in binding energies between $i^* = 1$ (where $E_1 = 0$) and $i^* = 2$ is large, and consequently the scaling regimes for $i^* = 1$ and $i^* = 2$ could be clearly identified. Direct STM measurements prove the absence of step atom detachment at 450 K, hence the distinction between stable and unstable nuclei is unambiguous and meaningful. Relying on the results obtained from the transition temperature T_t, we can therefore confidently state that the dimer binding energy on Pt(111) is $E_2 \approx 0.63\,{\rm eV}$ with an error margin well below 0.10 eV.

This finding is in sharp contrast to measurements by FIM. With the FIM, the dimer dissociation barrier is determined by defined heating of a dimer adsorbed to the apex plane of a field emitter to successively higher temperatures, until the dimer disappears. With the relation $E_{\rm diss,2} = E_2 + E_{\rm D}$ the experimentally determined dissociation barrier $E_{\rm diss,2} = 0.49 \pm 0.01\,{\rm eV}$ results in a dimer binding energy of $E_2 = 0.23 \pm 0.01\,{\rm eV}$ [83]. This value is lower by almost a factor of three compared to the estimate obtained from the analysis of island densities. The contradiction could hardly be more pronounced. To emphasize the discrepancy even further, island number densities were calculated on the basis of the dissociation and diffusion energies measured by FIM for clusters containing up to 7 atoms [83]. The densities were obtained from rate equations of the type (2.9), including loss terms due to dissociation and coalescence terms due to cluster diffusion [101]. The results are included in Fig. 2.9. The dashed curve takes cluster diffusion into account (no dissociation), the dotted curve includes dimer dissociation only (no diffusion) and the dash-dotted curve additionally includes cluster dissociation of larger clusters. Apparently the experimental island number densities are incompatible with dimer dissociation as well as additional dissociation of larger clusters below 400 K. According to the FIM measurements (dotted curve), dimer dissociation would lead to $T_t \approx 200\,\mathrm{K}$ instead of $T_t = 430\,\mathrm{K}$ as observed experimentally. The experimental island densities are consistently modeled by rate equations assuming the dimer to be stable up to 400 K and allowing for cluster diffusion as measured by FIM (dashed line). The agreement between the dashed line and the experimental data would be nearly perfect, if the value $\nu_0 = 5 \times 10^{12}\,\mathrm{s}^{-1}$ as determined by the onset method would have been used in solving the rate equations instead of the value $\nu_0 = 1 \times 10^{13}\,\mathrm{s}^{-1}$ as determined by FIM.

Ab initio DFT calculations for the dimer binding energy on Pt(111) agree far better with the estimate obtained from the island density analysis than with the FIM measurement: two independent calculations resulted in $E_2 = 0.48\,\mathrm{eV}$ [134] and $E_2 = 0.50\,\mathrm{eV}$ [130]. A critical inspection of the FIM dissociation experiments indicates that in these experiments cluster dissociation took place at the step edge, rather than on the central part of the small apex plane of the field emitter tip [130, 135, 136]. Once the dimer has surmounted the escape barrier of 0.46 eV [83] to reach the step edge, DFT calculations find an easy pathway for dissociation at the step edge with $E_{\mathrm{diss},2} = 0.41\,\mathrm{eV}$. This pathway of dimer dissociation is compatible with the FIM value $E_{\mathrm{diss},2} = 0.49\,\mathrm{eV}$ resulting in the small apparent $E_2 = 0.23\,\mathrm{eV}$. It appears therefore that FIM measures dimer dissociation at the step edges, a process not relevant to cluster nucleation, rather than dimer dissociation on the terrace. This explains the large discrepancy between the FIM result and the estimate for E_2 based on the analysis of the island densities.

Since the dimer binding energy defines the bond strength in the surface layer, one expects it to scale with the cohesive energy E_{coh}. Application of the time scale approach based on the transition temperature T_t to island density data for Ir(111) and Al(111) [137] yields dimer binding energies of 0.73 eV and 0.37 eV, respectively. Scaled to E_{coh}, the dimer binding energy for Ir(111), Al(111) and Pt(111) amounts in all three cases to $E_2 = (0.11 \pm 0.01) \times E_{\mathrm{coh}}$. Interestingly also the dimer binding energy for Rh(111) is $0.1\,E_{\mathrm{coh}}$ [121].

For dispersing an fcc crystal into its atomic constituents, e.g. by moving atoms from the repetitive kink position into the vacuum, in a pair binding model six bonds have to be broken. Thus for bonds of equal strength one would expect $E_2 = 1/6\, E_{\text{coh}} = 0.17\, E_{\text{coh}}$, which is much larger than the value found experimentally. This feature can be understood on the basis of a bond order – bond length argument: The fewer bonds an atom has, the shorter and the stronger these bonds are. An atom compensates the loss of coordination by strengthening the remaining bonds [138, 139] (see also Sect. 6.4.2).

The first stage in the transfer of a kink atom into the vacuum is the detachment from the kink to the step, which reduces the coordination number from 6 to 5 (process 8 in Table 3.1). As this amounts to creating two additional kinks, the corresponding energy is approximately twice the kink energy ϵ. Using the experimental estimate for ϵ obtained from step fluctuation measurements (see Table 3.2) we find a value of 0.33 eV, while DFT calculations in the LDA yield 0.42 eV for the creation of a step adatom at the A-step, and 0.51 eV at the B-step, respectively [140]. As expected from the bond order – bond length argument, these numbers for breaking a bond by reducing the coordination number from 6 to 5 are smaller than our estimate of the energy E_2 required to break the dimer bond, which reduces the coordination number from 4 to 3. Also embedded atom method (EAM) calculations of Pt cluster binding energetics on Pt(111) exhibit a pronounced decrease of the in-plane bond strength with increasing coordination of the cluster atoms from about 0.87 eV for the dimer to 0.55 eV for a terrace atom (coordination number 9) [141].

Assuming a monotonic increase of the bond strength with decreasing coordination, the experimentally determined Pt–dimer bond strength thus provides an upper bound for the in-plane bond strength of Pt atoms in higher coordinated situations on Pt(111). Since the strengths of all the six bonds of the kink atom have to add up to the cohesive energy, it follows, moreover, that the adsorption energy required to break the remaining three bonds of an isolated adatom is bounded from below by $E_{\text{ad}} \geq E_{\text{coh}} - 3\, E_2 \approx 4.0$ eV. This value is of a similar magnitude as the bond strength 3.70 eV of a Pt dimer in the gas phase [142].

2.4 Supplementary Topics

2.4.1 Condensation on Clusters and at Steps

The condensation of atoms on a flat, step-free fcc(111) surface is largely localized at the lattice site closest to the point of impact; practically no atoms reach a site more than two nearest neighbor spacings a away (see Sect. 2.1). Condensation close to step edges involves a number of interesting phenomena which are also localized within distances of $1 - 2\,a$ around the step, but which may be relevant for growth in situations where the step concentration is high.

Condensation Close to Ascending Steps. In a FIM experiment [143] iridium atoms were condensed at 20 K on an Ir(111) apex plane, which car-

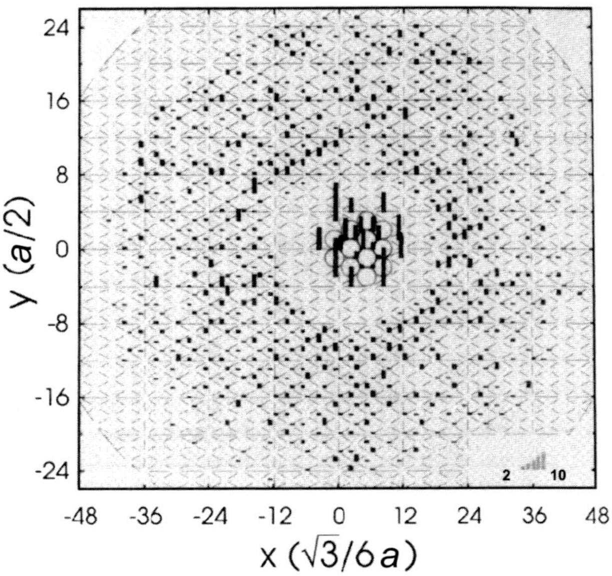

Fig. 2.10. Distribution of iridium adatoms condensed at $T \approx 20\,\mathrm{K}$ onto an Ir(111) surface with a central cluster of 12 iridium atoms. Height of bars indicates population on sites; the calibration is given at the lower right. The maximum plane diameter is $26\,a$ (Fig. 2 of [143])

ried a small adatom island (Fig. 2.10). Surprisingly a zone of about $2a$ around the ascending step was completely free of adatoms, while an excess of atoms was found attached to the step as well as at the outer boundary of the empty zone around the island. Although the interpretation of the data is not unambiguous, it must be concluded that the potential energy surface around the ascending step is severely distorted, allowing atoms that impact in the empty zone either to skitter out of the empty zone (towards the terrace or the step) or to leave it by thermal migration[19] even at 20 K. In agreement with these findings, in FIM observations of Ir adatom diffusion on Ir(111) an Ir adatom was never observed in a zone of about $2a$ around the ascending step of an Ir cluster [145]. The Ir adatom was either found outside the empty zone or attached to the cluster. From statistical analysis, it could be concluded that the kinetic barrier for entering the empty zone from the terrace side, followed by attachment to the cluster edge, is quite similar to the activation energy for Ir adatom hopping. It follows that, in general, the accommodation of an atom within a distance of about $2a$ from an ascending step may result in direct incorporation into the step[20].

[19] Ir adatom diffusion on a flat Ir(111) surface becomes relevant on the same time scale only at about 100 K [144].

[20] The consequences of this effect for multilayer growth are described in Sect. 4.9.7.

Condensation on Small Clusters and Close to Descending Steps. Thermal energy Pt–atom accommodation at Pt(111) on heptamers and close to descending steps was investigated by EAM molecular dynamics simulations [146]. Atoms arriving on top of a heptamer were occasionally (in 17% of the events) found to disrupt the cluster and to accommodate *in* the cluster by pushing out an atom of the cluster. Similar effects have also been observed in simulations of Cu–deposition on Cu(111) [147]. Even when the atom impacted in the vicinity of a descending step close to a stable adsorption site, in a few cases Pt–atom accommodation in the Pt(111) surface layer by step atom push-out could be observed. However, the effect relied crucially on atom alignment and vanished at finite temperatures. Switching to deposition of energetic Pt–atoms with kinetic energies of 10 eV or 20 eV greatly enhanced the frequency of such push-out events at heptamers (78% of the events) and at steps. About 60% of the energetic Pt–atoms impacting in a strip of one nearest neighbor spacing around the stable adsorption sites next to a step were accommodated in the surface layer and caused a push–out event. Thus, although accommodation in a lower layer upon arrival close to a descending step is not relevant in thermal energy deposition, it may be quite significant in energetic deposition. Finally, even on a flat surface ballistic exchange, i.e. the incorporation of the arriving atom in the surface layer, was observed in FIM experiments [148]: A small fraction of Ir atoms impinging on Rh(001) were found to exchange ballistically at 77 K. It should be noted that Ir on Rh(001) performs also thermal exchange, but only at a much higher temperature of 330 K.

Funneling. In an attempt to explain sustained reflection high energy electron diffraction (RHEED) oscillations during epitaxial growth at low temperatures, the theoretical concept of *funneling* was introduced [149, 150]. On a fcc(001) surface it means that if the vector connecting the impact point and the nearest adsorption site in the lattice plane of impact has a component along the dense packed directions, whose magnitude is larger than $a/2$, the atom moves downwards, until a stable adsorption site is reached (Fig. 2.11). A stable adsorption site is, in this context, defined as a fourfold hollow site, i.e. a site with four nearest neighbors. Accordingly, atoms arriving to the left of the dashed line in Fig. 2.11b in the upper layer will be funneled downwards towards the lower layer, since no adsorption site with more than 2 nearest neighbors is available in the layer of arrival. Figure 2.11c displays an example where funneling occurs, even though threefold coordinated adsorption sites are present: According to the funneling rule, threefold coordinated sites are not stable and therefore atoms arriving at a (111)-facet of the pyramid will descend all the way down to the base of the pyramid.

The concept of funneling partly agrees, but partly contradicts experimental findings. Onefold and twofold coordinated sites can indeed not be assumed to be stable adsorption sites, and there is neither theoretical nor experimental evidence for a metal atom on a metal surface being in a potential energy minimum in such a configuration. Twofold coordinated sites are typical transition states for adatom hopping on fcc(111) and fcc(001) surfaces. Thus funneling of an atom impacting close to an only onefold or twofold coordinated site must be assumed. For example, atoms impacting on an adatom or a dimer will always be funneled to the terrace.

However, threefold coordinated sites are stable adsorption sites at which condensation can occur. In the condensation experiments with Ir–adatoms on Ir(111) described in this section and in Sect. 2.1 [47, 89], the Ir–atoms arrived at an angle of about 45° towards the (111) surface normal, similar to the angle formed by atoms deposited at normal incidence on a fcc(001) surface with a (111) pyramid facet (35°, see Fig. 2.11d). The absence of transient mobility found in the experiments on condensation at Ir(111) thus conceptually contradicts the funneling of atoms arriving on large pyramids at the (001)–surface, which expose threefold coordinated adsorption sites. Even for small pyramids of only three layers height molecular dynamics simulations demonstrate that condensation at the sides of the pyramids is possible [44]. Therefore, the assumption of the deposited atom to be adsorbed at the nearest threefold *or* fourfold coordinated site appears to be more realistic.

Recent modeling of low temperature Ag/Ag(100) growth using KMC [151] and temperature–accelerated MD [152] has indeed revealed the importance of the trapping of atoms in threefold sites at the sides of nanoprotrusions, and has led to the extended notion of *restricted* downward funneling. In MD simulations of multilayer growth of fcc(100) films, the overhangs created by

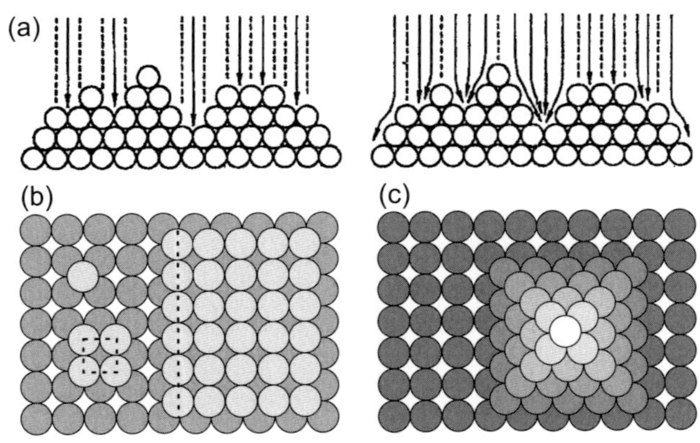

Fig. 2.11. (**a**) Schematic side view of random immobile deposition (left) versus downward funneling dynamics (right) (Fig. 1 of [149]). (**b**) Top view ball model of an fcc(001) surface with step, tetramer and adatom. Atoms arriving to the left of the dashed line at the step on the upper layer will be funneled to the lower layer. Atoms arriving outside of the dashed line on the tetramer will be funneled to the lower layer. Atoms arriving on the adatom will always be funneled to the lower layer. (**c**) Top view ball model of a pyramid on a fcc(001) surface with (111) facets. All atoms arriving on the pyramid facets will be funneled downwards according to [149], as the facets offer only threefold coordinated adsorption sites

atoms adsorbed in threefold coordinated sites were observed to lead to the incorporation of bulk vacancies into the growing film [153] (see also Sect. 4.8.2).

Steering. The strong attractive force that is felt by an atom approaching the surface leads to a considerable deflection towards the surface normal for oblique incidence trajectories [44]. While this *steering effect* is without consequences for deposition onto a perfect flat substrate, in the presence of steps and islands it induces lateral inhomogeneities in the deposition flux that can affect the growth morphology [154]. Steering as a mechanism for growth instability appears to have been invoked for the first time in a numerical study of one-dimensional deposition models [155]. It may play a role in the formation of mounds on amorphous films [156], and it appears to provide an important roughening mechanism in metal homoepitaxy at very low temperatures, where thermal mobility is turned off [152, 157]. The amount of deflection, and hence the roughness induced by steering, decreases with increasing kinetic energy of the deposited atoms.

Evidence of steering was obtained in deposition experiments of Cu on Cu(001) at grazing incidence (i.e. an angle of 80° from the surface normal), where the formation of elongated adatom islands as well as ripples in the direction perpendicular to the incidence plane was observed [158, 159] (see also Sect. 4.9.2). The situation is visualized in Fig. 2.12. Based on trajectory calculations for a simple Lennard–Jones potential, the distance between the target point (the dashed line in Fig. 2.12a) and the actual impact point is found to be considerably shorter for a trajectory terminating on an island close to an up–step in the beam direction, than for a trajectory terminating

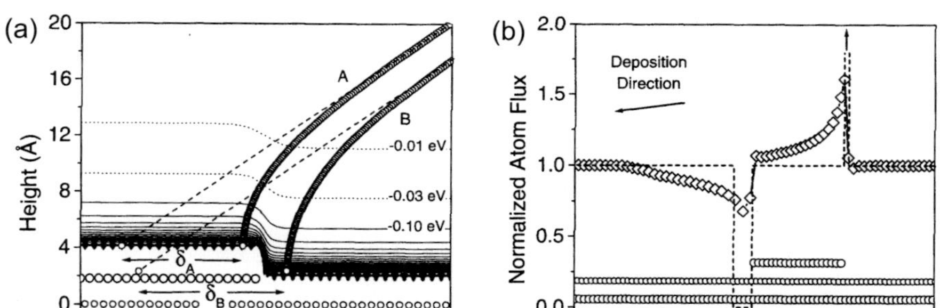

Fig. 2.12. (**a**) Calculated equipotential energy contours and two atom trajectories for a surface with a step in the direction perpendicular to the plane of incidence (note the different length scales on the two axes). The increase in the attractive potential is 0.1 eV between subsequent solid contour lines. The trajectory calculations for a deposition angle of 80° with respect to the surface normal start at 20 Å above the surface. (**b**) Calculated adatom flux at the surface, normalized to a homogeneous atom flux far above the surface, in the presence of an adatom island of monolayer height. The dashed line shows schematically the adatom flux in the case of classical shadowing without steering (Fig. 2 of [158])

on a step free terrace. The average flux onto the top of an adatom island near an up–step is therefore enhanced, while it is diminished on the lower terrace behind the adatom island. As the adatom flux arriving on top of the island is distributed also to its side edges, the net speed of step advancement in the direction of the beam is slower than perpendicular to it, leading to the observed rectangular island shapes. Steering is different from a purely geometric shadow effect, since in the latter case the flux enhancement would occur on the lower terrace in front of an up–step, and not on top of the island.

2.4.2 Exchange Diffusion

So far we have envisioned the migration of adatoms as a hopping motion between the adsorption sites of the surface lattice. Although this picture of diffusion through single particle motion applies in many cases to a first approximation (including the motion of Pt–adatoms on Pt(111)), it is not complete. In general, the positions of the substrate atoms respond to the presence of an adatom in order to minimize the total energy of the system. The relaxed atom positions and bond lengths around the adatom may differ only by a few percent from those of the unrelaxed substrate, but nevertheless relaxation of the substrate is crucial in ab initio calculations for obtaining correct values of adsorption energies and kinetic barriers. As an example, the activation energy for Pt–adatom hopping on Pt(111) decreases from 0.41 eV to 0.33 eV upon substrate relaxation in the calculations of [86].

The single particle picture definitely breaks down for a class of diffusion processes known as exchange diffusion or concerted motion. Exchange diffusion is sketched for the example of a fcc(100) surface in Fig. 2.13. Instead of performing a hop of length a along one of the $\langle 110 \rangle$-directions, the adatom be-

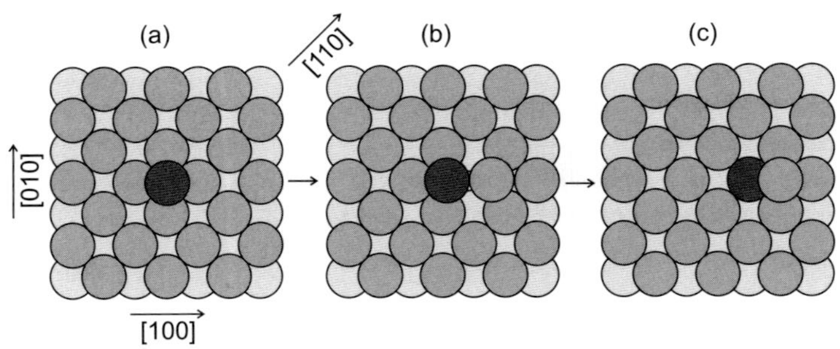

Fig. 2.13. Schematic representation of the exchange diffusion process on a fcc(100) surface. (**a**) An adatom in a fourfold hollow. To begin displacement, the adatom moves down while a substrate atom moves up in a concerted motion. (**b**) In the saddle–point configuration the two atoms are above the surface at equivalent positions with a surface vacancy below. (**c**) To complete displacement, the original adatom fills the vacancy and the displaced surface atom becomes a new adatom. In exchange diffusion displacements occur along $\langle 100 \rangle$ directions and not along $\langle 110 \rangle$ directions as in ordinary hopping diffusion (redrawn after Fig. 15 of [52])

comes a surface atom, and another atom initially located in the surface layer becomes an adatom displaced by a distance of $\sqrt{2}a$ from the initial adatom in one of the ⟨100⟩-directions. For exchange diffusion the substrate takes active part in the diffusion process and it is not just a passive checkerboard defining the allowed lattice sites.

Exchange diffusion on a fcc(100)-surface was predicted theoretically for Al–adatom diffusion on Al(100) [160]. It was confirmed experimentally by FIM for Pt–adatom diffusion on Pt(100) [161] and Ir–adatom diffusion on Ir(100) [162]. The telltale signature of exchange diffusion in these systems is the c(2×2) site visitation map, which comes about because the adatom proceeds only by displacements along the ⟨100⟩-directions. Exchange diffusion allows all atoms to maintain a higher coordination during the migration step, e.g. in the case of interest here the coordination of the saddle–point configuration is three, instead of two for hopping. On the other hand, in the saddle–point configuration *two* atoms are in low coordination for exchange diffusion, in contrast to only one atom for hopping. For Al–adatom diffusion on Al(100) it was argued that the exchange configuration is energetically preferred, because Al–atoms are trivalent [160].

Historically, exchange diffusion was first proposed on the basis of FIM measurements for the cross channel motion of Pt–adatoms on Pt(110) [88]. The first clear evidence for exchange diffusion came from the observation of W–adatom diffusion on Ir(110) [163]. Currently, whenever the activation energy for an atomic diffusion process is considered, it is also tested whether the process is a single particle hopping process or a multi–particle exchange process. To give an example of a complicated exchange process: the descent by exchange of a Pt–adatom at a straight A-type step on Pt(111) from the upper terrace to a position in the step (compare Fig. 2.1) has – according to ab initio calculations [87] – an activation energy of only 0.31 eV. Descent by ordinary hopping has an activation energy of 0.53 eV and is thus irrelevant compared to the exchange process. Similarly, effective medium calculations show that the descent of a Pt–adatom near a kink position is strongly favored to occur by exchange rather than hopping [90]. Experimentally, descent by exchange has been demonstrated for a W–adatom at a B-type step on Ir(111) [164]. We will return to the microscopic mechanisms involved in the descent from steps below in Sec. 4.3.4.

2.4.3 Nucleation and Random Walks

It should have become clear from the discussion in Section 2.3.1 that the celebrated expression (2.18) for the number density of stable islands relies on a rather large number of simplifying assumptions. Most fundamentally, the rate equation nucleation theory by construction ignores the fluctuations and correlations in the spatial arrangement of the clusters, a deficiency which can only partly be cured by a judicious choice of capture numbers. It is therefore important to supplement the classical theory by more direct, microscopic approaches which take into account the stochastic nature of the nucleation process [133].

As an illustration, we present here a simple random walk estimate of the island number density N for the case $i^* = 1$. Starting from the observation that most nucleation occurs in the steady state regime, we consider a situation where adatoms are predominantly captured by islands. The life–time of a freshly deposited adatom is then of the order of the diffusion time $\tau_D \sim 1/(ND)$, and the average adatom density is $n_1 = F\tau_D \sim F/ND$, compare to (2.17). The probability p_{nuc} that an adatom participates in a nucleation event, that is, that it encounters another adatom during its lifetime, is given by the product of Ωn_1 and the number $\mathcal{N}(\tau_D)$ of *distinct* adsorption sites[21] visited in time τ_D. For the latter quantity, random walk theory provides the expression $\mathcal{N} \approx \pi\nu\tau_D/\ln(\nu\tau_D)$ for $\nu\tau_D = 4D\tau_D/a^2 \gg 1$ [165]. The actual value of N is fixed by the requirement that, out of the approximately $1/(\Omega N)$ adatoms deposited in the area occupied by one island, only one (or two!) forms a nucleus, i.e. that $p_{\text{nuc}} \sim \Omega n_1 \mathcal{N}(\tau_D) \sim \Omega N$. This yields finally

$$\hat{N}^3 \ln(1/\hat{N}) \approx \frac{\Omega^2 F}{D}, \tag{2.29}$$

which coincides with the scaling law (2.19) only up to a logarithmic correction; the leading behavior of (2.29) for $D/\Omega^2 F \gg 1$ is

$$\hat{N} \sim (\Omega^2 F/D)^{1/3} [\log(D/\Omega^2 F)]^{-1/3}. \tag{2.30}$$

In a power law fit of N versus F/D this will tend to produce scaling exponents which are smaller than 1/3. For example, for $D/\Omega^2 F = 10^8$, an effective exponent of 0.30 is to be expected [107]. Correspondingly, taking into account the logarithmic factor in an Arrhenius plot of the island density vs. $1/T$ yields a somewhat smaller estimate for the activation energy of surface diffusion; for the data displayed in Fig. 2.8 one finds $E_{\text{D}} = 0.25\,\text{eV}$ instead of $0.26\,\text{eV}$, which is well within the experimental uncertainties.

The logarithmic factor can be reproduced within rate equation theory by using the expression

$$\sigma_1 \approx \frac{4\pi}{\ln[(D/F\Omega)n_1]} \sim -\frac{1}{\ln(\Omega N)} \tag{2.31}$$

for the capture number of adatoms, which is obtained from the uniform depletion approximation [104]. Inserting (2.31) into the prefactor η on the right hand side of (2.18) yields precisely a relation of the form (2.29). Reduced rate equations of the form (2.13), (2.16) which incorporate the logarithmic correction are able to quantitatively reproduce the results of kinetic Monte Carlo simulations [166].

[21] The validity of this statement becomes evident by assuming that all other adatoms are immobile: Then it is clear that repeated visits to the same unoccupied site do not increase the chance for nucleation.

Random walk arguments have been particularly helpful in clarifying the case of one-dimensional diffusion, which is relevant to nucleation on anisotropic surfaces such as fcc(110) and hex-reconstructed fcc(100) [48]. In one dimension the distance between islands is N^{-1}, and correspondingly the adatom diffusion time in the steady state regime is of the order of $\tau_D \sim 1/(N^2 D)$ and the adatom density is $n_1 \sim F/(N^2 D)$. The number of distinct site visited by a one-dimensional random walk grows as $\mathcal{N}(\tau_D) \sim \sqrt{\nu \tau_D}$. Inserting these expressions into the condition $p_{\text{nuc}} \sim N$ yields the estimate $N \sim (F/D)^{1/4}$ for $i^* = 1$ [107, 133].

Similar considerations can be employed to derive a useful general formula for the exponent χ in the scaling law (2.19), which covers all cases *except* $i^* = 1$ in one dimension [167, 168]. It reads

$$\chi = \frac{i^* d_D}{2(i^* + 1) + d_{\text{f}}}. \tag{2.32}$$

Here $d_D = 1, 2$ denotes the dimensionality of diffusion and d_{f} the (fractal or integer) dimensionality of the islands. Logarithmic corrections similar to (2.29) arise whenever $i^* d_D = 2$, because $i^* d_D$ is the effective dimensionality of the random walk in configuration space which describes the nucleation process, and two-dimensional random walks are well known to be marginally space filling [165]. Equation (2.32) implies a larger scaling exponent for fractal islands ($d_{\text{f}} < d_D$) compared to compact islands, which agrees with KMC simulations [100]. To some extent this increase may be compensated by the logarithmic correction in (2.29).

2.4.4 Nucleation with Mobile Clusters

In the derivation of the basic results of nucleation theory in Sect. 2.3.1 it was assumed that all clusters of sizes $s \geq 2$ are immobile. In fact the mobility of small clusters such as dimers is often quite important; it is clear that at temperatures where step atoms are mobile also dimers must be assumed to migrate, since the number of lateral bonds of an atom in a dimer is smaller or equal to the number of lateral bonds of a step atom.

Cluster mobility leads to additional terms of the form $\sigma_{ss'}(D_s + D_{s'}) n_s n_{s'}$ in the rate equations, where D_s is the diffusion coefficient of a cluster of size s, with $D_1 \equiv D$, and $\sigma_{ss'}$ is the capture number associated with the collision between clusters of sizes s and s' [169]. As a simple and practically relevant example (see Sect. 2.3.4), we consider here the case where dimers are mobile but do not dissociate, and clusters of sizes $s \geq 3$ are counted as immobile, stable islands [133]. Then the number density of stable islands grows due to adatom-dimer and dimer-dimer collisions,

$$\frac{\mathrm{d}N}{\mathrm{d}t} = \sigma_{12}(D + D_2) n_1 n_2 + \sigma_{22} D_2 n_2^2. \tag{2.33}$$

Provided $D_2 < D$, as would usually be the case, it can be shown that the second term on the right hand side is negligible compared to the first one [133]. Then (2.33) essentially reduces to the standard nucleation equation (2.13) with $i^* = 2$. To evaluate it, we note that the adatom density n_1 is still given by (2.17) in the steady state regime, where adatoms are captured primarily by stable islands. However, in contrast to the case of unstable dimers ($i^* = 2$), the dimer density n_2 is no longer given by the Walton relation (2.14). Instead, it is determined by the following rate equation:

$$\frac{\mathrm{d}n_2}{\mathrm{d}t} = \sigma_1 D n_1^2 - \sigma_{12}(D + D_2) n_1 n_2 - \sigma_{22} D_2 n_2^2 - \bar{\sigma}_2 D_2 n_2 N. \tag{2.34}$$

Here the first term on the right hand side is the dimer creation rate, the second and third terms describe adatom-dimer and dimer-dimer collisions, and the last term describes the capture of dimers by stable islands, with $\bar{\sigma}_2$ denoting the corresponding mean capture number. In the steady state regime $n_1, n_2 \ll N$, and therefore capture at stable islands is the dominant loss mechanism also for dimers. Setting $\mathrm{d}n_2/\mathrm{d}t = 0$ and equating the first and last terms on the right hand side of (2.34) yields the estimate $n_2 \approx (\sigma_1 D/\bar{\sigma}_2 D_2) n_1^2/N$ for the dimer density. Inserting this into the nucleation equation (2.33) and integrating over time, one then obtains the expression [133]

$$\hat{N} \approx \tilde{\eta}_2 \left(\frac{\Omega^4 F^2}{D_2 D} \right)^{1/5} \Theta^{1/5} \tag{2.35}$$

for the (dimensionless) island number density, where the prefactor is given by $\tilde{\eta}_2 = (5\sigma_{12}\sigma_1/\bar{\sigma}^3\bar{\sigma}_2)^{1/5}$.

These considerations can be generalized to the case where clusters up to and including a critical size i^*_{mob} are mobile, while larger clusters are immobile [170]. If the diffusivities of the mobile clusters are all equal, one obtains a scaling law of the form (2.19) with exponent

$$\chi = \frac{i^*_{\mathrm{mob}}}{2i^*_{\mathrm{mob}} + 1}. \tag{2.36}$$

On the other hand, if clusters of arbitrary size are mobile, with a mobility that decreases with cluster size more rapidly than $D_s \sim 1/s$, then the scaling exponent takes the value $\chi = 1/2$ up to logarithmic corrections [169]. This is consistent with the limit $i^*_{\mathrm{mob}} \to \infty$ of the expression (2.36). According to continuum fluctuation theory, the diffusivity of large islands decays with island size as $D_s \sim s^{-\alpha}$, where $\alpha = 3/2$, 1 and 1/2 for mass transport dominated by edge diffusion, terrace diffusion and attachment/detachment kinetics, respectively [171, 172]. For $\alpha < 1$ the island size distribution is predicted to reach a time-independent power law shape at long times [169].

2.4.5 Adatom–Adatom Interactions and Nucleation Theory

A central assumption of conventional nucleation theory is that adatoms interact only by forming bonds when they are located at neighboring adsorption sites, but move independently otherwise. This is only approximately true, since each adatom perturbs the spatial and electronic degrees of freedom of the substrate, and therefore *substrate mediated adatom–adatom interactions* should be the rule rather than an exception. Electrostatic dipole and elastic interactions are generically present, which decay asymptotically as r^{-3} with the distance r between the adatoms. For metals, interactions of even longer range occur if the substrate possesses a partially filled surface state. Each adatom causes an oscillatory, Friedel–type perturbation in the electron density, and the interference of these perturbations gives rise to an interaction energy of the form [173, 174]

$$E(r) \approx -A\left(\frac{4\epsilon_{\mathrm{F}}}{\pi^2}\right)\frac{\sin(2q_{\mathrm{F}}r + 2\delta_{\mathrm{F}})}{(q_{\mathrm{F}}r)^2}. \tag{2.37}$$

Here ϵ_{F} and q_{F} denote the Fermi energy and Fermi wavevector of the surface band, δ_{F} is the Fermi-level phase shift for the scattering of surface electrons off the adatom, and the dimensionless amplitude $0 < A < 1$ depends on δ_{F} and on the amount of bulk scattering.

Experimentally, the determination of adatom–adatom interactions is based on the probability $P(\boldsymbol{r})$ to find two adatoms in a vector separation $\boldsymbol{r}$ on a surface in thermal equilibrium at a temperature T:

$$P(\boldsymbol{r}) = CP_0(\boldsymbol{r})\mathrm{e}^{-E(\boldsymbol{r})/k_{\mathrm{B}}T}, \tag{2.38}$$

where $P_0(\boldsymbol{r})$ is the probability to find the two adatoms at separation $\boldsymbol{r}$ in the absence of interactions, and C is a normalization constant which accounts for the limited range of distances entering the analysis. Adatom–adatom interactions have long been known in field ion microscopy [52, 175]. For FIM the analysis was limited and complicated by the small size of the apex planes as well as by the influence of the bounding step. Statistics could be improved by measuring heteropairs, if one of the adatoms was immobile and situated at the center of the apex plane. For such heteropairs on W(110) oscillatory, orientation dependent interaction energies were obtained [176]. For example, the interaction between a Pd–W heteropair is strongly repulsive at next nearest neighbor sites with about 40 meV repulsion energy, but rapidly decays to a few meV beyond distances of $3a$.

The same methodology was applied for the determination of the interaction energies between Cu adatoms on a Cu(111) surface by STM [177]. Due to the larger field of view and the absence of steps, much better statistics was achieved and the long-range behavior could be determined. As can be inferred from Fig. 2.14, the adatom–adatom interaction energy oscillates with a wavelength of about half the Fermi wavelength of the electrons in

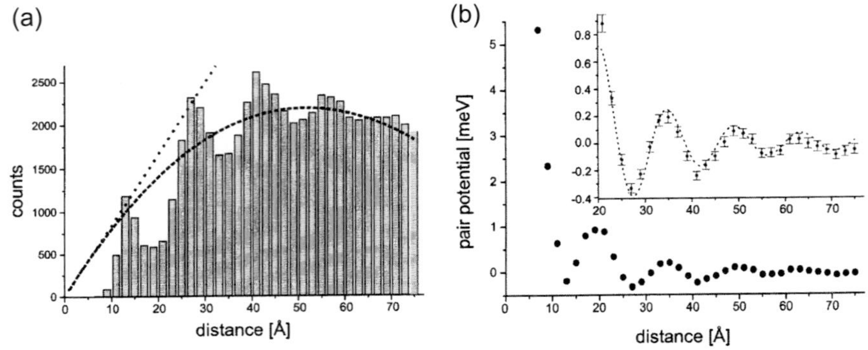

Fig. 2.14. (**a**) Histogram of the distribution of adatom separations determined from a series of STM topographs. The dotted curve displays the expected behavior due to the linear increase of the number of adsorption sites with distance r, and the dashed curve includes the effect of the finite image size. In the absence of adatom–adatom interactions, the measured data would follow the dashed curve. (**b**) Potential energy between two Cu adatoms. The inset shows the behavior for larger separations. The dotted line is a fit based on equation (2.37) (Fig. 2 of [177])

the surface–state band and decays quadratically with r, in agreement with (2.37). The inset in Fig. 2.14 shows a fit to (2.37) for larger distances. While the long range electronic interaction is relatively weak, the repulsion at short distances is dominated by the rapidly decaying electrostatic dipole and elastic interactions. The maximum interaction energy is found at a distance of $2a$ (at a distance of a a dimer is formed) and amounts to about 15 meV. This value was confirmed in recent STM experiments, which in addition found oscillatory adatom–adatom interactions also for Co atoms adsorbed on Ag(111) and Cu(111) [178].

Ab initio calculations for the systems Al/Al(111), Cu/Cu(111) [179] as well as Ag on strained and unstrained Ag(111) [180] indicate the general presence of adatom–adatom interactions for homoepitaxial or quasi–homoepitaxial growth systems. In the calculations the interaction could only be investigated up to a few nearest neighbor distances. For Cu/Cu(111) the maximum of the adatom–adatom repulsion was calculated to be at a distance of $2a$ as in the experiment. However, the calculated magnitude of the repulsion of 46 meV overestimates the experimental finding by a factor of three. For Al/Al(111) the calculated repulsion amounts to 17 meV.

While the long–ranged oscillatory Friedel–type interactions are evidently too small to affect the nucleation behavior noticeably, the effect of the repulsive short range interaction at a distance of $2a$ will depend on the interaction strength in relation to the diffusion barrier for individual adatoms. For the Pt/Pt(111) and Ir/Ir(111) systems, the excellent agreement between the results of the nucleation theory analysis [82, 137] and FIM measurements [67, 181] a posteriori justifies the assumption of vanishing adatom–adatom interactions: at least for these systems the energy scale of the interactions must be very small compared to E_D. In general, however, this need not be correct. In particular for situations with a small adatom diffusion barrier, the additional adatom–adatom repulsion may lead to errors in E_D when standard nucleation theory is applied.

KMC simulations of nucleation in the presence of repulsive adatom–adatom interactions indicate, in addition to an overall increase of the island density due to the suppression of dimer formation, a qualitative change in the temperature dependence of N [180, 183]. At low temperatures just above the onset of adatom diffusion the island density becomes almost independent of T (Fig. 2.15). The simulated island densities are well reproduced by a modified rate equation nucleation theory, in which the forces induced by the long-ranged interactions are included in the self-consistent calculation of the capture numbers [182, 184]. The main effect can be understood within a simplified model which retains a repulsive ring of interaction strength ϵ_R and radius R_{int} around each adatom and leaves E_D unchanged. At temperatures below $(\epsilon_R + E_D)/k_B$ the repulsion cannot be overcome, and no nucleation takes place until the coverage has reached a value of the order of $1/(\Omega R_{int}^2)$, when the deposition of one adatom within the interaction range of another becomes likely. If the adatom mobility is sufficiently high, and the flux is sufficiently low, then one island forms within each interaction zone and the island

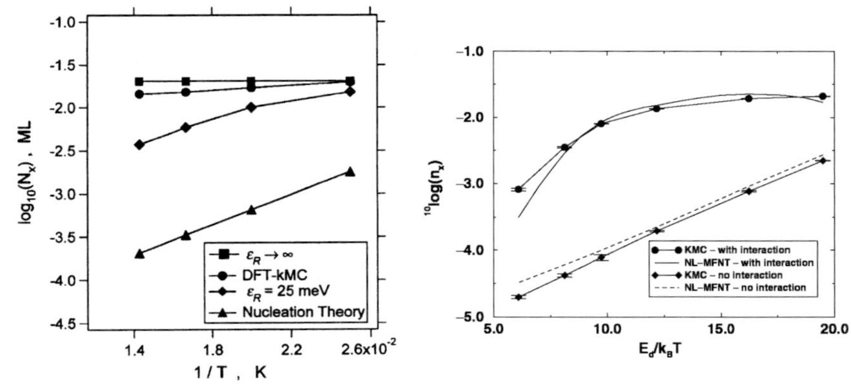

Fig. 2.15. *Left panel*: Arrhenius plot of the island number density obtained from KMC simulations of Ag on strained Ag(111). The figure shows data from simulations including different kinds of long ranged adatom–adatom interactions: Interactions obtained from a full DFT calculation (*circles*), an impermeable repulsive ring around each adatom (*squares*), and a repulsive ring with an energy barrier $\epsilon_R = 25$meV (*diamonds*). The ring radius is assumed to be $5a$. The triangles show the island density expected from nucleation theory in the absence of repulsive interactions (Fig. 2 of [180]). *Right panel:* Comparison of KMC simulations and self-consistent rate equations with (upper data sets) and without (lower data sets) long ranged interactions. The interaction potential is assumed to be dominated by elastic contributions, and decays asymptotically as r^{-4}. The island number density is shown as a function of E_D/k_BT, at a coverage of 0.05 ML and a flux of 0.03 ML/s (Fig. 2 of [182])

density is of the order of $1/R_{\text{int}}^2$ independent of T. This regime is bounded towards high temperatures by the onset of hopping across the repulsive barrier.

The repulsive ring scenario, and the resulting change in the island density curve, is similar to nucleation in the presence of trapping impurities, see Sect. 2.4.6. If the underlying mechanism is not recognized, in both cases the naive interpretation of the flat region in the Arrhenius plot of N vs. $1/T$ yields an anomalously small estimate for the diffusion barrier, accompanied by an anomalously small preexponential factor (see Fig. 2.19). Indeed several authors [179, 180] have suggested that the anomalous attempt frequencies obtained from nucleation experiments for Al on Al(111) [185] and Ag on strained Ag(111) [48] with a small diffusion barrier E_{D} may be attributed to adatom–adatom interactions. At least for the case of Al on Al(111), however, interaction with impurities was instead demonstrated to be the origin of the measured anomalously low attempt frequency [186, 187].

While the adatom–adatom interactions discussed above mostly affect the early time nucleation regime, where dimer formation is the dominant aggregation mode, in the late time saturation regime possible interactions between the adatoms and the edges of stable islands become important. The effect of repulsive energy barriers suppressing the attachment of adatoms to stable islands has been analyzed within rate equation theory [184, 188]. If the attachment probability is so small, that each adatom visits many islands before being incorporated [189], a new scaling regime with an island density exponent $\chi = 2i^*/(i^* + 3)$ in (2.19) is found. In this case the failure to recognize such an effect in a nucleation theory analysis would lead to an *overestimate* of both the adatom diffusion barrier, and of the preexponential factor.

2.4.6 Adsorbate Influenced Diffusion and Nucleation

In the early phase of epitaxial growth experiments under high or ultra high vacuum conditions one was well aware of the disturbing effects of impurities and adsorbates [190]. Varying impurity concentrations from one experiment to another may obviously create serious reproducibility problems. A few years ago the concern about the role of trace amounts of impurities, which were regarded hitherto as irrelevant, was raised again, and earlier experiments that had been interpreted as being representative of clean growth were questioned [131]. The experimental and theoretical discussion on the influence of adsorbates on adatom diffusion and nucleation is still moving rapidly, and here only two established effects will be presented.

The Sky Hook Mechanism. This phenomenon was first considered as a possible explanation for the drastic lowering of the onset temperature of Ni adatom diffusion on Ni(110) from 150 K to about 80 K in the presence of hydrogen, as measured by FIM [192]: A hydrogen atom bound to the adatom is likely to lower the adatom bonding to the substrate, thereby reducing also the corrugation of the potential energy surface and the adatom diffusion barrier. In an ab initio calculation of H adsorption on Be(0001) the existence of such a mechanism of adatom diffusion enhancement was confirmed [193]. Hydrogen binds strongly to the top of a Be adatom, which is the preferred binding site at the adatom. This bond weakens the bonds of the Be adatom to its Be(0001) surface neighbors, as signaled by a bond length increase of 10% and a lowering of the adatom diffusion barrier by a factor of three, from 0.06 eV to 0.02 eV. However, only H–coverages above 1/15 ML will make this mechanism operative, as only then the H adsorption on the adatom becomes energetically favorable compared to the terrace.

Experimentally the sky hook mechanism was observed for room temperature diffusion of Pt adatoms on Pt(110) [191]. In sequences of STM images tracing adatom diffusion along the troughs of the reconstructed surface, occasionally a Pt adatom was observed to light up for some time interval, during which its diffusion coefficient increased by a factor of 500 (Fig. 2.16). Based on the correlation between the hydrogen background pressure and the frequency of appearance of bright atoms, and in agreement with ab initio calculations, it was concluded that hydrogen adsorbed on top of the Pt adatom reduces the activation energy for diffusion by 0.16 eV.

Speculating, one may go one step further and consider the effects of adsorbates on top of small clusters. One possibility is a weakening of the bonds between the cluster and the substrate, which enhances cluster mobility. Another conceivable effect would be a weakening of the bonds *within* the cluster, giving rise to a reduced cluster binding energy. Ab initio calculations show that, for a Pt dimer on Pt(110), a CO molecule at its favorite adsorption site on-top of one of the dimer adatoms reduces the dimer binding energy by a factor of three from 0.21 eV to 0.07 eV [124]. Preliminary experimental results for island size distributions on Pt(110) support these ideas [123]. All three effects – the adatom sky hook mechanism, its analogue for clusters, and the adsorption-induced weakening of cluster binding energies – will tend to

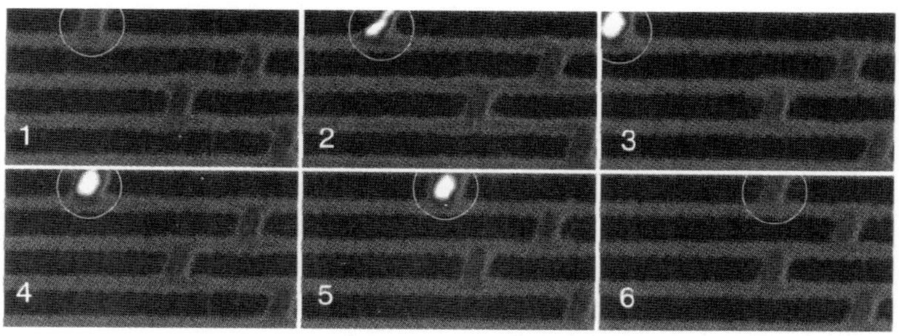

Fig. 2.16. Six STM images (56 Å × 31 Å) extracted from an STM movie. The sequence shows the transformation of a normal Pt adatom into a bright Pt–H intermediate complex, and back to a normal Pt adatom (the atom of interest is marked by a white circle). The STM images were obtained with a tunneling resistance above 100 MΩ, for which the tip influence on diffusion was found to be negligible (Fig. 2 of [191] reprinted by permission from Nature Macmillan Publishers Ltd.)

reduce the saturated island number density N, as is evident from (2.18). The observation that a small CO background pressure (1×10^{-9} mbar) during Pt deposition on Pt(111) gives rise to a slight decrease in N by a factor of about 1.5 [131] thus suggests that one of these mechanisms may be operative in the Pt/Pt(111) system[22].

Impurity Trapping. Foreign substitutional atoms in the surface layer or adsorbate molecules on the surface may act as trapping centers for adatoms, if (i) the impurity has a lower mobility than the adatom and (ii) if the adatom binds to the impurity. An example for a substitutional trap is an Ir atom in the surface layer of Rh(001), which, at 260 K, confines the random walk of a Rh adatom to the four lattice sites adjacent to the impurity [195]. By a statistical analysis the additional binding energy of the adatom to the impurity was found to be $E_\mathrm{t} = 0.12\,\mathrm{eV}$.

An example for the effect of trapping on the temperature dependence of island number densities is given in Fig. 2.17 for the deposition of Pt on Pt(111). The island density obtained under nominally clean deposition conditions (full squares, data shown also in Fig. 2.9) is compared to a contaminated series (open circles), where deposition was performed with a low rate and at a high background pressure [70]. The higher island densities in the contaminated case are attributed to codeposition of trapping impurities[23]. The contaminated series exhibits initially a much slower decrease of island density with temperature due to efficient trapping, followed by a high temperature regime with a large Arrhenius slope, which may reflect either detrapping or the onset of adsorbate mobility (see below).

If the small slope data for the contaminated case are used to determine the adatom diffusion parameters, both the activation barrier and the attempt frequency are severely underestimated: One obtains $E_\mathrm{D} = 0.11\,\mathrm{eV}$ and $\nu_0 = 10^4\,\mathrm{s}^{-1}$ compared to $E_\mathrm{D} = 0.26\,\mathrm{eV}$ and $\nu_0 = 5 \times 10^{12}\,\mathrm{s}^{-1}$ in the clean case. A similar situation is encountered for Al/Al(111), where the anomalously small preexponential factor of $\nu_0 = 8 \times 10^6\,\mathrm{s}^{-1}$ [185] was attributed to adsorbate trapping under poor deposition conditions [186]. In fact, as is explained below in Fig. 2.19, the occurrence of anomalously small barriers due to impurity trapping is always coupled to anomalously small prefactors. The island number density is generally sensitive to trapping when E_D is small. Then at moderate temperatures extremely small saturated island number

---

22 There is a strong correlation between the adsorption energy of a CO molecule on a Pt surface and the coordination number of the adsorption site, with low coordinated sites being most attractive [194], see Sect. 3.5. This supports the idea of CO-decorated Pt adatoms and clusters.

23 The adsorbate species leading to trapping was not be identified unambiguously, but CO is a likely candidate, as it is mainly responsible for the pressure rise during Pt deposition, adsorbs readily and becomes mobile between 160 K and 190 K [196]. At the temperatures of interest it could act as a trap, as long as the adsorption place exchange between the positions on the terrace and on top of the adatom is still suppressed.

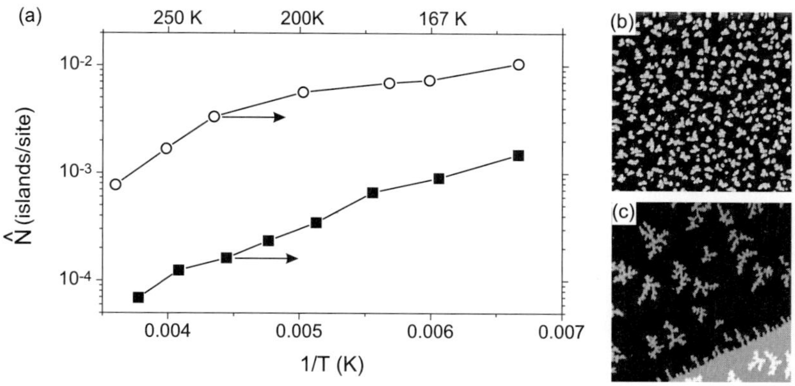

Fig. 2.17. (**a**) Platinum island number densities on Pt(111). Full squares: STM data [82] represented in Fig. 2.8, $F = 6.6 \times 10^{-4}\,\mathrm{ML\,s^{-1}}$. Open circles: STM data from a deposition sequence with high background pressure during deposition ($p \approx 1 \times 10^{-9}\,\mathrm{mbar}$) and $F = 3.3 \times 10^{-4}\,\mathrm{ML\,s^{-1}}$ [70]. Island densities normalized to $F = 6.6 \times 10^{-4}\,\mathrm{ML\,s^{-1}}$. (**b**) STM topograph after deposition at 230 K, image size 960 Å × 960 Å, contaminated deposition.(**c**) STM topograph after deposition at 225 K, image size 960 Å × 960 Å, nominally clean deposition. The arrows at the data points in (**a**) point to the corresponding STM topographs

densities result under clean deposition conditions (e.g. $N < 10^{-4}$ islands/site for Pt/Pt(111) at 400 K), and adsorbates with a similar concentration which are able to trap adatoms will significantly alter the nucleation behavior.

More detailed insights into the effect of impurity trapping can be gained from kinetic Monte Carlo simulations, in which the adsorbate atoms are included as a second (mobile or immobile) surface species [198]. For comparison with Fig. 2.17, Fig. 2.18 shows the temperature dependence of the island density for the case of predeposited, immobile adsorbates. Varying the binding energy E_t and the coverage Θ_t of the impurities, one finds generically[24] an island density plateau in a temperature range $T_1 < T < T_2$, where N is independent of T and roughly equal to the trap concentration Θ_t. In the plateau regime a freshly deposited adatom is sufficiently mobile to find the nearest trap prior to the arrival of the next adatom within the capture zone of the trap. In addition the trapping is strong enough to keep the first adatom fixed until the second arrives, such that one stable nucleus forms around each impurity. For $T < T_1$ the first condition is no longer satisfied; most nucleation then occurs away from the impurities, and the island density takes on a value $N > \Theta_\mathrm{t}$ characteristic of clean growth. On the other hand, for $T > T_2$ detachment from impurities becomes relevant. Each adatom then visits many traps before joining a cluster, and the effect of the impurities is limited to a decrease of the adatom diffusivity due to temporary trapping. In this regime standard

[24] The existence of a plateau depends on deposition flux and impurity coverage through the condition $(\nu_0/\Omega F)^\mu > \Theta_\mathrm{t}^{-(1+2\mu)}$, where $\mu = E_\mathrm{t}/E_\mathrm{D}$ [197].

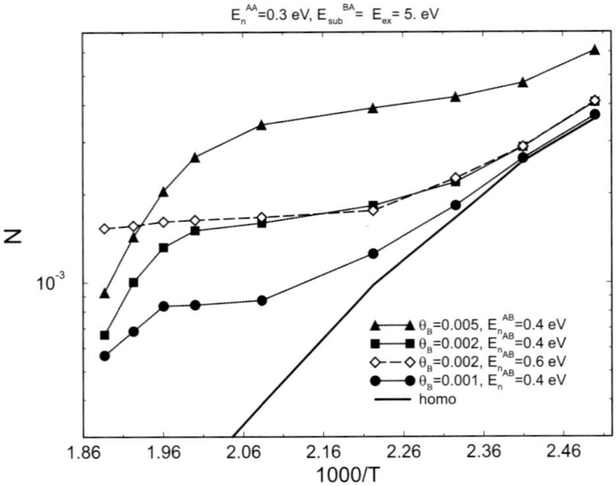

Fig. 2.18. Island number density as a function of temperature, obtained from KMC simulations of a two–species growth model. Atoms are deposited onto a square lattice, where immobile adsorbates have been predeposited at a coverage Θ_{t}. Adatoms interact through a bond–counting energy expression, in which each lateral nearest neighbor bond contributes a pair bond energy of 0.3 eV. The diffusion barrier of an isolated adatom is $E_{\mathrm{D}} = 0.8\,\mathrm{eV}$, the attempt frequency is set to $10^{13}\,\mathrm{s}^{-1}$, and the flux is 0.001 ML/s. Full symbols show data for a trapping energy $E_{\mathrm{t}} = 0.4\,\mathrm{eV}$ and different impurity coverages, while the open diamonds show data for $E_{\mathrm{t}} = 0.6\,\mathrm{eV}$. The bold line depicts the island density for clean growth ($\Theta_{\mathrm{t}} = 0$) (Fig. 2 of [197] reprinted with kind permission of Kluwer Academic Publishers)

nucleation theory applies with a modified effective diffusion coefficient of the form [102, 198]

$$D_{\mathrm{eff}} = \frac{D}{1 - \Theta_{\mathrm{t}} + \Theta_{\mathrm{t}} e^{E_{\mathrm{t}}/k_{\mathrm{B}}T}}. \tag{2.39}$$

This expression defines a third characteristic temperature $T_3 > T_2$ through $\Theta_{\mathrm{t}} e^{E_{\mathrm{t}}/k_{\mathrm{B}}T_3} \approx 1$. For $T < T_3$ the effective diffusion coefficient displays Arrhenius behavior with an activation energy $E_{\mathrm{t}} + E_{\mathrm{D}} > E_{\mathrm{D}}$, which accounts for the larger slope in the island density plots in Figs. 2.17 and 2.18, whereas for $T > T_3$ the traps become irrelevant and the island density reverts to the clean growth behavior. The three regimes are summarized in Fig. 2.19, which also illustrates a general coupling between anomalous apparent energy barriers and anomalous apparent attempt frequencies.

Simulations with mobile impurities show a similar but less pronounced increase of the island density; in particular no clear plateau regime seems to develop [198, 199]. In general, mobile impurities are less efficient in promoting nucleation, because they are depleted by capture at stable islands in the later stages of growth. The modeling of impurity trapping effects within rate equation theory is possible, if the areal concentration of free traps and the concentration of trapped critical nuclei are included in addition to the densities of adatoms and stable islands [197, 200, 201].

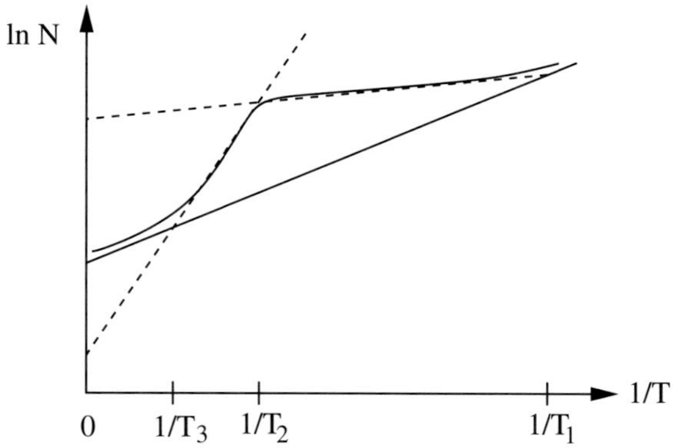

Fig. 2.19. Schematic of the effect of trapping impurities on the island number density N. Upper and lower full curves show the island density in the presence and in the absence of impurities, respectively. The three temperatures T_1, T_2 and T_3 delimit the regimes described in the text. The dashed lines illustrate the erroneous interpretation of the temperature dependence of the island density in terms of standard nucleation theory: The slope of the Arrhenius plot is underestimated for $T_1 < T < T_2$ and overestimated for $T_2 < T < T_3$. An underestimation of the slope is always accompanied by an overestimation of the intercept, and vice versa. Consequently small (large) apparent barriers imply small (large) apparent preexponential factors

Finally also repulsive impurities, i.e. adsorbates that have a repulsive interaction with the deposited adatoms, lead to enhanced island number densities and different scaling regimes for N [125]. Examples for systems with repulsive impurities are homoepitaxial growth on Al(111) in the presence of O adsorbates [202] and homoepitaxial growth on Ag(111) in the presence of Sb adsorbates [203, 204]. These examples are briefly discussed in Sect. 3.5.

3. Island Shapes

Haben Sie schon gesehn in was für Figuren
die Schwämme auf dem Boden wachsen?
Wer das lesen könnt.[1]
Georg Büchner, Woyzeck

In the last chapter we explained how the adatom diffusion barrier and cluster binding energies determine the number density of adatom islands. To understand the *shape* of an island it is not sufficient to consider these quantities. Two additional factors come into play: the atomic processes occuring at the island edge, and the effects of the local environment of the island on the diffusion field of the adatoms.

3.1 Island Shapes and the Hierarchy of Diffusion Processes

The formation of homoepitaxial adatom islands can be described as an essentially two dimensional (2D) problem. All islands shown in Fig. 3.1 are of single atomic layer height. For an island formed after deposition of 0.15 ML, on average only 7% of the island atoms arrived on top of it and were incorporated by an inter-layer process at the island edge, while all other atoms arrived on the substrate layer and were attached when the adatom random walk encountered the island boundary. The fraction of island atoms originating from the island top may be further reduced by reducing the deposited amount, without affecting the qualitative features of the island shape at a given temperature[2]. Therefore, to a good approximation the adatom supply from the top of the island may be neglected as a shape forming factor.

The two dimensional island shape problem is a 2D analog to the three dimensional (3D) crystal growth problem. In both cases the crystals grow by the transport of particles from the surrounding phase (fluid or gas), which is

[1] "Have you ever noticed how the toadstools grow in patterns? If we could only read them!" [205]

[2] For example, island shapes after deposition of 0.04 ML at 600 K (where less than 2% of the atoms arrive on the island top) are qualitatively identical to those after deposition of 0.15 ML (compare Fig. 1d and Fig. 4a of [206]).

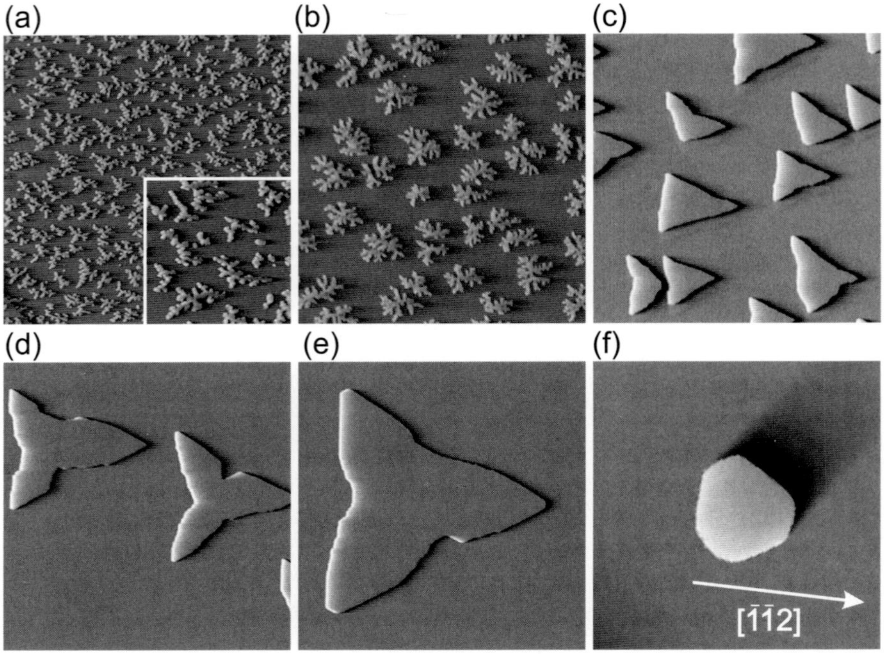

Fig. 3.1. STM topographs of adatom islands on Pt(111) after deposition of Pt at (**a**) 200 K, (**b**) 300 K, (**c**) 400 K, (**d**) 500 K, (**e**) 600 K and (**f**) 700 K. The deposited amount is 0.15 ML in (**a**)–(**e**) and 0.08 ML in (**f**), the deposition rate is 7×10^{-3} ML/s in (**a**)–(**e**) and 2.7×10^{-3} ML/s in (**f**). The topograph size is always 1560 Å× 1560 Å

driven by concentration gradients around the crystal. The crystal shapes reflect the symmetry of the 2D or 3D crystal lattice and the resulting energetics at the phase boundaries (1D steps or 2D surfaces). In addition, the conditions of supersaturation (or undercooling) and the diffusion processes at the phase boundaries are encoded in the shape, be it a snow flake or a platinum adatom island. In this sense "a snowflake is a letter to us from the sky" [207] and an adatom island shape is a letter from the atoms. The purpose of this chapter is to learn to decipher their handwriting.

Indeed, the shapes of Pt adatom islands grown in the temperature range between 200 K and 700 K on Pt(111) exhibit a dramatic variability (Fig. 3.1). At 700 K and above the 2D equilibrium shape, a hexagon with threefold symmetry, is realized (Fig. 3.1f). This shape is determined entirely by the energetics of the phase boundary between the crystal and the surrounding 2D adatom lattice gas. The island boundary consists predominantly of B-steps, which form almost straight 1D facets, and shorter A-steps[3]. In Sect. 3.2 the conditions for equilibrium shape formation will be discussed, and the question

[3] See Fig. 2.1 to recall the definition of A- and B-steps.

of how quantitative information on the step energetics can be obtained from a shape analysis will be answered.

At the low temperature end of the shape manifold, at 200 K, fractal-dendritic islands are formed (Fig. 3.1a and b, compare also Fig. 2.7). These islands have the branched ramified structure commonly associated with fractals. Their formation is dominated by noise, and their shape and Hausdorff dimension are similar those of the structures obtained by simulations of diffusion limited aggregation (DLA); in this simple growth model, particles deposited one–by–one far from the aggregate perform a lattice random walk until they stick irreversibly to the aggregate boundary at their point of impact [36, 208, 209]. The ramified structure results because the protruding branches screen the inner parts of the aggregate from the diffusion field, thus capturing the majority of the particles. In contrast to the (largely isotropic) DLA fractals, however, the Pt islands display preferential directions of branch growth in the $[\bar{1}\bar{1}2]$, $[\bar{1}2\bar{1}]$ and $[2\bar{1}\bar{1}]$ directions. Such preferential growth directions reflecting the crystal symmetry are a well-known feature of *dendrites*, the tree-like structures developing in solidification, with branches strictly aligned along the crystal axes of most efficient heat transport [210]. Details of the formation and growth of fractal-dendritic islands are discussed in Sect. 3.3.

In the intermediate temperature range between 400 K and 600 K (Figs. 3.1c–e) compact, triangular islands form. On the one hand, this island shape may be viewed as a kinetic exaggeration of the equilibrium shape: The facets which are most prominent in the equilibrium shape are kinetically preferred and suppress other facets and step orientations[4]. Alternatively, the triangular shapes may be viewed as a high temperature reminiscence of the fractal-dendritic islands: They still have the same three preferential growth directions, but the branches are thickened to such an extent that the islands become compact. From this point of view the concave parts on the sides of the triangles are remains of the fjords in the branched low temperature island shape. It is interesting to note that the concavity is more pronounced in the islands grown at 500 K (Fig. 3.1d) than at 400 K (Fig. 3.1c); a possible mechanism for this "reentrant" shape evolution is described in Sect. 3.7.3. The formation of compact islands, the origin of the concave shapes, and the sensitivity of compact islands to the growth conditions are the topics of Sects. 3.4, 3.5 and 3.6.

To a first approximation, the variation of island shapes with temperature can be related to the hierarchy of the diffusion processes at the island edge. The islands presented in Fig. 3.1 are grown on the time scale of seconds. Thermally activated processes at the step edge may be considered to be relevant for the shape formation, if they occur on this time scale, i.e. if $\nu_{\text{process}} \geq 1\,\text{s}^{-1}$. Assuming a process independent effective attempt frequency

[4] The kinetic exaggeration of the equilibrium shape by growth shapes is also observed in 3D crystal growth [211, 212]. In fact both kinds of shapes can be described within a similar formalism, see Sect. 3.4.

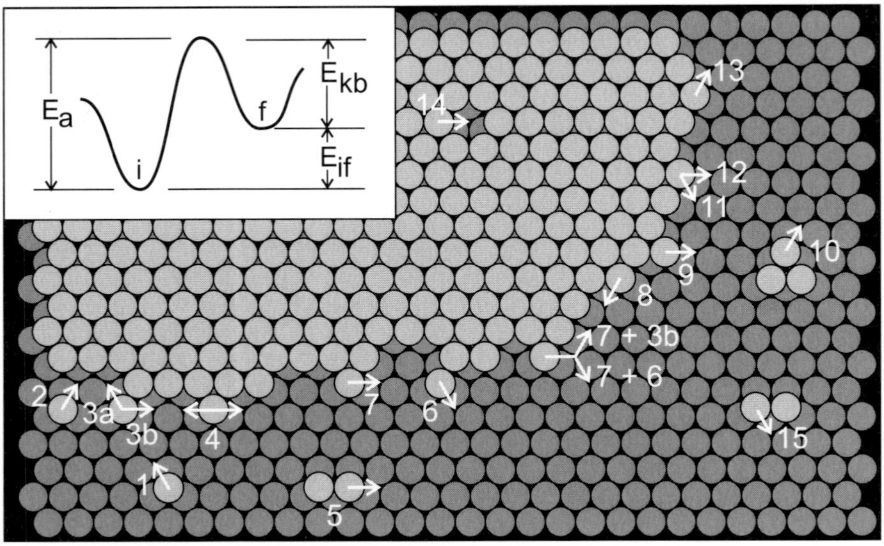

Fig. 3.2. Ball model sketch of in-layer atomic processes on a fcc(111) surface. **1** adatom diffusion, **2** attachment of adatom to step edge, **3a** diffusion of atom at a corner into the kink position, **3b** diffusion of atom at a corner to step adatom position, **4** diffusion of step adatom along step edge, **5** dimer dissociation, **6** detachment of atom at corner, **7** diffusion of step adatom to position at a corner, **7+6** detachment of step adatom via corner, **7+3b** diffusion of step adatom around corner, **8** step adatom generation, **9** detachment of step adatom, **10** trimer dissociation, **11** diffusion of kink atom to corner site , **12** kink detachment, **13** corner break-up, **14** surface vacancy diffusion, **15** dimer diffusion. Inset: Potential energy diagram for a diffusion process between nn-sites. The total activation energy E_{a} is the sum of the energy difference E_{if} between the initial position i and the final position f, and the kinetic barrier E_{kb}

$\nu_0 = 5\times10^{12}\,\mathrm{s}^{-1}$, the scale of activation energies of the different processes can be related to the temperature scale via the universal Arrhenius law (2.4). The onset temperature for a process with activation energy E_{a} is then estimated to be

$$T_{\mathrm{onset}} = 400\,\mathrm{K/eV} \times E_{\mathrm{a}}. \tag{3.1}$$

The hierarchy of kinetic processes at the island edge can be conveniently organized within a simple nearest neighbor model (nn-model) for the activation energies. As illustrated in the inset of Fig. 3.2, the total activation barrier E_{a} of a process is subdivided into two parts: the energy difference E_{if} between the initial state i and the final state f, and the kinetic barrier E_{kb}. If one assumes the kinetic barrier to be proportional to the state of minimal (either final or initial) nn-coordination and the energy difference to be proportional to the difference in nn-coordination of inital and final state, one arrives at

$$E_{\mathrm{a}} = n_f E_{\mathrm{k-nn}} + (n_i - n_f) E_{\mathrm{nn}} \text{ if } n_i \geq n_f \tag{3.2}$$

Table 3.1. Table of activation energies E_a of in-layer diffusion processes on Pt(111) given in eV. The processes are visualized in Fig. 3.2 and numbered accordingly. The change in coordination due to the jump process is indicated in the second column. The third column values (model) are calculated according to the nn-model discussed in the text, the DFT and EMT values are the results of density functional theory (DFT) calculations in the local density approximation, and effective medium theory (EMT) calculations, respectively. The experimental values given are determined by FIM or STM.

Process	i→f	Model	DFT (LDA) A	DFT (LDA) B	EMT [213] A	EMT [213] B	Experiments A	Experiments B
adatom diffusion, **1**	3→3	0.3	0.29[87]		0.16		0.26[67, 82]	
step attachment, **2**	3→5	0.3	$(0.29)^{2}$		0.12–0.15			
corner to kink, **3a**	4→6	0.4	0.45	0.39[140]	0.24^{1}	0.19^{1}		
corner to step, **3b**	4→5	0.4	0.40	0.49[140]	0.24^{1}	0.19^{1}		
step edge diffusion, **4**	5→5	0.5	0.84	0.90[140]	0.45	0.4	0.69 0.60[214]	0.84 [88]
dimer dissociation, **5**	4→3	0.8	0.81[134] 0.79[130]				0.49[83] 0.88[130]	
detachment from corner, **6**	4→3	0.8	$(0.79)^{2}$		0.5–0.6			
step to corner, **7**	5→4	0.9	0.99	0.90[140]	0.4–0.45			
around corner, **7+3b**	5→4 →5	0.9	0.99		0.4–0.45			
step adatom generation, **8**	6→5	1.0	(1.26	$1.40)^{2}$	0.63–0.65		$\leq 1.50^{3}$ [215]	
step adatom detachment, **9**	5→3	1.3	$(1.28)^{2}$		0.71–0.74			
step adatom detachment (corner), **7+6**	5→4 →3	1.3	$(1.28)^{2}$		0.71–0.74			
trimer break-up, **10**	5→3	1.3	$(1.28)^{2}$				$\geq$ 1.19[130]	
kink to corner, **11**	6→4	1.4	1.32	1.39[140]	0.63–0.65			
kink detachment, **12**	6→3	1.8	$(1.72)^{1,2}$		$\geq$0.96			
corner break-up, **13**	6→4	1.4	(1.32	$1.39)^{2}$	0.63–0.65			
vacancy diffusion, **14**	8→8	0.8						
dimer diffusion, **15**	4→4	0.4	0.37[134]				0.37[67]	

[1] Averaged value.
[2] Estimated on the basis of references [87, 130, 140] assuming energies and barriers to be due only to the nearest neighbor environment.
[3] From time-dependent step edge fluctuations, see Sect. 3.7.1.

$$E_a = n_i E_{k-nn} \text{ if } n_i \leq n_f, \tag{3.3}$$

where E_{k-nn} is the energy per coordination in the kinetic barrier, E_{nn} is the nn-bond strength and n_i and n_f are the number of initial and final state nearest neighbors, respectively. Table 3.1 collects the activation energies obtained from this model using the values $E_{k-nn} = 0.1\,\text{eV}$ and $E_{nn} = 0.5\,\text{eV}$. They are compared to available experimental activation energies, as well as to estimates calculated by DFT in the local density approximation and by effective medium theory (EMT). The processes are visualized in Fig. 3.2. It is apparent that the simple nn–model adequately reproduces the hierarchy of diffusion processes and the activation energies as obtained from experiments and density functional theory. While the nn–model, ab initio calculations and experiments roughly match, the EMT–energies and onset temperatures are smaller by about a factor of two.

Summarizing the content of the table, we may conclude that at low temperatures (below 200K/350K according to the nn–model/DFT–calculations) step edge diffusion is suppressed, causing branched shapes. Full mobility along the step edge, including motion around corners, is possible at intermediate temperatures above 360 K (400 K) according to the nn–model (DFT). Thermally activated shape changes in the *absence* of adatom supply during growth will set in with the formation of mobile species at the step edge, i.e. with step adatom creation at kinks (process 8 in Fig. 3.2). Phase equilibrium between the adatom lattice gas and the solid island phase is eventually reached, if atoms are effectively exchanged between the two phases. Direct detachment from kinks (process 13) guarantees this equilibrium at an onset temperature of 720 K in the nn–model.

Whereas the nn–model produces identical energy values and onset temperatures for the processes at the two different step edges, in the DFT– and EMT–calculations it becomes apparent that the energetics at the two types of dense packed steps is slightly different. These small energy differences explain the breaking of the sixfold symmetry of the surface layer to the reduced, threefold symmetry of the island shapes. In the next section these, obviously relevant, energy differences for atomic processes at the two types of dense packed steps will be traced back to a difference in their step free energies.

3.2 The Equilibrium Shape of Adatom and Vacancy Islands

3.2.1 Definition and the Wulff Construction

The shapes of objects in our surrounding everyday world are usually trapped in metastable non-equilibrium states. Upon equilibration, amorphous objects would turn into spheres and polycrystalline objects would recrystallize into single crystals, taking on polyhedral shapes with rounded parts between the high symmetry facets. Generally, the surface energies governing shape changes are small compared to the other energies involved in the formation of useful objects, thus guaranteeing the stability of man-made artifacts. It is only when the objects become small and the fraction of surface atoms becomes significant, that surface energies start to play an important role. In fact adhesion phenomena are some of the most cumbersome problems in micro-mechanics [216].

For small particles and under conditions allowing sufficient material transport, the equilibrium shape may be reached experimentally. An example are small metal crystallites close to their melting temperature in equilibrium with their vapor, which rest on a non-wetting support [211, 217, 218]. The equilibrium shape is defined as the shape of minimum total surface free energy under the condition of fixed particle number and volume [219]:

$$\int_{\mathcal{S}} \gamma(\theta, \phi) \mathrm{d}s = \min! \tag{3.4}$$

In this equation $\gamma(\theta, \phi)$ is the specific surface free energy, i.e. the surface free energy per unit surface area, for a surface orientation parametrized by the angles θ and ϕ. The integral is taken over the entire surface $\mathcal{S}$ of the crystal. The solution of the variational problem (3.4) is the equilibrium shape attained in the limit of large particle numbers, as only then shape fluctuations become negligible and finite size effects disappear [220]. It is uniquely determined by the state variables p, T and the material parameter $\gamma(\theta, \phi)$, and is independent of history. As a consequence of uniqueness, the equilibrium crystal shape is identical to that of a pore in the same material (the hollow shape).

In perfect analogy to the 3D case, the 2D equilibrium shape of an adatom or vacancy island is defined by the minimum of the total *step free energy* for an island of fixed particle number and area:

$$\int_{\mathcal{L}} \delta(\phi) \mathrm{d}l = \min! \tag{3.5}$$

Here, $\delta(\phi)$ is the step free energy per unit length, and the integral is taken over the entire edge length $\mathcal{L}$.

Equilibrium shapes are of great interest, because they provide direct access to surface (3D case) or step edge (2D case) energetics. The link between shape and energetics is established by the Wulff theorem [221]. It was first proved by Burton, Carbrera and Frank [22] for the 2D case of interest here; a short derivation is given in [222]. The geometric interpretation of the Wulff theorem is the Wulff construction, which is defined as follows: At each point of the polar plot of the step free energy $\delta(\phi)$ (*delta*-plot), a straight line perpendicular to the radius vector to that point is drawn; the inner envelope of the resulting family of lines is then geometrically similar to the equilibrium shape. The Wulff construction is equivalent to the analytic expression [222]

$$r(\phi) = \min_{\vartheta} \frac{\varrho\, \delta(\vartheta)}{\cos(\vartheta - \phi)} \tag{3.6}$$

for the shape in 2D polar coordinates (r, ϕ), where the scale factor ϱ sets the size of the island, and the minimization over the auxiliary angular variable ϑ

corresponds to the construction of the inner envelope. From (3.6) the simple relationship

$$\tilde{\delta}(\phi)\kappa_{\rm st}(\phi) = 1/\varrho = \text{const.} \tag{3.7}$$

can be derived, where $\kappa_{\rm st}(\phi)$ is the local curvature of the shape, and the quantity

$$\tilde{\delta}(\phi) = \delta(\phi) + \frac{d^2\delta}{d\phi^2} \tag{3.8}$$

is the *stiffness* of the step edge [172]. Equation (3.7) is a local equilibrium condition which states that the step chemical potential $\mu_{\rm st}$ is constant along the island edge. Indeed, in complete analogy to the Gibbs–Thomson relation for curved droplets [222], the chemical potential at a curved step deviates from its equilibrium value for a straight step by the amount

$$\Delta\mu_{\rm st} = \Omega\tilde{\delta}\kappa_{\rm st}. \tag{3.9}$$

This identifies the scale factor in (3.6) as $\varrho = \Omega/\Delta\mu_{\rm st}$.

If the equilibrium shape is smooth, without any sharp corners, the Wulff construction can be inverted and the entire orientation dependence of the step free energy can be obtained from the shape. In fact, from general arguments on the absence of long range order in one dimension, it may be argued that steps form no true facets and that 2D equilibrium shapes should always

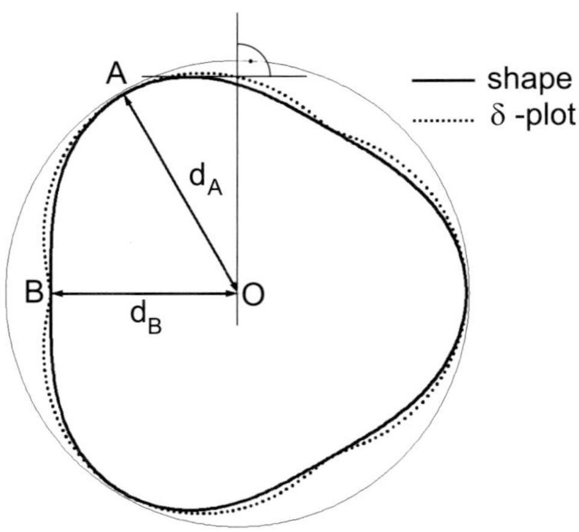

Fig. 3.3. Schematic of the inverse Wulff construction. From the equilibrium island shape (*full line*), a curve geometrically similar to the polar plot of the step free energy δ is constructed (*dotted line*). The δ-plot is obtained by connecting all points, where radial lines through the origin O intersect at right angles with tangents to the equilibrium shape. At directions corresponding to mirror axes of the shape either minima (B) or maxima (A) of δ are found (based on Fig. 3 of [223])

be smoothly curved at any finite temperature[5] [219]. The inverse Wulff construction, i.e. the reconstruction of the δ-plot from the equilibrium shape, is visualized in Fig. 3.3 for a threefold symmetric equilibrium island shape, which is the expected symmetry for an island on a (111)-surface[6]. The intersection at right angles of a tangent to the equilibrium shape and a radial line constitutes a point of the δ-plot. It is evident from the figure that a direction where a mirror axis intersects the threefold symmetric equilibrium shape corresponds either to a minimum (indicated by B in Fig. 3.3) or a maximum (indicated by A) of the step free energy. The step free energy ratio for these specific directions is obviously given by the ratio of the distances $d_{\mathrm{A,B}}$ of the step positions from the island center,

$$\delta_{\mathrm{B}}/\delta_{\mathrm{A}} = d_{\mathrm{B}}/d_{\mathrm{A}}. \tag{3.10}$$

For the example shown in Fig. 3.3 this ratio is 0.82.

3.2.2 Experimental Realization of 2D Equilibrium Shapes

Two different approaches exist for the experimental determination of equilibrium island shapes. The *quench method* relies on the analysis of the "frozen" morphology of a surface with 2D islands. The surface is subjected to a heat treatment at a temperature T allowing equilibration by surface diffusion, and is subsequently quenched to a lower temperature T' for imaging, at which morphological changes are absent on the time scale of observation. While this method was first applied to Si islands on Si(001) [226], the first demonstration of equilibration together with quantitative results on step free energies was achieved for Pt–islands on Pt(111) [227]. The second approach is the *fluctuation method*, which is based on the observation of a single island for a long time by taking a time lapse sequence of topographs at the equilibration temperature. The equilibrium shape is then obtained by averaging over a large number of images. In the following the two approaches will be discussed and compared.

Figures 3.4b, c, e and f exemplify vacancy and adatom islands equilibrated at temperatures at or above 700 K and quenched to room temperature. In order to lend credibility to the claim that the quenched morphologies actually represent equilibrium shapes, a number of conditions were checked.

1. *History independence*: As a thermodynamic state, the equilibrium shape must be independent of the kinetic pathway leading to its formation. Specifically, islands formed at a low temperature and annealed to T must be identical in shape to islands formed directly at T. This is explicitly demonstrated by the identity of the shapes in Figs. 3.4b, c, e and f.

---

[5] Experimentally, there is a remarkable exception for islands on Au(110), which have a large forbidden angular range of step orientations due to the complications of a surface reconstruction [224].

[6] For an analytic formulation of the inverse Wulff construction, see [225].

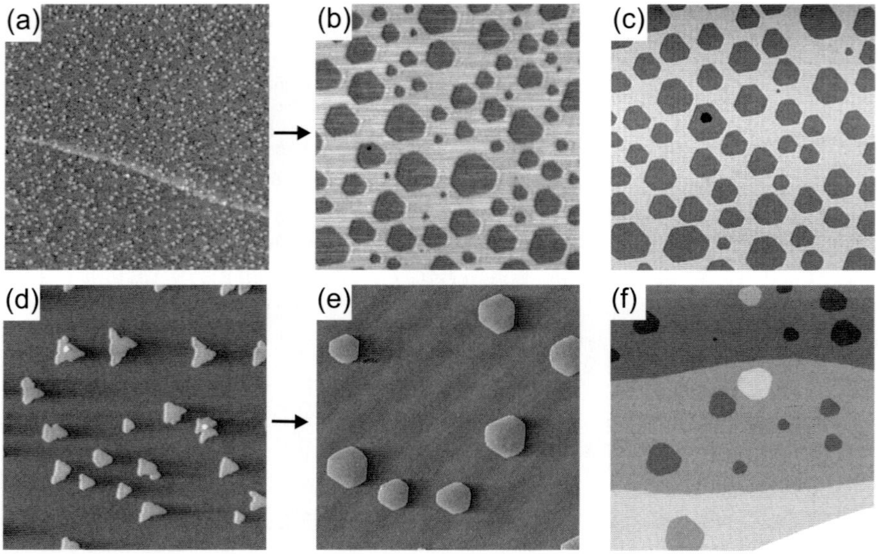

Fig. 3.4. Vacancy and adatom islands generated by ion bombardment (600 eV Ar^+, 2.8×10^{16} ions/m^2s) and by Pt–deposition on Pt(111). (**a**) 0.45 ML removed during 120 s bombardment at $T = 245$ K. (**b**) Same as (**a**), but with subsequent 120 s annealing at $T = 760$ K. (**c**) 0.45 ML removed during 120 s bombardment at $T = 760$ K. (**d**) 0.1 ML deposited at T = 425 K. (**e**) Same as (**d**), but with subsequent 60 s annealing at $T = 700$ K. (**f**) 0.45 ML removed during 120 s bombardment at 625 K and subsequent 120 s annealing at $T = 850$ K. Topograph sizes: (**a**), (**b**), (**c**): 1650 Å × 1650 Å; (**d**) and (**e**): 2000 Å × 2000 Å; (**f**): 3300 Å × 3300 Å

2. *Size independence*: Irrespective of their size (provided it exceeds, say, 100 Å in diameter) the island shapes are identical after annealing to a given temperature (see Figs. 3.4b, c, e and f).
3. *Time independence*: The quenched shape does not depend on the duration of the annealing time.
4. *Identity of crystal shape and hollow shape*: The shapes of adatom and vacancy islands are identical. This is particularly clear in Fig. 3.4f, which contains both types of islands. Adatom and vacancy islands are rotated relative to each other by 180°. This is due to the fact that the A- and B-step edges are interchanged in adatom and vacancy islands (Fig. 3.5).

Averaging over a large number of adatom and vacancy islands on Pt(111), which had been equilibrated between 700 K and 800 K, the ratio $\delta_B/\delta_A = 0.87 \pm 0.02$ between the free energies of the two types of close-packed step edges was determined [227]. Although no systematic investigation of the temperature dependence of the equilibrium shape could be performed, it was noticed that after equilibration between 700 K and 800 K the A-step orientations are orientations of local *minima* in the δ-plot, while after equilibration at 1270 K the A-step orientations are orientations of absolute *maxima*, as in

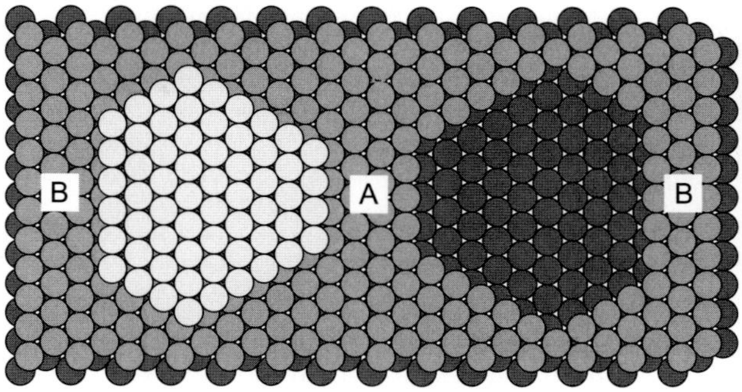

Fig. 3.5. Ball model of islands with a step free energy ratio for the dense packed A- and B-steps similar to the one determined experimentally for Pt(111). Due to their small size, these islands deviate slightly from the equilibrium shape, which is a threefold symmetric hexagonal shape with rounded corners. The reduced symmetry of the island shape results from the breaking of the sixfold symmetry of the (111)-surface by the underlying atomic layer. The origin of the different orientations of the adatom and vacancy island is obvious in the ball model

Fig. 3.3 [228]. In addition, the step free energy ratio tends towards 1 with increasing T, growing from 0.87 to 0.89 after equilibration at 1270 K. We will come back to these findings in the next subsection.

The main drawback of the quench method is obvious already from the topographs in Fig. 3.4: In general, different islands slightly deviate in shape from each other. At the equilibration temperature the islands perform entropic fluctuations around their minimum free energy shape, which are frozen in during the quench. The islands thus do not represent the exact equilibrium shape. Averaging over many islands partly removes the contribution of shape fluctuations, but they cannot be removed completely due to the different sizes of the islands and the slight spatial variations in the image distortions caused by the piezo–elements of the STM.

The problem can be alleviated by equilibrating at higher temperatures for longer times in order to produce large islands, for which the effect of frozen fluctuations becomes negligible (compare e.g. Fig. 5b of [228]). For large equilibrated Pb–islands the precise determination of the δ-plot could be achieved in this way [223]. Nevertheless, two difficulties remain: One is the uncertainty introduced by the quenching procedure, the other is the difficulty to identify contamination effects.

The fluctuation method largely avoids these disadvantages, because the elimination of the fluctuation-induced deviations from the equilibrium shape is much more straightforward and efficient when the shape is averaged over a large number of images of one and the same island. This method has been applied successfully for the determination of the equilibrium shapes of islands on Cu(111), Cu(001) and Ag(111) [229–231].

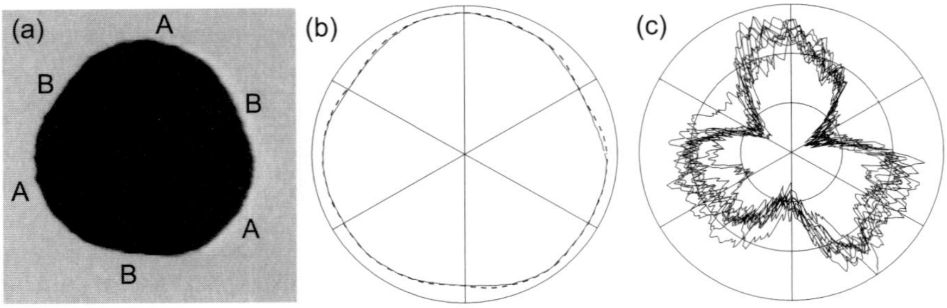

Fig. 3.6. (**a**) Aluminum vacancy island created by 1 keV Ne^+ ion bombardment at 200 K followed by annealing for 300 s at 425 K. Imaging temperature 300 K, topograph size 700 Å × 700 Å. (**b**) Average island shape (*thin line*) and corresponding δ-plot (*dashed line*), after averaging over 10 STM topographs of the same island. (**c**) Superposition of the corresponding single island boundaries of the time lapse sequence on an exaggerated scale. The outer circle corresponds to the maximum radius measured for all island boundaries, while the center corresponds to 85% of this value. Note the large fluctuations of the A-steps

The method is illustrated here for the case of Al vacancy islands on Al(111) [232]. Figure 3.6a shows a snapshot of an island. The island boundary averaged over a time lapse sequence of such topographs exhibits the very regular, threefold symmetric shape shown in Fig. 3.6b. Application of the inverse Wulff construction yields the δ-plot also shown in Fig. 3.6b. The resulting step free energy ratio $\delta_B/\delta_A = 0.951 \pm 0.005$ is in reasonable agreement with an early prediction of $\delta_B/\delta_A = 0.935$ derived from density functional theory [233].

In Fig. 3.6c the single snapshots of the island boundary are superimposed on a magnified radial scale. It is apparent that the island fluctuates freely along its perimeter. For the reactive Al(111) surface this representation is of particular importance, since it allows unambiguously to identify step edge pinning due to contamination. Such step edge pinning evidently prevents the formation of the equilibrium island shape and is found quite frequently on Al(111). A step edge that is pinned in one or several points would be characterized in the representation of Fig. 3.6c by one or several points where *all* island boundary snapshots intersect. Another interesting observation from Fig. 3.6c is that the fluctuation amplitude of steps close to the A-step orientation appears generally to be larger than close to the B-step orientation. We will return to this point in the next section.

3.2.3 The Ising Model and the Awning Approximation

In the present subsection a very simple microscopic picture of the equilibrium shape, its temperature dependence, and the relation between the step free energy δ and the kink energy ϵ needed to excite a kink in an otherwise dense packed step will be developed. This picture yields information on the energetics of atomic processes at the step edges, which is necessary to obtain a clear understanding of island growth kinetics.

The simplest description of the equilibrium island shapes on a (111) surface is provided by the Ising model on the triangular lattice[7]. In this model the energy E of an arbitrary configuration of adatoms is given by

$$E = -E_{\mathrm{nn}} \sum_{\langle \boldsymbol{r}\boldsymbol{r}' \rangle} \eta_{\boldsymbol{r}} \eta_{\boldsymbol{r}'}, \tag{3.11}$$

where the sum runs over all nearest neighbor pairs of sites $\boldsymbol{r}$, $\boldsymbol{r}'$, the occupation number $\eta_{\boldsymbol{r}} = 1$ ($\eta_{\boldsymbol{r}} = 0$) when site $\boldsymbol{r}$ is occupied (vacant), and E_{nn} is the bond energy[8]. A step is the boundary between an occupied region and an empty region, and its energy is equal to the number of broken bonds along the step times $E_{\mathrm{nn}}/2$; each broken bond increases the energy by $E_{\mathrm{nn}}/2$, because breaking apart a pair of atoms costs an energy E_{nn} and creates two broken bonds.

Consider first the equilibrium shape at 0 K. In this case the step free energy $\delta(T) = \delta_0 - TS$ reduces to the energy δ_0. As a natural unit of length we use the nearest neighbor distance a, such that $\delta(T), \delta_0$ and S denote the step free energy, step energy and entropy per step atom, respectively. The step energy of a dense packed step oriented parallel to a $\langle 1\bar{1}0 \rangle$-direction (see Fig. 2.1) is then simply given by $\delta_{0,\langle 1\bar{1}0 \rangle} = E_{\mathrm{nn}}$, as each step atom has two additional broken bonds compared to a terrace atom. The kink energy ϵ is

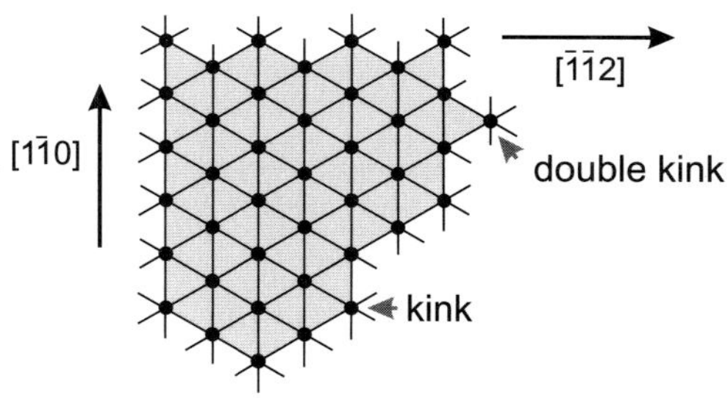

Fig. 3.7. Schematic sketch of atoms (*full dots*) on a triangular lattice. In-plane bonds and broken bonds are indicated by lines. The island area is shaded. The expressions given in the text for the step energy in the dense packed and open directions, as well as for the kink energy, in terms of the bond strength E_{nn}, can be read off from the figure

[7] Recent work on Cu(100) shows, however, that the Ising model may fail to quantitatively account for the orientation dependence of the step free energy [234].

[8] In its original formulation the Ising model is a model of magnetism which involves spin variables $2\eta_{\boldsymbol{r}} - 1 = \pm 1$. The present version is known as the Ising lattice gas. The Ising model is discussed extensively in many textbooks of statistical mechanics, see e.g. [235].

given by $\epsilon = \delta_{0,\langle 1\bar{1}0\rangle}/2 = E_{\mathrm{nn}}/2$ and the step energy of a completely kinked step oriented parallel to a $\langle\bar{1}\bar{1}2\rangle$ direction is given by $\delta_{0,\langle\bar{1}\bar{1}2\rangle} = (2/\sqrt{3})\delta_{0,\langle 1\bar{1}0\rangle}$. These simple relations follow immediately from the geometry of Fig. 3.7.

In the Ising model the zero temperature equilibrium shape of an island is an equal-sided hexagon with sharp corners. This is easily seen by starting from a circular island (the equilibrium shape in the isotropic case). To such an island one may add atoms to the kink positions until all kinks have disappeared and an equal-sided hexagon emerges, while keeping the total number of broken bonds (the total edge energy) constant. This proves that the hexagon is the shape of maximum area for a given edge energy.

Increasing the temperature above 0 K turns on the effects of entropy. For a dense packed step edge the partition function per step atom reads, to leading order in the kink concentration $\mathrm{e}^{-\epsilon/k_{\mathrm{B}}T}$ [22]

$$Z = \mathrm{e}^{-\delta_{0,\langle 1\bar{1}0\rangle}/k_{\mathrm{B}}T} + 2\mathrm{e}^{-(\delta_{0,\langle 1\bar{1}0\rangle}+\epsilon)/k_{\mathrm{B}}T}, \tag{3.12}$$

expressing the possibilities that the step may contain no kink, a forward kink, or a backward kink at a given position. To the same order the step free energy $\delta = -k_{\mathrm{B}}T \ln Z$ is then given by

$$\delta_{\langle 1\bar{1}0\rangle}(T) = \delta_{0,\langle 1\bar{1}0\rangle} - 2k_{\mathrm{B}}T\mathrm{e}^{-\epsilon/k_{\mathrm{B}}T}. \tag{3.13}$$

It is clear from (3.13) that for a step oriented along $\langle 1\bar{1}0\rangle$, configurational entropy relies upon kink excitation. It is negligible at low temperatures and becomes relevant only when kink excitation occurs with significant probability.

The situation is different for the opposite extreme case of a step along $\langle\bar{1}\bar{1}2\rangle$. The number of paths of equal length connecting two points on such a step is exponentially large in the distance between the points. Since all these paths are energetically degenerate, the configurational entropy is temperature independent. As long as longer paths do not contribute significantly to the entropy, the step free energy is [230]

$$\delta_{\langle\bar{1}\bar{1}2\rangle} = \delta_{0,\langle\bar{1}\bar{1}2\rangle} - \frac{2\ln 2\, k_{\mathrm{B}}T}{\sqrt{3}}. \tag{3.14}$$

The expressions (3.13) and (3.14) are low temperature approximations to the full Ising step free energy, which has been worked out in [246, 247]. The corresponding equilibrium shapes show a gradual transition from a hexagon to a circle at the Ising critical point.

We can use (3.13) and (3.14) to assess the temperature dependence of the step free energy ratio $\delta_{\mathrm{B}}/\delta_{\mathrm{A}}$ determined experimentally in Sect. 3.2.2. Assuming $E_{\mathrm{nn}} = 0.5\,\mathrm{eV}$ as in Sect. 3.1, the step free energy in the $\langle 1\bar{1}0\rangle$ direction decreases only marginally from $\delta_{0,\langle 1\bar{1}0\rangle} = 0.500\,\mathrm{eV}$ at 0 K to $\delta_{\langle 1\bar{1}0\rangle} = 0.497\,\mathrm{eV}$ at 750 K, while in the open $\langle\bar{1}\bar{1}2\rangle$ direction it drops significantly from $\delta_{0,\langle\bar{1}\bar{1}2\rangle} = 0.577\,\text{eV}$ to $\delta_{\langle\bar{1}\bar{1}2\rangle}(750\,\text{K}) = 0.526\,\text{eV}$. Thus, at a typical equilibration temperature only the free energy of the open kinked steps deviates significantly from the zero temperature step energy. The ratio $\delta_\text{B}/\delta_\text{A} = 0.87$ determined at 750 K is therefore representative of the zero temperature step energy ratio to a very good approximation. This makes this ratio well suited for comparison with ab initio density functional theories, which evaluate step free energies only at 0 K. As can be seen from Table 3.2, the recent DFT results for $\delta_\text{B}/\delta_\text{A}$ on Pt(111) in the local density approximation (LDA) agree quite well with the experimental results. Earlier calculations [243] disagreed with the experiment, giving $\delta_\text{B}/\delta_\text{A} = 0.98$. The DFT calculations in the generalized gradient approximation (GGA) [86] also yield the correct step edge energy ratio, but it has been argued that the re-

Table 3.2. Step and kink formation energies, as well as ratios of step and kink energies for A- and B-steps, on fcc(111) metal surfaces. The density functional theory calculations are either in the local density approximation (LDA) or in the generalized gradient approximation (GGA). The energies are given in eV/atom. Numbers in parentheses indicate the error of the last or the last two digits.

Surface	Method	Year	δ_A	δ_B	$\delta_\text{B}/\delta_\text{A}$	ϵ_A	ϵ_B	$\epsilon_\text{B}/\epsilon_\text{A}$
Pb(111)	STM [223]	1999	0.091	0.082	0.901	0.061	0.087	1.43
	GGA [236]	2000	0.095	0.078	0.82	0.041	0.060	1.46
	STM [237]	2002	0.131	0.117	0.89	0.0400	0.0603	1.51
	STM [238]	2003	0.1283(3)	0.1157(58)	0.90	0.0425	0.0606	1.51
Al(111)	LDA [233]	1994	0.248	0.232	0.935			
	STM [232]	2003			0.951(5)			
Ag(111)	STM[239]	1998	0.22(4)					
	GGA [240]	1999		0.216			0.074	
	STM [231]	2001	0.25(3)		1.004(3)	0.101(5)		0.99(1)
Cu(111)	STM [241]	1998	0.45					
	STM [229]	1999	0.22(2)		1.02(2)			
	STM [230]	1999	0.31(4)		1.01(1)	0.110(5)		1
	GGA [242]	1999	0.27	0.26	0.96	0.09	0.121	1.33
	STM [231]	2001	0.27(3)		0.989(7)	0.113(7)	0.121(7)	1.06(2)
Pt(111)	STM [227]	1991			0.87(2)			
	LDA [243]	1995	0.48	0.47	0.98			
	STM [215]	1996				0.167(5)		1
	LDA [86]	1998	0.43(2)	0.38(2)	0.88(7)			
	GGA [86]	1998	0.29(2)	0.25(2)	0.86(10)			
	GGA [240]	1999		0.67		0.316		
	LDA [244]	2000	0.40	0.35	0.88	0.18	0.25	1.39
Ir(111)	STM [245]	1998			0.93(1)			
	GGA [240]	1999		0.795				

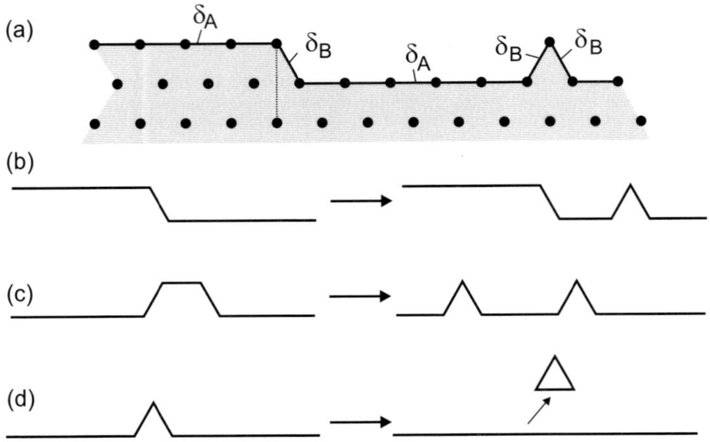

Fig. 3.8. (**a**) In the awning approximation, a step configuration is decomposed into microscopic segments of A-step and B-step, and the energy of the configuration is assigned accordingly (see text). (**b**)–(**d**) Schematic sketches for the determination of energy differences of atomic processes in the awning approximation (see text).(**b**): step adatom generation; (**c**): step dimer dissociation; (**d**): step adatom detachment

sulting step energies are less reliable[9] [86]. The absolute values for the step edge energy and kink energy from DFT also correspond reasonably well to the predictions of the nearest neighbor model. Assuming a bond strength of 0.5 eV as suggested by DFT results for the dimer binding energy [130], the nn–model predicts 0.25 eV for the kink energy and 0.5 eV for the step energy per atom, which is consistent with the LDA calculations (compare to Table 3.1).

We now turn to a slightly modified model which takes into account the different step energies of the two types of dense packed steps. This cannot be accomplished simply by an orientation dependent bond strength, because the difference between the two types of steps relies on the position of one triangular lattice with respect to a second, substrate lattice (see Fig. 2.1). The most simple way to account for this feature is the awning approximation [249]. This is a zero temperature approximation which attributes to each segment of a step edge that has unit length and is oriented along the A-step or B-step direction the corresponding step edge energy δ_A or δ_B. Although this model is not based directly on atomic bond energies, it allows to analyze the effect of step energetics on atomic processes at the step edge. For example, it is clear from Fig. 3.8a that the awning approximation treats a kink on the

[9] In our discussion the possible influence of vibronic entropy on the temperature dependence of the step free energy has been neglected. It can be argued that, although vibronic entropy lowers the step free energy at the equilibration temperature, its effect is identical for the different steps and thus to first approximation not relevant for the free energy ratio [230, 248].

A-step as a piece of a B-step, which replaces half a unit length segment of the A-step. The kink energy at the A-step is therefore

$$\epsilon_{\mathrm{A}} = \delta_{\mathrm{B}} - \frac{1}{2}\delta_{\mathrm{A}}. \tag{3.15}$$

Together with the corresponding relation for the B-step we find

$$\epsilon_{\mathrm{A}} - \epsilon_{\mathrm{B}} = \frac{3}{2}(\delta_{\mathrm{B}} - \delta_{\mathrm{A}}). \tag{3.16}$$

The kink formation energy at the step edge with the *lower* step energy is *higher* than at the step with higher step energy. Comparison with Table 3.2 shows that this prediction of the awning approximation holds true for all calculated or measured systems (with the exception of [215] for Pt(111)). It simply reflects the fact that kink excitation at the energetically more favorable B-step involves formation of a small piece of the energetically disfavored A-step, while kink excitation at the less favorable A-step involves formation of a small piece of the energetically favorable B-step. This explains why thermal step fluctuations are more pronounced at the A-step, as was observed in Sect. 3.2.2 [for a quantitative relation between the step fluctuation amplitude and the kink energy see (3.21)].

Similarly the ratio of the kink energies is given by

$$\frac{\epsilon_{\mathrm{B}}}{\epsilon_{\mathrm{A}}} = \frac{2 - \delta_{\mathrm{B}}/\delta_{\mathrm{A}}}{2\delta_{\mathrm{B}}/\delta_{\mathrm{A}} - 1}. \tag{3.17}$$

It is interesting to analyze Table 3.2 using (3.17). For Pt(111) with an input value of $\delta_{\mathrm{B}}/\delta_{\mathrm{A}} = 0.88$ obtained from DFT calculations [86, 244], the predicted kink energy ratio is $\epsilon_{\mathrm{B}}/\epsilon_{\mathrm{A}} = 1.47$ in decent agreement with the calculated ratio of $\epsilon_{\mathrm{B}}/\epsilon_{\mathrm{A}} = 1.39$ [244]. From this point of view the kink energy ratio $\epsilon_{\mathrm{B}}/\epsilon_{\mathrm{A}} = 1$ obtained in [215] from measurements of step fluctuations appears questionable. For the case of Pb(111), the measured step energy ratio $\delta_{\mathrm{B}}/\delta_{\mathrm{A}} = 0.901$ predicts $\epsilon_{\mathrm{B}}/\epsilon_{\mathrm{A}} = 1.37$ in good agreement with the experimental estimate 1.43 [223]. Qualitatively, the following rule holds: The anisotropy of the kink energies is larger and inverted with respect to the anisotropy of the step energies.

A second test of the awning approximation against experiments is possible by noting that, within this approximation, a step adatom is a double kink (Fig. 3.8b). The energy cost for the formation of a step adatom at an A-step is therefore

$$E_{\mathrm{stad,A}} = 2\delta_{\mathrm{B}} - \delta_{\mathrm{A}}. \tag{3.18}$$

The transfer of a step adatom from an A-step to a B-step then gives rise to an energy increase of

$$\Delta E_{\mathrm{stad}} = E_{\mathrm{stad,B}} - E_{\mathrm{stad,A}} = 3(\delta_{\mathrm{A}} - \delta_{\mathrm{B}}). \tag{3.19}$$

In other words, a step adatom is more strongly bound to the step of larger step energy. Intuitively this result is clear, since the step of higher step energy has less saturated bonds and is thus in larger need of bond saturation. Using DFT–values for δ_A and δ_B taken from [86,244], we estimate $\Delta E_{stad} \approx 0.15\,\text{eV}$. Indeed, DFT calculations using LDA (GGA) show that a step adatom binds more strongly to the A- than to the B-step by about 0.11 eV (0.13 eV) [140], in excellent agreement with the prediction based on (3.19).

There is further independent experimental evidence for the relation between higher step energy and stronger step adatom binding suggested by (3.19). In FIM measurements it was found that an Ir step adatom on Ir(111) is bound more strongly to the A-step than to the B-step by about 0.2 eV [250–252]. Based on the step energy ratio $\delta_B/\delta_A = 0.93$ obtained experimentally [245], and using a total step energy $\delta_B \approx 0.5\,\text{eV}$ (suggested by the value for Pt(111) scaled by the ratio of the cohesive energies) one ends up with an estimate of 0.11 eV for the binding energy difference of a step adatom, which is qualitatively consistent with the FIM result.

Based on the preceding discussion, the awning approximation can be expected to yield qualitatively correct predictions for the energy differences involved in atomic processes at the two types of dense packed step edges. Apart from the formation of kinks and step adatoms, there are many other processes in which the energetic differences between the two steps become relevant. Two further examples are illustrated in Fig. 3.8c and d. Within the awning approximation, the binding energy of a step dimer at the A-step is $2\delta_B - \delta_A$, which is lower by $3(\delta_A - \delta_B)$ than the binding energy of a step dimer at a B-step. Finally, step adatom detachment is energetically more expensive from an A-step than from a B-step by $3(\delta_A - \delta_B)$.

3.2.4 Absolute Energy Values for Steps and Kinks

Experimental methods for the determination of absolute values for kink and step energies are highly desirable within the framework of an atomic scale understanding of epitaxial growth. A variety of methods were developed in recent years [215, 225, 229–231, 237, 238, 253, 254] and several reviews on this subject exist [172, 255, 256].

Absolute values for the kink energy can be obtained from the analysis of step fluctuations. Consider a vicinal surface with steps oriented along the dense packed directions, and with a step separation that is sufficiently large so that step–step interactions can be neglected on the length scale of interest. Due to thermal fluctuations, typical step conformations display random deviations from the dense packed orientation. We use a coordinate system in which the steps are oriented along the y-axis, and describe the step conformation by the function $x = \zeta(y)$ (compare to Fig. 4.37). Since kink excitations at different points of the step are independent, the step position $\zeta(y)$ performs a random walk in "time" y, which can be characterized by the mean square displacement

$$\langle(\zeta(y) - \zeta(y'))^2\rangle = b^2(T)\frac{|y - y'|}{a}. \tag{3.20}$$

The right hand side of (3.20) is the mean square displacement per site b^2, also known as the *step diffusivity*[10], multiplied by the number of sites where a step displacement could occur. The step diffusivity is easily calculated within the low temperature approximation of Sect. 3.2.3. Each kink induces a displacement $\pm a_\perp$ perpendicular to the step, so the mean square displacement per kink is $a_\perp^2$. The probability that a segment of the step contains a kink of either sign is proportional to the Boltzmann factor $\mathrm{e}^{-\epsilon/k_\mathrm{B}T}$, normalized by the partition function $1 + 2\mathrm{e}^{-\epsilon/k_\mathrm{B}T}$ which sums over the three possibilities of no kink, a forward kink or a backward kink. This yields[11]

$$b^2(T) = \frac{2a_\perp^2\,\mathrm{e}^{-\epsilon/k_\mathrm{B}T}}{1 + 2\mathrm{e}^{-\epsilon/k_\mathrm{B}T}} \approx 2a_\perp^2\,\mathrm{e}^{-\epsilon/k_\mathrm{B}T}. \tag{3.21}$$

Equations (3.20) and (3.21) form the basis of the determination of kink energies from the spatial fluctuations of the steps. In practice, it is often easier to measure the mean square displacement between two steps rather than the displacement of a single step. This method has been used to obtain kink energies for Pt(111) [215] (see also Table 3.2).

Another method for the determination of the kink energy is based on measuring the temperature dependence of the curvature in the equilibrium island shapes at the points of dense packed step orientations [231]. The connection to the previously described method becomes clear by noting that the continuum theory of step fluctuations yields the relationship [172, 255]

$$b^2(T)/a = k_\mathrm{B}T/\tilde{\delta} \tag{3.22}$$

between the step diffusivity and the step stiffness. According to the local form (3.7) of the Wulff construction, the curvature of the equilibrium shape is inversely proportional to the step stiffness, hence it follows from (3.22) that the curvature is also proportional to b^2. This relation is intuitively plausible, since one would expect the radius of curvature to be of the order of the kink distance $\sim \mathrm{e}^{\epsilon/k_\mathrm{B}T}$. The method based on the equilibrium shape curvature was used to determine the kink energy for dense packed steps on Cu(111) and Ag(111) [231].

[10] It is worth emphasizing that the term *diffusivity* refers here to *spatial* fluctuations of a temporally frozen step configuration. Some aspects of *temporal* step fluctuations are discussed in Sect. 3.7.1.

[11] Strictly speaking, (3.21) is valid only for surfaces with square symmetry. It has to be modified for surfaces with hexagonal symmetry, because there each kink causes not only a displacement in the x-direction, but also in the y-direction. However, the resulting correction term is of the order of $\mathrm{e}^{-2\epsilon/k_\mathrm{B}T}$ and may thus be neglected at the level of accuracy of (3.21).

A related approach combines (3.6) and (3.7) to obtain absolute values of kink and step energies. For the high symmetry directions ϕ_{A} and ϕ_{B} corresponding to the A- and B-step orientations in the equilibrium shape, (3.6) implies that $r(\phi_i) = \varrho\delta(\phi_i)$, where i stands for A or B [compare to (3.10)]. Together with (3.7) it follows that the (dimensionless) product of island radius and island step curvature equals the ratio of step free energy to step edge stiffness [231, 238],

$$r(\phi_i)\kappa_{\mathrm{st}}(\phi_i) = \delta(\phi_i)/\tilde{\delta}(\phi_i), \quad i = \mathrm{A, B}. \tag{3.23}$$

Temperature-dependent measurements of the left hand side of (3.23) for Pb islands on Pb(111) compared to analytical expressions for $\delta(\phi_i)/\tilde{\delta}(\phi_i)$ in terms of $\delta_{0,i}$ and ϵ_i allow to fit the latter quantities with good precision [237, 238].

When the step free energy is isotropic, the step stiffness and the step free energy coincide according to (3.8). Through the relation (3.22) the analysis of step fluctuations then also gives access to the absolute value of the step free energy. This idea has been applied in the analysis of island shape fluctuations [229]. Sequences of STM–topographs allow one to determine the radially averaged mean square displacement $\langle(\Delta r)^2\rangle$ of the position of the island boundary from its average position. A fluctuation theory for islands with an isotropic edge free energy $\bar{\delta}$ then yields the relation

$$\langle(\Delta r)^2\rangle = \frac{3k_{\mathrm{B}}TR}{4\pi\bar{\delta}}, \tag{3.24}$$

where R is the radius of the circular equilibrium shape. Equation (3.24) is the analog of (3.20) for a closed, isotropic step. It could be shown that the error introduced by approximating the six-fold symmetric equilibrium island shape of, e.g., Cu(111) islands by an isotropic circular shape amounts to only a few percent. From an analysis of the island size dependence of G the average step energy on Cu(111) was obtained [229]. The analysis of island shape fluctuations was recently extended to yield absolute step energies for highly anisotropic islands on TiN(111) [225].

Finally, step energies may be estimated by comparing the temperature dependence of the step free energy for dense packed steps to that of maximally open steps which consist only of kinks [231]. The analysis is based on an equation obtained by dividing (3.13) by (3.14). In an iterative procedure the temperature dependence of $\delta_{\langle 1\bar{1}0\rangle}(T)/\delta_{\langle\bar{1}\bar{1}2\rangle}(T)$ is used to determine the absolute value of the step energy $\delta_{\langle 1\bar{1}0\rangle}(0)$, relying, however, on known kink energies. The method yields step energies for Cu(111) in reasonable agreement with the step fluctuation method (see Table 3.2).

3.3 Fractal-Dendritic Islands

The formation of fractal-dendritic islands is a common feature in low temperature homoepitaxial growth on fcc(111) metal surfaces. So far it has been observed on Pt(111) [206, 257, 258], Ag(111) [259], Ir(111) [260] and Al(111) [202] at temperatures ranging from the onset of island formation up to about 15% of the melting temperature T_m. Figure 3.9 shows examples of fractal-dendritic growth on three of these surfaces at around $0.1\,T_m$. The same kind of morphology has been found also in heteroepitaxial growth of fcc metals on surfaces with triangular symmetry [48, 259, 261]. The fractal-dendritic growth mode discussed here is characterized by (i) anisotropy in the branched island shapes due to preferential growth into the $\langle\bar{1}\bar{1}2\rangle$-directions, and (ii) island branch widths of only a few atomic diameters. Fractal growth with thick branches and without anisotropy [262] can be traced to a different mechanism and is treated in Sect. 3.6.

Starting from the very first observations of fractal [262] and fractal-dendritic growth [257], the phenomenon was linked to diffusion limited aggregation (DLA). It is evident from a look into Table 3.1 that a considerable activation energy gap separates the onset of adatom diffusion ($E_D = 0.26\,\mathrm{eV}$) from the onset of step edge diffusion ($E_a \geq 0.6\,\mathrm{eV}$) on Pt(111). Since step atom detachment requires even higher activation energies, in the temperature range between the onset of adatom diffusion and the onset of edge diffusion, deposited adatoms migrate until they stick *irreversibly* at the point of their first impact onto an aggregate of adatoms. The only difference in comparison to the DLA model studied in computer simulations [36, 208, 209] lies in the deposition rule: While in DLA the particles are deposited one–by–one far away from the aggregate, in epitaxial growth the adatoms are deposited at random in space and time. As a consequence several islands grow simul-

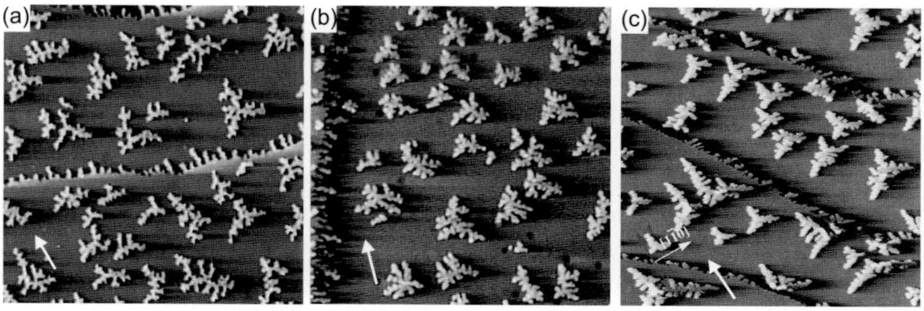

Fig. 3.9. Fractal-dendritic islands created by homoepitaxial growth on (**a**) Pt(111), (**b**) Al(111) and (**c**) Ag(111). The deposition temperatures are (**a**) 10.2% of T_m, (**b**) 9.8% of T_m and (**c**) 8.9% of T_m. The topograph size is always 1200 Å × 1200 Å. The $[\bar{1}\bar{1}2]$ direction is indicated by a white arrow in the lower left corner of each topograph ((**c**) is Fig. 32a of [48] reprinted with permission from Elsevier Science)

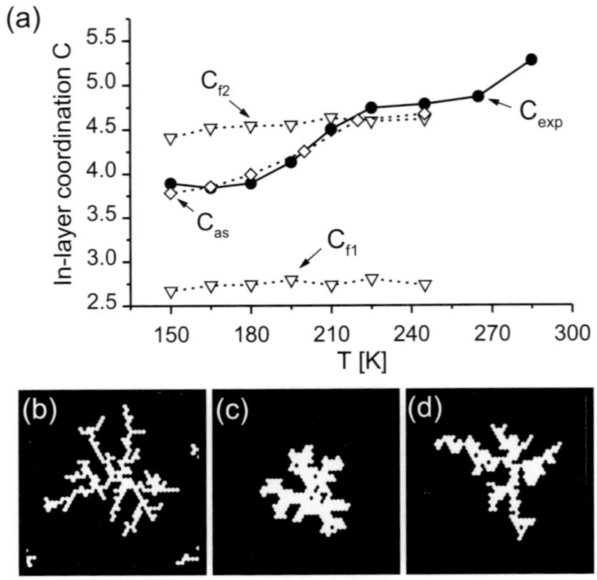

Fig. 3.10. (**a**) In-layer coordination numbers $C_{\rm exp}, C_{\rm f1}, C_{\rm f2}$ and $C_{\rm asy}$ resulting from experiment, from KMC simulations for fractals type-I, KMC simulations for fractals type-II and from KMC simulations with the corner asymmetry assumption, respectively (see text). (**b**), (**c**) and (**d**) exhibit example islands created by KMC simulations for fractals type-I, fractals type-II and fractals with the corner asymmetry assumption (based on Fig. 2 of [258])

taneously, there is a finite density of adatoms surrounding the aggregates, and adatoms can be deposited close to the islands, also in the fjords between branches. All three features become negligible in the limit of large island spacing and small coverage, hence the analogy to DLA should become exact when $\Theta \ll 1$ and $\Omega N \ll 1$.

However, closer inspection of the experimentally observed island shapes reveals important qualitative deviations from the basic DLA model. First, the shape anisotropy of the islands (which was overlooked in the first publications [206,257]) is much more pronounced than in DLA. Although diffusion-limited aggregation may be sensitive to the lattice anisotropy, this anisotropy is reflected in the aggregate shape only for very large aggregates and lattices of sufficiently low symmetry (triangular lattices appear not to cause anisotropic shapes) [36,263]. Second, the average in-layer coordination number C of the atoms forming the islands is considerably higher than expected from DLA simulations. This finding contains the key to an atomistic understanding of fractal-dendritic growth, as it is outlined below.

In the following the evolution of fractal-dendritic growth on Pt(111) with temperature is analyzed in the range between 150 K and 265 K [258]. The

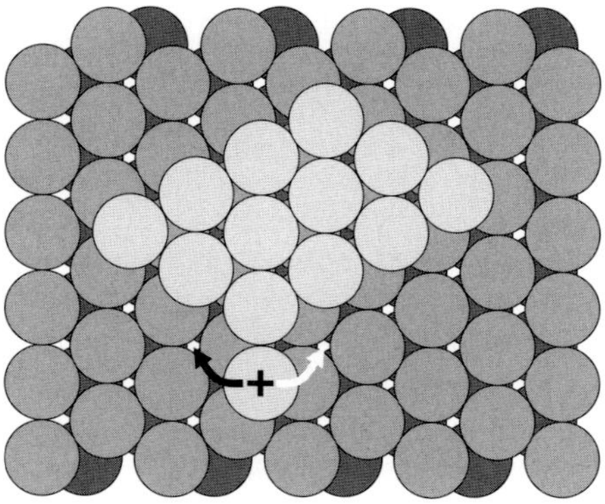

Fig. 3.11. Ball model of a small aggregate on an fcc(111) surface. The atom marked by a + has an in-layer coordination of one. The black and white arrow indicate nearest neighbor hops towards higher coordinated sites via different transition states. These hops are named bridge and atop hop, respectively

corresponding island shapes are shown in Fig. 2.7. In order to obtain a number characterizing the island morphology, the average branch thickness of the islands was measured, carefully corrected for STM–tip effects and recalculated into an average in-layer coordination number C_{exp} of the aggregate atoms. The data for C_{exp} shown in Fig. 3.10 exhibit a clear temperature dependence. The coordination number is slightly below 4 up to 180 K, increases between 180 K and 225 K to about 4.75 and remains on this plateau value up to 265 K. Further insight into the growth mechanism was obtained from KMC simulations performed on a triangular grid with a deposition rate as used in the growth experiments, and the experimentally determined adatom diffusion parameters (compare Table 2.1 and [82]). Assuming the adatoms to stick to the aggregate at the lattice site of first encounter – irrespective of the number of lateral nearest neighbors, provided it is larger than zero – results in the coordination numbers labeled C_{f1} in Fig. 3.10. Islands grown according to this model are classified as type-I fractals [264]. An example of such an island is shown in Fig. 3.10b.

In contrast to the experimental coordination number C_{exp}, C_{f1} is temperature independent. Moreover, the magnitude of C_{f1} is much too small to match C_{exp}. In order to increase C one may assume that laterally onefold coordinated atoms are unstable and increase their coordination number by a single hop towards a two- or threefold coordinated site. Each atom marked by a cross in Fig. 3.11 will then relax by a hop, as indicated by the arrows, to a higher coordinated site prior to the arrival of another adatom. Islands grown according to this model are classified as type-II fractals. An example is shown in Fig. 3.10c. The resulting in-layer coordination C_{f2} matches C_{exp} fairly well for temperatures above 210 K. However, in the low temperature range C_{f2} clearly overestimates C_{exp}. Moreover, the resulting island shape still does not show any anisotropy. Evidently more complicated assumptions on the nearest neighbor hops of laterally onefold coordinated atoms to sites with higher coordination are necessary, involving in particular temperature-dependent jump probabilities. Two possible scenarios are analyzed in the following, which explain not only the magnitude and temperature dependence of C_{exp}, but also produce the required island anisotropy.

Corner Asymmetry. The first scenario relies on the fact that there are two different paths for laterally onefold coordinated atoms towards higher coordinated sites, which differ in their geometry relative to the substrate lattice. The two distinct nearest neighbor hops may be characterized as "atop" or "bridge" hops, depending on whether the diffusion path proceeds via an atop position or a bridge position. In view of Fig. 3.11 it appears plausible that the onefold coordinated atom is able to maintain a larger coordination with respect to the substrate by performing a bridge hop instead of an atop hop. Consequently one might speculate on a smaller activation energy for the bridge jump[12]. There is no direct experimental evidence for this assumption on Pt(111), but FIM measurements have determined a smaller activation energy for bridge hops of onefold coordinated Ir atoms at an Ir island corner on Ir(111) [252].

Assuming an energy barrier of 0.33 eV for bridge and 0.51 eV for atop hops (corresponding to onset temperatures of 130 K and 200 K, respectively), the coordination number C_{asy} obtained from the KMC simulations reproduces precisely the experimental values (Fig. 3.10a). The corner asymmetry assumption also reproduces the observed anisotropy of the island shape in an adequate fashion. The simulated island in Fig. 3.10d exhibits preferential growth in the $\langle\bar{1}\bar{1}2\rangle$-directions, as observed in the experiment (compare to Fig. 2.7). The agreement between experiment and simulation is strikingly confirmed also at higher temperatures. This is illustrated in Fig. 3.12, which confronts the shapes of islands grown at 245 K in the experiment (a) and in the KMC simulation (b). In both cases islands with triangular envelopes and preferential branch growth in the $\langle\bar{1}\bar{1}2\rangle$-directions form, along with less regular shapes.

Further evidence in support of the corner asymmetry hypothesis is provided by the pattern of aggregation of Pt adatoms at the two types of dense packed steps. The STM image in Fig. 3.13b shows a B-step that is relatively smooth, while the A-step is decorated by asperities. The KMC simulation (Fig. 3.13c) performed at the same temperature of 180 K and including the corner asymmetry shows the same difference in growth behavior at the two steps. The underlying reason can be recognized in Fig. 3.13a. The asperity formation is controlled by the fate of the atoms marked by a +, which are

[12] The actual diffusion paths of onefold coordinated atoms to higher coordinated sites do not necessarily follow the arrows in Fig. 3.11. For example, the jump indicated by the white arrow could also proceed via two or more bridge and hcp positions. However, such a path implies that the atom moves further away from the step and looses in-layer coordination.

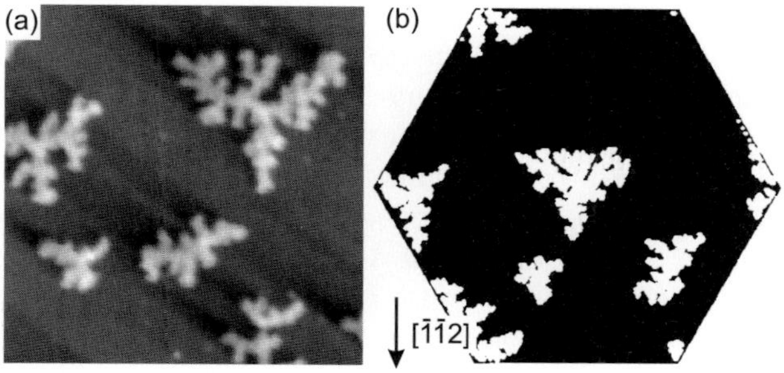

Fig. 3.12. (**a**) Experimental (500 Å× 500 Å) and (**b**) simulated (482 Å× 556 Å) island shapes at 245 K. In order to facilitate the comparison a thickening of 2 Å has been added to the branches of the simulated islands, which mimics the thickening of the topography induced by the STM tip (Fig. 18 of [115])

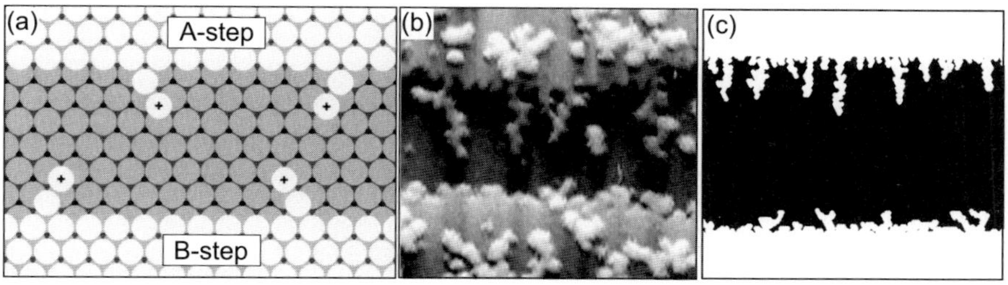

Fig. 3.13. (**a**) Based on the corner asymmetry assumption (see text), for a onefold coordinated atom attached to a step adatom at the B-step a nearest neighbor hop towards the step edge is possible via bridge positions, whereas no such easy path exists for an atom attached to a step adatom at the A-step. In (**b**) experiments and (**c**) simulations at 180 K, asperity growth is observed at A-step edges, while the B-step edges remain relatively smooth (based on Fig. 19 of [115])

laterally onefold coordinated to step adatoms: While those near the B-step can reach the step by a jump via bridge positions, the atoms at the A-step can relax only through the unfavorable atop sites. Accordingly the latter will remain longer in the onefold coordinated position, and have a higher chance of becoming stabilized by another arriving atom. Once initiated, the tip of the asperity will receive more adatom supply than the sides and its growth will accelerate (compare also Fig. 3.14b below and Sect. 3.6).

In summary, the corner asymmetry effect explains satisfactorily the temperature dependence of the in-layer coordination of island atoms, the island shape, and the aggregation behavior at step edges on Pt(111). The same mechanism would be expected to apply also to other fcc(111) systems which exhibit fractal-dendritic island growth. Indeed, DFT calculations for a onefold coordinated atom at an island corner on Al(111) yield a much smaller activation energy for a hop towards an A-step (0.04 eV) than towards a B-step (0.17 eV) [265], i.e. a lower activation energy for bridge hops, consistent with the low temperature fractal-dendritic Al island shape on Al(111) (compare Fig. 3.9). Likewise, the potential energy landscape for a onefold coordinated Ir atom at the corner of an Ir island is consistent with the preference for bridge hops [252], explaining the orientation and fractal-dendritic appearance of the low temperature Ir islands on Ir(111) [137]. Also for Ag(111), EMT calculations yield a smaller activation barrier for the hop of a onefold coordinated corner atom towards the A-step compared to the B-step [259].

For Pt(111), however, the ab initio DFT calculations are in contradiction with the picture developed above: The activation energy for a hop towards the A-step is 0.49 eV, whereas the hop to the B-step requires only 0.40 eV [140] (process 3b in Fig. 3.2, compare to Table 3.1). The same ordering of the activation barriers for a onefold coordinated atom was also found earlier in EMT calculations [213], although the EMT barriers are more than a factor of two smaller than the DFT values (see Table 3.1). Since the accuracy of the DFT calculations is comparable to the activation energy difference of 0.09 eV, one might be tempted to dismiss these results in favor of the nice agreement obtained between KMC simulations and experiments for fractal-dendritic islands on Pt(111). Nevertheless, in the following we show that it is possible to develop an alternative scenario for fractal-dendritic growth, which removes the discrepancy with theory (at the price of simplicity and definiteness, however).

The hcp-site Effect. It was noted in Sect. 2.4.1 that the potential energy landscape of an adatom is likely to be influenced by the presence of a step in its vicinity. This leads to effects which cannot be captured in simple nearest neighbor lattice models, and which may be important for island growth at low temperatures. Figure 3.14a shows the diffusion paths of an adatom towards the corner of an adatom island. As mentioned already in Sect. 2.2.3, the minimum energy path proceeds via intermediate hcp-sites, which are local minima of the potential energy [50, 68, 86]. When the adatom reaches the hcp-site at the corner marked by a full black circle, it is in close vicinity to the island boundary. The difference in coordination between the two possible final (fcc) adatom positions at the step edge is then likely to affect the adatom potential energy at the saddle point. The diffusion path from the hcp-site towards the laterally twofold coordinated A-step site will have a smaller activation energy than the one towards the laterally onefold coordinated corner site. For adatoms approaching the island corner this implies a preferential attachment to laterally twofold coordinated sites at the A-step, mimicking at least partially the corner asymmetry effect discussed above (compare to Fig. 3.14b). Since moves from hcp-sites towards fcc-sites set in at 45 K for Pt(111) [50], the effect will be operative already at the onset of adatom diffusion and nucleation. The hcp-site effect, which was first pointed out on the basis of EMT calculations for Ag islands on Pt(111) [259], will thus contribute to the formation of fractal-dendritic islands with preferential growth in the

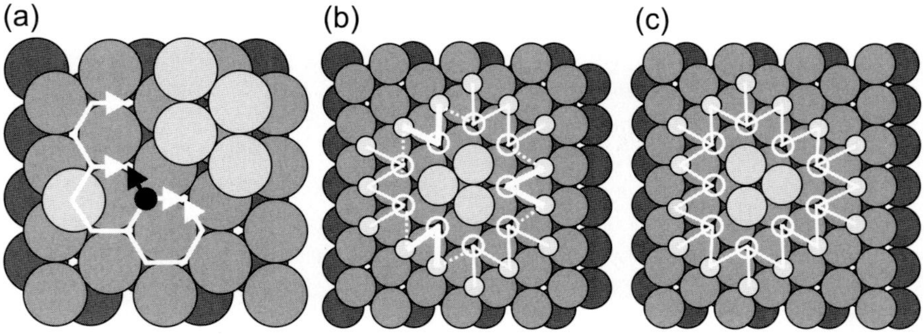

Fig. 3.14. (**a**) Schematic sketch of the hcp-site effect as suggested in [259]. Diffusion paths for the adatom towards the adatom cluster are indicated. At the hcp-site next to the island corner (marked by a full black dot) the influence of the nearby aggregate atoms makes the activation energy for motion towards the twofold coordinated site smaller than to the laterally onefold coordinated site. Consequently, the diffusion path indicated by the white arrow will be forbidden. (**b**) Schematic sketch of the possible nearest neighbor hops of adatoms (small balls) towards a trimer exhibiting A-steps. Final positions are indicated by white circles. Due to the hcp-site effect the nn–hops indicated by the dotted lines are absent, leading to a doubling of the probability for nn–hops indicated by the fat lines. Therefore adatoms will attach most likely to the A-step sites, transforming the trimer into a cluster dominated by B-steps. (**c**) Schematic sketch of the possible nearest neighbor hops of adatoms towards a trimer exhibiting B-steps. Adatoms preferentially attach to laterally onefold coordinated sites, initiating branch growth in the absence of jumps from onefold to twofold coordinated sites

$\langle\bar{1}\bar{1}2\rangle$-directions, irrespective of the corner asymmetry[13] (compare Figs. 3.14b and c).

Based on the hcp-site effect a consistent picture for the formation of fractal-dendritic islands on Pt(111), which avoids the discrepancy with the DFT calculations for the corner asymmetry, can be sketched as follows: At temperatures below the onset of jumps from laterally onefold to higher coordinated sites, the hcp-site effect is solely responsible for the increase of the coordination number C beyond C_{f1} for type-I fractals (see Fig. 3.10a), as well as for the experimentally observed shape anisotropy with preferential growth into the $\langle\bar{1}\bar{1}2\rangle$-directions. With the onset of thermally activated hops of laterally onefold coordinated atoms to twofold coordinated sites between 160 K (towards B-steps) and 200 K (towards A-steps), as deduced from the ab initio calculations [87] using (3.1), C increases further to the value C_{f2}. Since the hcp-site effect continues to favor adatom attachment to A-steps near corners even above 200 K, the growth anisotropy is maintained at higher temperatures.

[13] Adatoms on Ir(111) are observed to attach to island corners, i.e. to onefold coordinated sites [252]. Therefore on Ir(111) the hcp-site effect appears not to be effective.

Whether the incorporation of the hcp-site effect in KMC simulations of fractal-dendritic island growth on Pt(111) improves the agreement between simulations and experiments still needs to be tested. It appears, however, that while this effect should yield at least slightly different island shapes and temperature dependencies of C for Pt(111), its possible consequences are obscured for those systems in which both the corner asymmetry and the hcp-site effect favor attachment to A-steps near corners. This must be assumed to be the case for Al islands on Al(111). It is therefore not surprising that KMC simulations [266] using the corner asymmetry energies calculated by DFT [265] adequately reproduce the fractal-dendritic island shapes on Al(111) shown in Fig. 3.9b [see Sect. 3.7.2 for examples of simulated island shapes on Al(111)].

***

For completeness we note that recently a third scenario based on a barrier for attachment of adatoms to B-steps has been developed [267]. While the corresponding KMC simulations for Pt on Pt(111) qualitatively reproduce the correct shape anisotropy, at 180 K the simulated islands are considerably more compact than those observed experimentally.

The increase in C_{exp} beyond the level set by C_{f2}, which is seen in Fig. 3.10a for $T > 265\,\mathrm{K}$, indicates the onset of a transition from fractal-dendritic growth to compact island growth. It is caused by the appearance of a new atomic process at the island boundary, the motion of step adatoms along dense packed steps (process 4 in Fig. 3.2). With an onset temperature of 275 K as suggested by Fig. 3.10a, (3.1) yields 0.7 eV as a crude estimate for the activation energy of this process. This number is in reasonable agreement with the experimental FIM values for the activation energy of A-step diffusion. Also DFT calculations, which give somewhat higher edge diffusion barriers, support the idea that the shape transition is triggered by the onset of A-step diffusion (see Table 3.1). The compact growth regime is treated in the next section, while further discussion on the morphological transition between fractal-dendritic and compact islands is provided in Sect. 3.7.2.

3.4 Compact, Triangular Islands

Homoepitaxial growth on Pt(111) and Al(111) in the temperature range between 20 % and 30 % of the melting temperature $\mathrm{T_m}$ produces compact triangular islands. Figure 3.15 exhibits STM topographs for these two materials at $0.20\,\mathrm{T_m}$ after deposition of about 0.15 ML. The resulting islands are mainly bounded by B-steps, although some step edge roughness is visible. The shape typically develops from smaller, more or less hexagonal islands to triangular islands. This is also suggested by the coexistence of small hexagonal islands and larger triangular islands on the same topograph (see Fig. 3.15b).

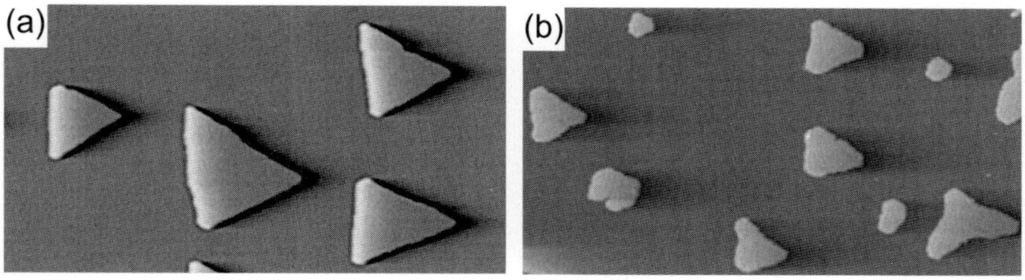

Fig. 3.15. STM topographs (1100 Å× 600 Å) after deposition at $0.2\,T_m$ with a rate of 7×10^{-3} ML/s exhibiting compact triangular islands on fcc(111) surfaces. (**a**) 0.15 ML Pt deposited on Pt(111) at 400 K. (**b**) 0.11 ML Al deposited on Al(111) at 183 K

These islands form in a temperature range where the motion of step adatoms along dense packed steps (process 4 in Fig. 3.2) and around island corners (process 7+3b) is thermally activated. On the other hand, the formation of compact triangular islands requires neither step adatom formation (process 8) nor step adatom detachment (process 9). This conclusion follows from the absence of observable thermal step fluctuations, and from the fact that the islands grown on Pt(111) at $0.2\,T_m$ do not change their size when the deposition flux is turned off.

In order to understand the growth shapes of compact islands, it is useful to define the advancement speed v of a step normal to itself[14]. Starting from a regular hexagon with two inequivalent step types and different speeds v_A and v_B, the geometric construction in Fig. 3.16a shows that the slow B-step edge expands at the expense of the faster A-step. A stationary hexagonal shape with a constant ratio of the lengths of A- and B-step segments develops if $v_A/v_B < 2$. Since the distances $d_{A,B}$ of the step edges from the island center grow as $d_{A,B} = v_{A,B}t$, the stationary shape is evidently characterized by

$$v_A/v_B = d_A/d_B. \tag{3.25}$$

This is the kinematic analog of the relation (3.10) derived from the Wulff construction for equilibrium shapes. In fact, the Wulff construction applies quite generally to compact growth shapes, with the normal step edge velocity v taking over the role of the step free energy in the equilibrium case [268]. This was known already to Wulff [221]; a general derivation can be found in [269] and [270]. For the two-dimensional islands of interest here, the kinematic Wulff construction implies that the asymptotic growth shape attained at long times from an arbitrarily shaped initial seed satisfies, in analogy to (3.6),

$$\lim_{t\to\infty} \frac{r(\phi,t)}{t} = \min_{\vartheta} \frac{v(\vartheta)}{\cos(\vartheta-\phi)}. \tag{3.26}$$

[14] The description used here is analogous to the description of facet growth for 3D crystals [212, 268].

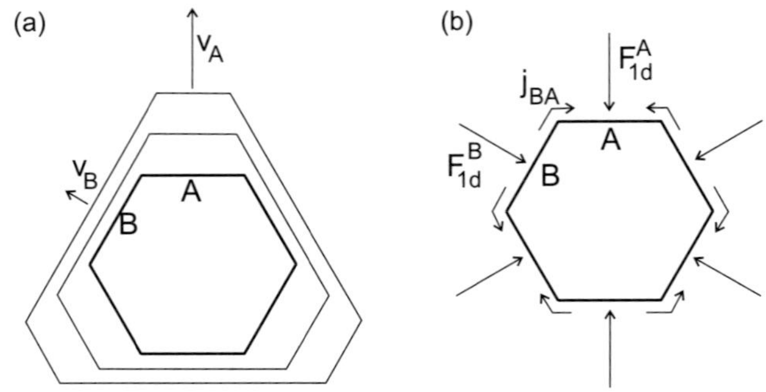

Fig. 3.16. (**a**) Starting with a small hexagonal adatom island bounded by dense packed A- and B-steps, a triangular B-step island eventually grows, if the step advancement speed for the B-step v_{B} is significantly larger than v_{A}. (**b**) Assuming identical fluxes of adatoms $F_{\mathrm{1d}}^{\mathrm{A}}$ and $F_{\mathrm{1d}}^{\mathrm{B}}$ to the A- and B-step sides of the hexagonal island, the condition $v_{\mathrm{A}} > v_{\mathrm{B}}$ may be realized by a current j_{BA} of material from the B-step to the A-step

In this equation $r(\phi, t)$ describes the time-dependent island shape in two-dimensional polar coordinates, and $v(\vartheta)$ is the normal step edge displacement speed along the direction ϑ.

Since the distance from the center of an equilateral triangle to one of the corners is twice the (minimal) distance to one of the edges, it follows from (3.25) that the fast A-steps disappear, and the growth shape degenerates to a triangle when $v_{\mathrm{A}}/v_{\mathrm{B}} > 2$. The experimental observation of triangular islands bounded by B-steps implies that this inequality is satisfied. For a regular, hexagonal island the supply of adatoms is approximately isotropic, i.e. the adatom fluxes $F_{\mathrm{1d}}^{\mathrm{A}}$ and $F_{\mathrm{1d}}^{\mathrm{B}}$ towards the two types of step edges are identical. To explain the difference in step edge velocities, one is thus forced to assume[15] a current j_{BA} of material from the B-step to the A-step (Fig. 3.16b). In fact, the corner asymmetry and the hcp-site effect discussed in the preceding section give rise to a current from the B-step to the A-step. However, these effects influence only the adatoms arriving right at an island corner. For the large islands of interest here, they are too weak to produce the required difference in step advancement speed.

One possible origin of a mass flux from B-steps to A-steps is a difference in binding energy of step adatoms at the two types of steps. A step adatom migrating along the island edge from the side with the smaller binding energy to the one with the larger binding energy is unlikely to return. The awning approximation discussed in Sect. 3.2.3 indeed predicts a binding energy difference of step adatoms to the two types of step edges, which is related through

[15] In principle different rates for the detachment of adatoms from step edges would lead to different net mass fluxes to the step, and hence to different speeds of step advancement. However, experimentally it can be ruled out that detachment is relevant at all on the time scale of the experiment.

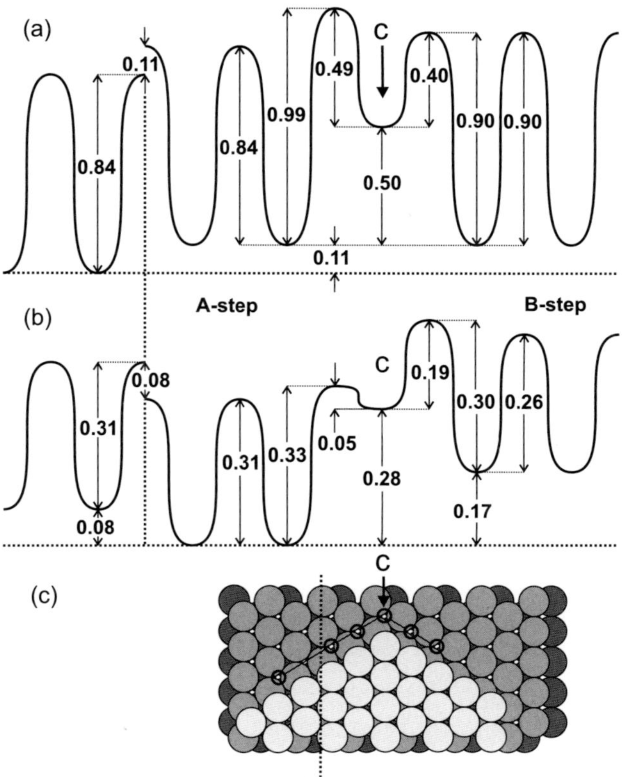

Fig. 3.17. Potential energy of an atom bound to an island edge as obtained by DFT within the LDA for (**a**) Pt(111) [140] and within the GGA for (**b**) Al(111) [265] along the sequence of positions indicated in (**c**). The jump in the potential energy curves indicated by a vertical dotted line results from the constraint of identical potential energies of the atom in the kink positions at the two step edges far away from the corner. It may reflect errors in the calculations as well as a gradual change of the potential energy landscape along the step edge. In any case, binding energy differences for adatoms at the two types of steps should be referred to positions far from the corner. Thus the binding energy of an adatom decreases by 0.11 eV and 0.09 eV if moved from the A- to the B-step on Pt(111) and Al(111), respectively

(3.19) to the difference in step energies. For Al(111) and Pt(111) $\delta_A > \delta_B$, and thus it follows immediately from (3.19) that an adatom is more strongly bound to the A-step.

A more detailed view is provided in Fig. 3.17, which shows the potential energy diagrams for an atom bound to the step edge on Pt(111) [140] and Al(111) [265] as obtained from DFT calculations. To the left of the dashed line in Figs. 3.17a and b the potential energy at the A-step far away from the corner is indicated, as obtained by matching the potential energy levels of an atom in the kink positions at the A- and B-step. These energies have to be the same at the two steps in order to allow for an island containing both types of steps to be in equilibrium with the terrace adatoms. The relative binding energy gain at the A-step amounts to 0.11 eV and 0.09 eV for Pt(111) and Al(111), respectively, which is consistent with the awning approximation (see Sect. 3.2.3).

The efficient transport of step adatoms from the B- to the A-step requires not only migration along the step edges, but also the rounding of the island corner. As can be inferred from the potential energy diagrams shown in Fig. 3.17, the activation energy for the motion of step adatoms around corners is in both cases only slightly larger than the one for motion along step. At $0.2\,T_\mathrm{m}$, where edge diffusion is activated, a significant transfer of material from the B-steps to the A-steps can thus plausibly be assumed to take place, which easily explains the formation of triangular compact islands.

Further support for this picture is provided by the fact that hexagonal, rather than triangular islands are observed in the corresponding temperature range during homoepitaxial growth on Ag(111) [203, 271]. For Ag(111) the two dense packed steps have identical energies (see Table 3.2), and thus no flux of adatoms between the two types of island edges which breaks the hexagonal symmetry would be expected. Our conclusion is also in agreement with KMC simulations, which show that even slightly stronger binding of step adatoms to one type of steps causes the disappearance of the corresponding island edges and leads to a pronounced triangular island shape [266, 272–274] (see also Fig. 3.18).

One may ask whether the different activation energies for motion of step adatoms along the two types of dense packed steps also affect the island growth shape. Initially it was argued that slower diffusion along one type of step will increase its roughness and kink density, thus inducing a larger speed of advancement and leading to the disappearance of the corresponding island edges [206]. Several independent KMC simulations explicitly testing the influence of different diffusion rates along the dense packed steps [266, 272, 273] find, however, the opposite result[16]: Slower diffusion along one type of steps tends to *extend* the size of the corresponding edges (Fig. 3.18).

This somewhat counterintuitive behavior can be explained as follows. In the temperature range of interest, kinks are generated primarily by the nucleation and growth of adatom rows (one-dimensional islands) along the island edge. The nucleation theory described in Sect. 2.4.3 provides the estimate[17]

$$l_\mathrm{1d} \approx (aD_\mathrm{e}/F_\mathrm{1d})^{1/4} \tag{3.27}$$

[16] The scenario of [206] suggesting faster growth of the edges with slower diffusion was supported by KMC simulations for island shapes on Al(111) [275–277]. However, as will be discussed below in Sect. 3.7.2, these simulations show significant deviations from the experimental island shapes, which indicate an incomplete treatment of the energetics of step edge atoms.

[17] See also Sect. 4.7.3.

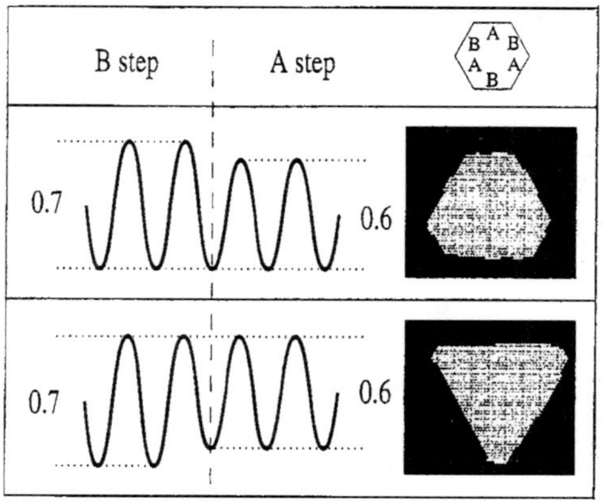

Fig. 3.18. Illustration of the opposing effects of the anisotropy of the step adatom binding energy and the edge diffusion barrier on simulated island shapes. In the upper panel the binding energies at the two step edges are the same, while in the lower panel the transition state energies have been aligned. The edge diffusion barriers at the two step edges are the same in both cases (Fig. 4 of [272] reprinted with permission from Elsevier Science)

for the typical spacing between kinks, where D_e is the edge diffusion coefficient and F_1d denotes the adatom flux from the terrace onto the edge, which in the present case depends on the deposition flux, the island number density and the island size (see Sect. 3.7.3). Thus the kink density is indeed larger at the step with the lower edge mobility. However, the higher density of sinks for the adsorption of step adatoms is overcompensated by the fact that the time required for a step adatom to reach one of the sinks is smaller at the step with faster edge diffusion [272]. To be quantitative, consider an adatom which reaches the island edge near the corner between the A-step and the B-step. The distance to the nearest kink along the A-step (B-step) is then l_1d^A (l_1d^B), and the times required to reach the kinks are

$$\tau^{\mathrm{A,B}} \approx (l_\mathrm{1d}^{\mathrm{A,B}})^2 / D_\mathrm{e}^{\mathrm{A,B}} \sim (D_\mathrm{e}^{\mathrm{A,B}})^{-1/2}. \tag{3.28}$$

Thus the kink at the step with faster diffusion, though further away, is reached first, and consequently this step captures the majority of the arriving atoms. This conclusion is confirmed by a more detailed diffusion calculation [278]. When both the binding energies of step adatoms and the activation energies for step adatom diffusion differ at the two steps, a net mass transfer to the A-step is predicted to occur if

$$\sqrt{D_\mathrm{e}^\mathrm{A}/D_\mathrm{e}^\mathrm{B}} \exp[\Delta E_\mathrm{stad}/k_\mathrm{B}T] > 1, \tag{3.29}$$

where ΔE_stad is the binding energy difference [see (3.19)]. Thus the difference in step edge diffusion barriers has to be at least twice the difference in binding energies in order for it to determine the island shape. In particular, if the difference in binding energies is equal to the difference in the edge diffusion barriers, as is the case in the lower panel of Fig. 3.18, the binding energy effect prevails and the islands are bounded primarily by A-steps. According to the energy landscapes in Fig. 3.17, the edge diffusion and binding energy anisotropies act in the same direction for Pt(111), both favoring islands bounded by B-steps, while the two effects compete for Al(111). However, since the difference in the edge diffusion barriers on Al(111) amounts to only 0.05 eV, according to (3.29) the binding energy effect dominates also in this case. Based on the comparison of experimentally observed and simulated island shapes, in Sect. 3.7.2 it will be shown that in fact also for Al(111) both anisotropies favor B-step formation.

3.5 Adsorbates and Island Shapes

In this section the influence of an adsorbed gas species on the shape of islands will be addressed. Depending on the concentration of the adsorbed species one may distinguish two regimes: a *surfactant regime* of large adsorbate concentration, in which gases are intentionally adsorbed to or close to saturation in order to modify the growth behavior, and a *contamination regime* of small adsorbate concentration, which may result from adsorption out of the residual gas pressure during or prior to deposition. In the surfactant regime island shape changes are expected, and have been observed in a number of cases [279–281]. An example of such an island shape change due to a preadsorbed layer will be discussed in the context of surfactant-induced layer-by-layer growth in Chap. 5. In the present section we are concerned with the contamination regime. The discussion is focused on the case of CO adsorbates in the homoepitaxial growth on Pt(111) [131], an example which has had historical implications for the understanding of island shapes on fcc(111) surfaces.

The strong influence that an extremely small amount of an adsorbed species may exert on the island shape is strikingly demonstrated by the effect of CO molecules on Pt island growth on Pt(111) at 400 K (Fig. 3.19) [131]. The five STM topographs show island morphologies after deposition of 0.15 ML at 400 K in the presence of different CO partial pressures. The CO pressure was set 100 s prior to deposition and was maintained during the 30 s deposition time. The Pt islands grown under nominally clean deposition conditions ($p_{\mathrm{CO}} < 5 \times 10^{-12}$ mbar with the total background pressure during deposition below 2×10^{-11} mbar) in Fig. 3.19a have compact, triangular shapes bounded by B-steps with a certain number of kinks, as discussed in Sect. 3.4. The triangles point to the right. A CO partial pressure of 1×10^{-10} mbar causes drastic changes in the island shape (Fig. 3.19b): The islands are no longer triangular but have compact irregular shapes which vary significantly and also exhibit portions of the dense packed A-steps. In

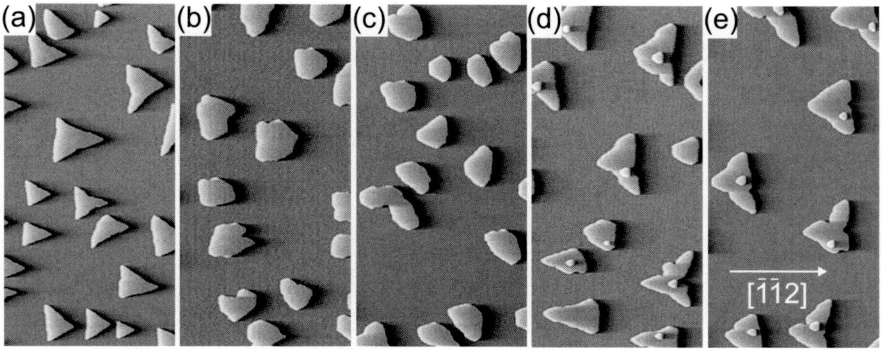

Fig. 3.19. STM topographs after deposition of 0.15 ML Pt on Pt(111) at 400 K with a deposition rate $F = 5.0 \pm 0.3 \times 10^{-3}$ ML s^{-1} and with different CO partial pressures during growth. (**a**) $p_{\mathrm{CO}} < 5 \times 10^{-12}$ mbar ("clean"); (**b**) $p_{\mathrm{CO}} = 1 \times 10^{-10}$ mbar; (**c**) $p_{\mathrm{CO}} = 4.7 \times 10^{-10}$ mbar; (**d**) $p_{\mathrm{CO}} = 9.5 \times 10^{-10}$ mbar; (**e**) $p_{\mathrm{CO}} = 1.9 \times 10^{-9}$ mbar. CO was admitted 100 s prior to and during deposition. The topograph size is always 1250 Å× 2500 Å

Fig. 3.19c at a CO pressure of 4.7×10^{-10} mbar, the island edges become dominated by A-steps and the islands start to develop a triangular shape which is now pointing to the left. At CO pressures of 9.5×10^{-10} mbar (Fig. 3.19d) and 1.9×10^{-9} mbar (Fig. 3.19e) fully triangular islands consisting almost exclusively of A-steps form. In addition, most of the islands in Figs. 3.19d and e support a second-layer island, in contrast to those grown at lower CO pressure. This shows that the presence of CO increases the nucleation probability on top of the islands, an effect which is of great importance for the morphological evolution of multilayer films, and which will be discussed in the next chapter.

How is it possible that small amounts of CO (to be quantified below) bring about such dramatic changes? To answer this question it is useful to recall the adsorption behavior of CO on Pt(111). On the flat terrace and at low coverage, CO adsorbs preferentially on-top [282–284]. In the presence of surface defects the binding of the molecule is stronger, the lower the coordination of the Pt–atom that it binds to [194]: In the limit of low coverage the adsorption energy is about 1.4 eV on the terrace [196, 285], about 1.7 eV at step atoms [282,283,286,287], and even higher for kink positions [287]. At the temperatures of interest here, the CO molecule is highly mobile [196, 288]. It rapidly exchanges between terrace and step edge sites [289], preferring the latter by far due to their stronger binding. The saturation coverage of CO at steps is one molecule per step atom [283]. An upper bound on the CO–coverage at the end of the 130 s exposure used for the experiments presented in Fig. 3.19 may be estimated assuming the sticking coefficient of 0.79 determined at 300 K [284], and an infinite lifetime of the adsorbed molecule.

This yields[18] $\Theta_{CO} \approx 1.9 \times 10^{-3}$ ML for Fig. 3.19b, $\Theta_{CO} \approx 9 \times 10^{-3}$ ML for Fig. 3.19c and $\Theta_{CO} \approx 1.8 \times 10^{-2}$ ML for Fig. 3.19d.

Using these estimates, the following picture emerges: At the beginning of the deposition process the highly mobile CO molecules adsorbed on the (111) terrace become captured by the newly formed step edges. Through their binding to the step atoms, these molecules will change the kinetics and energetics of adatom incorporation into steps, and thus affect the island growth. This implies that the change in island shape should saturate already for a small amount of adsorbed CO, which is sufficient to populate all step atoms. Saturation of the island–shape effect is reached somewhere between Fig. 3.19c and 3.19d (there is no further shape change between Fig. 3.19d and 3.19e), i.e. for a coverage between 9×10^{-3} ML and 1.8×10^{-2} ML. With the additional information that the ratio of the number of atoms in a step position to the number of atoms in a terrace position after deposition of 0.15 ML at 400 K is about $R_{s/t} \approx 9 \times 10^{-3}$, the upper bound coverages result in a CO concentration of 1-2 CO molecules per step atom, which is reasonable. The dramatic effects seen already in Fig. 3.19b correspond to only 0.2 CO molecules per step atom. Deviations from the clean growth behavior are observed for CO concentrations as low as 0.1 CO molecules per step atom, which may be realized e.g. by several 100 s CO exposure at a partial pressure in the low 10^{-11} mbar range.

While the selective adsorption of CO at step edges – the locations where growth takes place – explains the extreme sensitivity of growth to CO exposure, the CO-adsorption induced switch in island shape from triangles bounded by B-steps to triangles bounded by A-steps still needs to be analyzed. In the clean case discussed in Sect. 3.4 it was shown that triangular B-step islands develop, because step adatoms bind more strongly to A-steps. Building on this idea, if CO passivates A-steps more efficiently, so that step adatoms bind less well to CO–decorated A-steps than to B-steps, the observed switch in island shape is a consequence [131]. The differences in peak desorption temperatures between terrace– and step–CO for vicinals of Pt(111) consisting of B-steps or A-steps indeed suggest a stronger binding of CO to A-steps [282, 283, 286, 287, 289, 290], which lends credit to the above scenario. Kinetic Monte Carlo simulations using preferential binding of CO to A-steps as an input, implying thus a selective passivation of A-steps to step adatom binding, indeed reproduce the adsorption-induced switch in island shape from B-step to A-step islands [274].

For a number of years the understanding of the evolution of island shapes on fcc(111) metal surfaces with temperature was obscured by measurements which had been distorted by the effect of CO adsorption during growth [206]. In these early measurements electron–beam evaporation was used instead of

[18] More precise coverages may be calculated using data for the pressure dependence of the saturation coverage of CO on Pt(111) at 400 K [196]. For the short exposure times and low pressures under concern here these coverages are only 10%–20% lower than the infinite lifetime estimates.

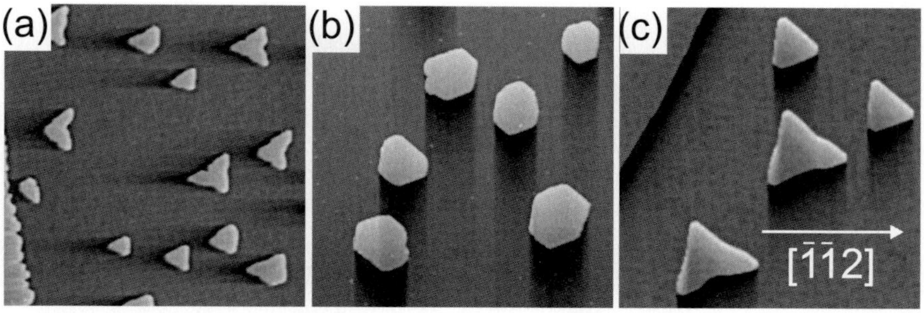

Fig. 3.20. Shapes of Pt islands on Pt(111) grown at various temperatures under the influence of a significant CO partial pressure (see text) and with a typical deposition rate of 1×10^{-2} ML/s. (**a**) $T = 400$ K, $\Theta = 0.2$ ML, topograph size 2200 Å× 2200 Å. (**b**) $T = 455$ K, $\Theta = 0.14$ ML, topograph size 1100 Å× 1100 Å. (**c**) $T = 640$ K, $\Theta = 0.15$ ML, topograph size 4000 Å× 4000 Å

resistive wire evaporation, which resulted in a measured CO partial pressure rise of a few 10^{-10} mbar due to deposition. The actual CO pressure during deposition at the sample, close to the evaporator, was probably larger than the pressure rise estimated from measurements with ion gauge and mass spectrometer far away from the evaporator. In the measurements at 400 K a triangular island shape bounded mainly by A–steps was found (Fig. 3.20a). In a narrow temperature interval between 450 K and 470 K a hexagonal transition shape was observed (Fig. 3.20b), while above 470 K triangular islands bounded by B–steps formed (Fig. 3.20c).

The apparent phenomenon of a temperature-dependent inversion of step advancement speed anisotropy triggered a considerable amount of theoretical work [272,273,291–293] and influenced predictions for homoepitaxial growth of other materials [275–277]. In most of these publications the difficulty to reproduce the observed island shape transition with reasonable assumptions about the atomic processes was explicitly stated. With hindsight, it is easy to relate the shape transition illustrated in Fig. 3.20 to the known adsorption and desorption behavior of CO on Pt(111) [294]: At 400 K, where the lifetime of the impinging CO molecules is still large (of the order of 1000 s) compared to the time scale of the experiment, the amount of CO accumulated during the heating up of the evaporator and the deposition process is sufficient to completely decorate the steps by CO. This leads to triangular islands bounded by A-steps, similar to the islands generated under the controlled conditions of Fig. 3.19d. As the growth temperature is raised, the lifetime of CO rapidly decreases and the CO concentration is no longer sufficient to fully decorate the steps during growth. Consequently in Fig. 3.20b hexagonal islands develop, which have boundaries partially free of CO (B-steps) and partially covered by CO (A-steps). At even higher temperatures CO adsorption is negligible and the islands grown in the presence of a small CO partial pressure are

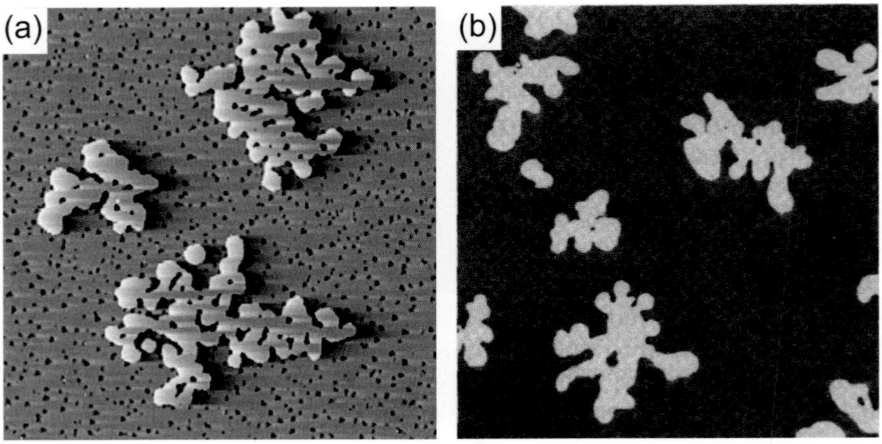

Fig. 3.21. Homoepitaxial growth on contaminated surfaces leading to increased island number densities and ramified (fractal) island shapes. (**a**) Islands formed after deposition of 0.09 ML Al on Al(111) at 300 K. Prior to deposition the surface was exposed to 0.027 Langmuir of oxygen (one Langmuir corresponds to an exposure of 1×10^{-6} Torr/s^{-1}), causing the formation of small O-islands (dark) with a coverage of about 0.1 ML. Topograph size 1500 Å× 1500 Å. (**b**) Islands formed after deposition of 1.1 ML of Ag on Ag(111) at room temperature. Prior to Ag deposition 0.08 ML Sb was adsorbed, which alloys in the first deposited Ag monolayer (see text). Topograph size 4800 Å× 4500 Å (Fig. 2b of [203])

identical to those grown under nominally clean conditions (compare Fig. 3.1e and Fig. 3.20c).

The effect of CO on the Pt island shapes on Pt(111) is not the only example for the extreme sensitivity of growth to small amounts of adsorbates. In general, any adsorbate present on the surface will tend to influence the shape of the growing islands, be it in heteroepitaxial or homoepitaxial growth. Figure 3.21 gives two examples, where ramified islands grow instead of the compact shapes expected under clean conditions at the given temperature. Figure 3.21a shows ramified Al islands, which were grown on Al(111) with small immobile oxygen islands (dark) formed prior to Al deposition by adsorption at 420 K. The Al adatoms and Al island steps are repelled by the oxygen islands [202]. The repulsion of the Al steps by the oxygen forces the advancing island edges to grow around O–islands, leaving holes in the Al islands and a ramified island boundary [202].

In Fig. 3.21b, 0.08 ML Sb was adsorbed on Ag(111) prior to deposition [203]. Subsequently 1.1 ML Ag was deposited. During the growth of the first monolayer, a Ag–Sb surface alloy is formed. Further growth on this dilute alloy layer differs from growth on the clean surface, because the incorporated Sb slows down the Ag diffusion (leading to a higher island number density) and apparently impedes the formation of compact islands [203]. Theoretically a repulsive interaction between the Sb patches in the first monolayer and the

Ag adatoms migrating in the second layer has been postulated [204]. As in the case of Al with preadsorbed oxygen, this would explain the formation of ramified islands.

3.6 Shape Instabilities and the Diffusion Field

As a complement to the atomic scale description of island shape evolution developed in the preceding sections, here a continuum approach to island growth will be sketched. This allows to make several qualitative remarks on the relation between the adatom diffusion field, the island shapes and the origin of shape instabilities. All the effects discussed rely on the inhomogeneous supply of adatoms to the island. The most important origin of such inhomogenities is the point effect of diffusion, which amplifies shape fluctuations and leads to shape instabilities.

The relevance of the point effect of diffusion for crystal growth was probably recognized for the first time by Lehmann in 1888 [295]. The idea is visualized in Fig. 3.22. Consider a three dimensional rhomboedric crystal growing from solution. While at the interface between crystal and liquid the solute concentration $c(\boldsymbol{x})$ is given by its equilibrium value c_0, far away from the crystal center it is kept at $c_\infty > c_0$. The flux j of solute particles onto a surface element of the crystal is given by

$$j = -D_\mathrm{s} \boldsymbol{n}_\mathrm{s} \cdot \nabla c, \tag{3.30}$$

where D_s is the diffusion coefficient of the solute particles in the solution and $\boldsymbol{n}_\mathrm{s}$ is the unit vector normal to the surface. The surface element then advances at the velocity

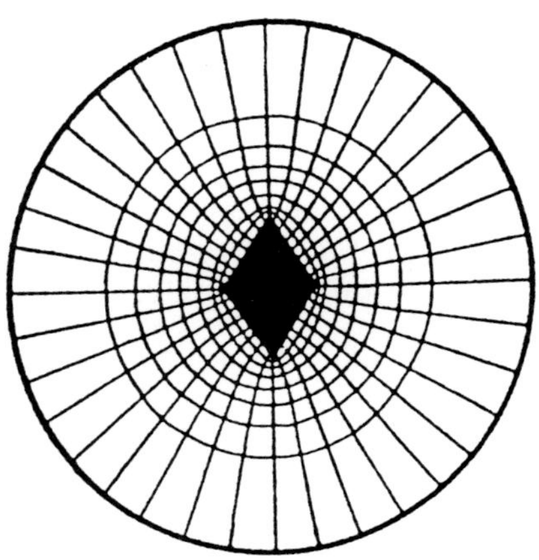

Fig. 3.22. Isoconcentration lines and rays of diffusion for the growth of a rhomboedric crystal from a supersaturated solution (Fig. 18 of [212])

$$V = \frac{D_s}{c_c - c_0} \boldsymbol{n}_s \cdot \nabla c, \tag{3.31}$$

where c_c is the concentration of the solute in the crystal. For slow advancement speeds the concentration field can be determined quasistatically[19] as the solution of the Laplace equation

$$D_s \nabla^2 c = 0. \tag{3.32}$$

This is evidently mathematically equivalent to an electrostatic problem with Dirichlet boundary conditions. Consequently, far away from the crystal surface the concentration behaves like the electrostatic potential of a sphere capacitor. However, close to the crystal surface the nonspherical shape of the crystal is felt. The "rays of diffusion" (corresponding to the electric field lines) meet the crystal surface at right angles, and the isoconcentration lines (equipotential lines in the electrostatic analogy) bunch at the crystal corners. Accordingly, at the crystal corners the concentration gradient is largest, and solute material is supplied preferentially to the corners[20]. The speed of advancement at the crystal corners will thus be significantly larger than at the crystal facets, and the crystal shape is unstable, i.e. it cannot grow in a self similar fashion. The concept described here was used by Lehmann to explain the formation of *skeletal crystals*, crystal shapes resulting from the coarsening of polygonal crystals by the preferential growth of corners and edges [295].

The point effect of diffusion would seem to imply that every polygonal crystal shape is unstable, and that any protrusion resulting from a shape fluctuation is amplified [296]. This is not so, because the instability is counteracted by the stabilizing effects of interface diffusion and transport through the surrounding phase. Both stabilizing effects rely on the fact that the formation of a protrusion or tip in the crystal shape is associated with an increase in the interface area, which causes in turn an increase in the total interfacial free energy. This implies a restoring force proportional to the specific interface free energy γ, which induces mass transport along the interface and through the solution. Because of the Gibbs–Thomson relation (4.65) (see also (3.9)), both transport channels are driven by the differences in curvature and are directed from locations with positive curvature (protrusions) to locations with negative curvature (indentations).

A quantitative description of the competing stabilizing and destabilizing effects is provided by linear stability analysis, which was first performed by

[19] See Sect. 4.3.1 for a discussion of the quasistatic approximation in a related context.

[20] From an atomistic perspective the point effect of diffusion arises because corners and protrusions have a greater chance of capturing incoming atoms, thus screening other parts of the crystal. This is the very same mechanism that is responsible for the growth of the fractal aggregates discussed above in Sect. 3.3. Fractal growth by diffusion limited aggregation (DLA) is the manifestation of the diffusional shape instability acting across length scales in a highly noisy environment.

Mullins and Sekerka for the case of isotropic γ (a spherical crystal) [296,297]. The most important result of this analysis is that the shape instability sets in when the crystal has grown beyond a *critical radius* R^*. The critical radius is proportional to $\gamma/(c_\infty - c_0)$, which clearly reflects the competition between surface free energy and supersaturation.

The same concepts – the destabilizing point effect of diffusion competing with stabilizing, curvature-driven mass transport along island edges and across terraces – apply to two-dimensional island growth on a crystalline substrate. Some differences arise because the diffusing species (the adatoms) is not conserved, but rather is constantly replenished by the deposition flux F. This implies that, in the quasistatic approximation, the Laplace equation (3.32) is replaced by the Poisson equation

$$D\nabla^2 n + F = 0 \tag{3.33}$$

(see Sect. 4.3.1). Moreover, in contrast to the surface of the solute-grown crystal, the edges of the two-dimensional islands are fed from both sides, from the terrace as well as by atoms deposited on top of the island, although the fraction of the latter is small at small coverages.

Avignon and Chakraverty [298] extended the Mullins–Sekerka analysis to a circular, two-dimensional island growing on a surface subject to a constant deposition flux[21]. We summarize the most important results. The distorted island shape is expanded in circular harmonics,

$$r(\phi, t) = R(t) + \sum_{k=1}^{\infty} \varepsilon_k(t) \mathrm{e}^{\mathrm{i}k\phi}, \tag{3.34}$$

and evolution equations for the shape are derived to linear order in the ε_k. The circular shape is unstable, if the perturbation grows more rapidly than the mean island radius, i.e. if the condition

$$\frac{\mathrm{d}}{\mathrm{d}t}\left(\frac{\varepsilon_k}{R}\right) > 0 \tag{3.35}$$

is satisfied for some k. As the island grows, the instability condition (3.35) is first met for $k = 3$, corresponding to a deformation of threefold symmetry. The radius at which this occurs defines the lower critical island radius R_1^*. In contrast to the conventional Mullins–Sekerka scenario, however, here the circular island is restabilized above an upper critical radius $R_2^* > R_1^*$. This is attributed to the increasing fraction of atoms deposited on top of the island, which feed the island edge from above and thus counteract the point effect of diffusion. Depending on the growth parameters it is also possible that the two critical radii merge, such that the circular island remains stable for all times.

[21] A clear example of this kind of single island instability was observed in Si growth on Si(111) [299].

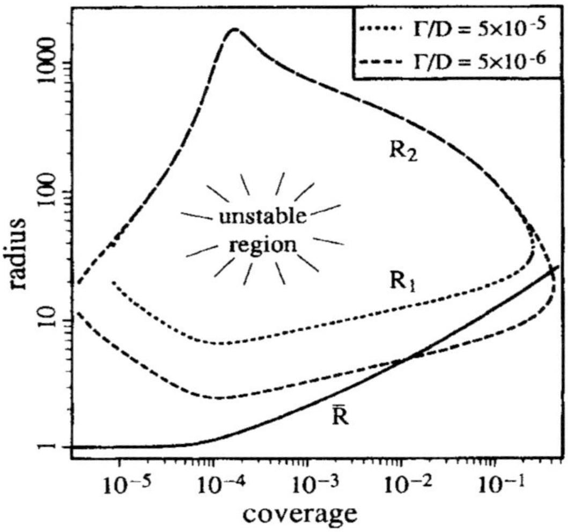

Fig. 3.23. Stability diagram for a circular island derived from self-consistent nucleation theory. R_1 and R_2 are the lower and upper critical island radii, and $\bar{R}$ denotes the mean radius of the ensemble of islands. The mean radius crosses the stability boundary for one set of conditions ($\Gamma/D = 5 \times 10^{-6}$, where $\Gamma = \Omega\sigma_{\rm st}\tilde{\delta}$), while for the second set of conditions ($\Gamma/D = 5 \times 10^{-5}$) the island remains stable (Fig. 2 of [300])

In [298] the growth of a single island was considered under conditions where a limiting adatom density is maintained far away from the island through the balance between deposition and evaporation. A refined treatment which takes into account the presence of other islands in a self-consistent manner[22] was presented by Bales and Chrzan [300]. Key features of the earlier analysis, such as the onset of the instability at $k = 3$, the restabilization at large island radius and the possibility of completely stable island growth, are recovered (Fig. 3.23). In addition, it is shown that the stability condition can be written in the form[23]

$$\frac{\sigma_{\rm st}\tilde{\delta}}{DN} > f_c(\Theta), \tag{3.36}$$

where f_c is a function of the coverage Θ only, independent of flux and temperature, and $\sigma_{\rm st}$ denotes the adatom mobility along the step edge (see Sect. 3.7.1). The theory assumes that island nucleation is irreversible, so that the scaling law (2.19) with $i^* = 1$ can be used.

For a qualitative discussion it suffices to consider the stability of a piece of a step edge, which may be part of an island boundary, or part of a step

[22] See also Sect. 2.3.1.

[23] A similar criterion, with the edge diffusion coefficient $D_{\rm e}$ replacing $\sigma_{\rm st}\tilde{\delta}$, can be derived from the estimate (3.49) in Sect. 3.7.3 for the critical size of a compact island with straight edges.

on a vicinal surface[24]. A protrusion of wavelength λ (wavenumber $q = 2\pi/\lambda$) can be written as $\varepsilon_q(y,t) = \varepsilon_0 \, \mathrm{e}^{\mathrm{i}qy+\omega(q)t}$, where the y-coordinate runs along the step (this is the analog of (3.34) for a straight step segment). The stability of the step is determined by the sign of the (relative) growth rate $\omega(q) = (\partial\varepsilon_q/\partial t)/\varepsilon_q$ of the perturbation. If $\omega(q) > 0$, the perturbation grows exponentially and the step is unstable with respect to fluctuations of wavelength $2\pi/q$, while for $\omega(q) < 0$ the perturbation is driven back and the straight step remains stable. Collecting order of magnitude estimates derived in [301, 302], one arrives at the following compact expression for the growth rate:

$$\omega(q) \approx \frac{F\Omega}{RN}|q| - \frac{Dn_1\tilde{\delta}\Omega^2}{k_{\mathrm{B}}T}|q|^3 - \sigma_{\mathrm{st}}\tilde{\delta}\Omega q^4. \tag{3.37}$$

Here R and N denote the average island size and island number density, and n_1 is the adatom density on the terrace. Equation (3.37) summarizes the main features of the diffusional shape instability. The first term is destabilizing (positive) and proportional to the deposition flux, while the second and third terms describe smoothening through terrace and edge diffusion, respectively; both stabilizing terms are proportional to the step edge stiffness $\tilde{\delta}$, representing the thermodynamic driving force for smoothening, and to the respective transport coefficients D and σ_{st}. The additional stabilizing effect of atoms that are incorporated to the step from the upper terrace is neglected in this analysis, which is therefore limited to small coverages.

According to (3.37), stabilizing effects dominate at small wavelengths (large q) and destabilizing effects at large wavelengths (small q). The possible range of wave numbers q of fluctuation-induced protrusions is limited from below by the island size, since wave numbers $q < 2\pi/R$ are impossible. Thus, as long as the islands are small, the stabilizing terms dominate the right hand side of (3.37) and all protrusions are driven back. However, as soon as the islands become larger than a critical value R^*, sufficiently small wave numbers (long wavelength protrusions) are possible such that island growth becomes unstable. The critical island size is found by balancing the destabilizing term with the dominant stabilizing term. Consequently R^* is generally expected to decrease with increasing flux and decreasing temperature[25].

***

In the following a number of examples for two-dimensional shape instabilities in epitaxial growth are discussed [228, 304]. The role of crystal anisotropy

[24] A detailed treatment of step instabilities on vicinal surfaces is provided in Sect. 4.7. For another example of a linear stability analysis see also Sect. 4.5.2.

[25] In fact the situation is more subtle because the destabilizing term in (3.37) depends on flux and temperature also through the island number density N, see Sect. 3.7.3 and [303].

in the destabilization, be it the anisotropy of the step free energy, the step edge diffusion coefficient or the rate of step adatom detachment, will be emphasized. It will turn out to be useful to distinguish two types of instabilities: those that may be understood within the framework of an isotropic linear stability analysis of the kind sketched above (instability-I), and those that crucially rely on anisotropies (instability-II). This distinction is similar to the classification of growth patterns into (isotropic) "seaweeds" and (anisotropic) dendrites that has been proposed in the theory of two-dimensional solidification [305]. So far the theoretical analysis of island instabilities of type II has been limited to simple scaling arguments of the kind described in Sect. 3.7.3.

Pt Islands on Pt(111) Grown Below 300 K. The fractal-dendritic Pt islands observed on Pt(111) below 300 K, as shown in Fig. 2.7 and in Fig. 3.1a and b, can be interpreted as a consequence of the diffusional growth instability. Indeed, at these low temperatures both step atom detachment and step edge diffusion are inactive (compare to Table 3.1). Accordingly, the absence of these stabilizing effects allows the point effect of diffusion to accelerate protrusion growth of wavelengths down to the atomic dimensions. The small characteristic length scale implies that individual attachment events are permanently frozen in, leading to a pattern that is strongly ramified and disordered. As discussed in Sect. 3.3, understanding the subtle effects of crystal anisotropy which are superimposed on the random structure requires a detailed analysis of atomic processes, and cannot be captured on the level of continuum stability analysis.

Growth of Fractal Au–islands on Ru(0001). Figure 3.24 shows the formation of monatomic high, strongly ramified, isotropic fractal islands after deposition of 0.3 ML Au on Ru(0001) at room temperature [262]. The typical branch widths are about 100 Å, considerably larger than the atomic scale. The island contour displays the same structure in all directions: The branch edges are rounded, exhibiting a waviness with radii of curvature of the same magnitude as the branch width. Thus, step edge diffusion with a diffusion coefficient largely independent of orientation appears to be effective. Island shape and step edge diffusion being isotropic, the standard linear stability scenario is applicable: As soon as the critical radius R^* is reached during growth, the stabilizing effect of step edge diffusion is outweighed by the point effect of diffusion and the protrusions grow rampantly.

An inspection of Figs. 3.24b and c points to an additional aspect. Annealing of the Au islands does not result in convex islands, but in meandering step edges. Although annealing at 650 K (Fig. 3.24b) does lead to more compact islands with larger radii of curvature, these islands are by no means convex. Upon annealing at 1100 K (Fig. 3.24c) the Au–adlayer is fully rearranged, but it still exhibits meandering steps and many open channels. The behavior is different from that of a homoepitaxial system, which always exhibits compact islands at high temperatures (compare Fig. 3.1f or Figs. 3.4d and e). Since the minimization of the total step free energy drives the generation of

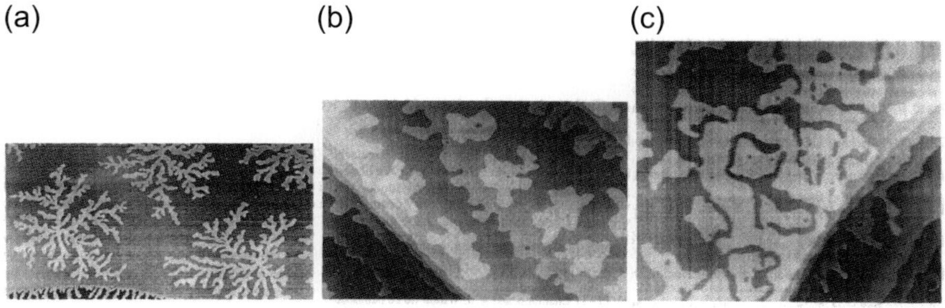

Fig. 3.24. (**a**) Island shapes formed upon deposition of 0.3 ML of Au on Ru(0001) at room temperature (scale 10000 Å × 6500 Å); similarly prepared surface (except for $\Theta_{\mathrm{Au}} = 0.37$ ML) but with additional annealing at (**b**) 650 K (scale: 8300 Å × 5400 Å) and (**c**) 1100 K (scale: 7600 Å × 6600 Å) (Figs. 2 and 3 of [262])

compact islands, one may conclude that the specific step free energy δ for the Au/Ru(0001) system is very small[26]. A small δ makes the stabilizing terms in (3.37) ineffective. Accordingly, the fractal shapes observed in Fig. 3.24a for room temperature growth of Au on Ru(0001) are a consequence of the small step free energy, rather than of limited step edge diffusion as is usually assumed. One may speculate that in well wetting systems the strong binding between deposit and substrate atoms could reduce the mutual binding within the adsorbed layer, and thus also the step free energy.

Stability of Triangular Pt Island Growth Shapes on Pt(111). Figures 3.25a to c show successive growth stages of islands in the temperature range between 440 K and 600 K. The islands appear to develop from small compact, either irregular or hexagonal (stage-1) shapes via the triangular (stage-2) shape to large dendritic-skeletal (stage-3) island shapes. The latter are named dendritic-skeletal, because the islands develop their preferential (dendritic) growth directions only after going through an intermediate stage of self-similar triangular island growth, like the skeletal 3D crystal shapes mentioned above [212, 295].

The instability of the polygonal Pt–island shapes cannot be properly described by isotropic linear stability analysis, since anisotropy effects are decisive for the destabilization. The key point is that the formation of the triangular stage-2 shape with its sharp corners is not the consequence of the amplification of an accidental fluctuation, but originates in the anisotropic growth kinetics discussed in Sect. 3.4. Once the corners of the triangle are developed, the point effect of diffusion leads to an increased material transport to these corners. As long as the side length L of the triangular shape does

[26] A possible influence of the lattice mismatch on the shape of the structures generated in *hetero*epitaxial systems such as this cannot be excluded a priori. Indeed, the resulting strain energy can be relaxed by breaking up large islands into smaller ones [306, 307]. However, the images in Fig. 3.24 show that the structures do not become smaller by annealing, giving no evidence for such relaxation effects.

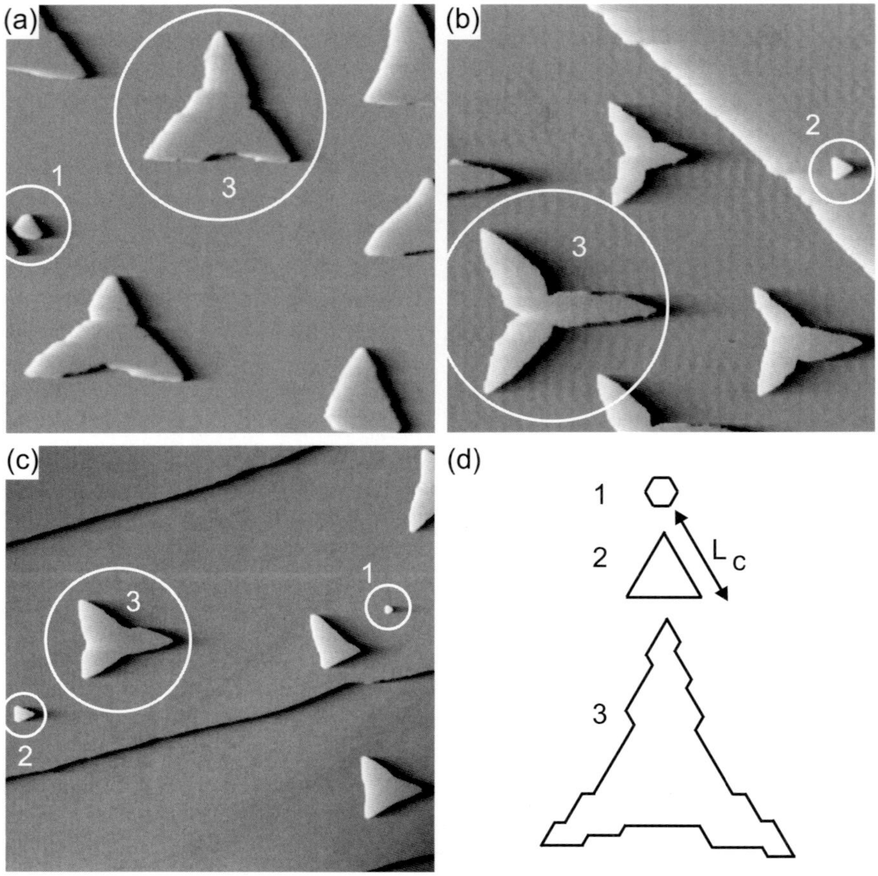

Fig. 3.25. Growth stages of islands imaged after deposition of Pt on Pt(111) at (**a**) $T = 440\,\mathrm{K}$, (**b**) $T = 460\,\mathrm{K}$ and (**c**) $T = 600\,\mathrm{K}$ and (**d**) schematic sketch of island shape evolution. The islands develop from an irregular or hexagonal (stage-1) shape via a well defined triangular (stage-2) shape to a dendritic-skeletal (stage-3) shape. The numbers placed near the circled islands indicate the stage in the shape evolution. The topograph sizes are (**a**) 1500 Å× 1500 Å, (**b**) 1800 Å× 1800 Å and (**c**) 6000 Å× 6000 Å

not exceed a critical length L_c, the excess material supplied to the corners by the diffusion field can be carried away by step edge diffusion; the triangular stage-2 shape remains stable, as evidenced by the smooth edges. For $L > L_c$ this is no longer the case (compare Fig. 3.25d). The material supplied to the corners is not carried away sufficiently fast to prevent enhanced growth at the corners, which – due to the point effect of diffusion – becomes self-accelerated. This leads to tip formation and eventually to the dendritic-skeletal shape of stage-3. Simple estimates of L_c based on an analysis of the mass transport along the island edge are derived in Sect. 3.7.3.

The important role of the diffusion field in the transition to dendritic-skeletal island shapes has been demonstrated in KMC simulations of Pt island growth on Pt(111), by comparing a full simulation (where atoms are deposited randomly on the terrace) with a restricted model in which atoms are placed directly at the island edge [308]. In the latter case the compact island shape remains stable, because the diffusion field is effectively turned off.

This discussion suggests to distinguish two types of island instabilities: the instability-I of a locally straight step against destabilization, as described by (3.37) under conditions of isotropic kinetics and energetics, and the instability-II of a polygonal island shape against tip formation at an island corner. Instability-II can only occur for a system with anisotropic growth kinetics. Examples of instability-I are the formation of Au–fractals on Ru(0001) or, as an extreme case, the fractal-dendritic islands on Pt(111) at low temperature, while the destabilization of the triangular island shape as shown in Fig. 3.25 is an example of instability-II. It is worth noting that, despite the qualitative differences between the two types of instabilities, the stability criteria (3.36) and (3.49) take a similar form, though they involve different transport coefficients (σ_{st} and D_{e}, respectively).

The Effect of Carbon on Pt Island Growth on Pt(111). Experiments on Pt(111) have shown that in the presence of carbon contamination the equilibrium shape of Pt–islands is no longer hexagonal with threefold symmetry, but becomes essentially circular [309]. This demonstrates that carbon adsorbs preferentially at steps and – of particular significance in the present context – that the specific free energy of the carbon-decorated steps is isotropic. The consequences for the onset of the diffusional growth instability are shown in Fig. 3.26. The three images have been taken after Pt deposition onto increasingly C–contaminated Pt(111) surfaces, under otherwise identical conditions. Figure 3.26a displays dendritic–skeletal islands similar to those shown in Fig. 3.25, while the islands in Fig. 3.26c are isotropic fractals; a care-

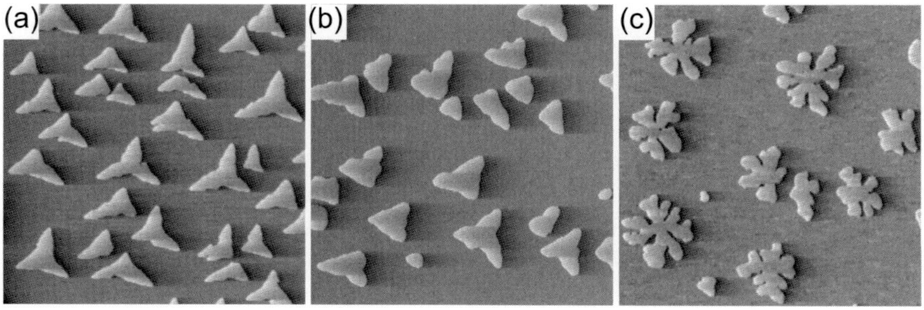

Fig. 3.26. Influence of adsorbed carbon on the island shape. Images taken after deposition of 0.2 ML Pt on a (**a**) clean, (**b**) slightly, and (**c**) heavily ($\approx$ 0.03 ML) C–contaminated Pt(111) surface under otherwise identical conditions (apart from a small difference in temperature: (**a**) 600 K, (**b**) and (**c**) 610 K). Topograph sizes 4400 Å × 4400 Å

ful look at Fig. 3.26c shows that tip splitting has taken place, which is a clear indication for the absence of anisotropy [305]. The observed shapes are obviously the results of instabilities-II and -I, respectively. Because carbon adsorption removes the anisotropy, instability-II, present in Fig. 3.26a, disappears for C–contaminated growth. At the same time carbon adsorption seems to lower the step free energy and thereby creates the conditions for the onset of instability-I.

The Influence of the Diffusion Field on Island Growth. Until now we have assumed that the diffusion field around an island is distorted only by the island itself, so that the adatom concentration remains homogeneous and isotropic sufficiently far from the island. This assumption is fairly well satisfied in the early stages of growth. However, as soon as the distances between edges of neighboring islands become comparable to the island size, the islands start to communicate through their overlapping diffusion fields. Around each island the diffusion field is depleted, which reduces the nucleation rate[27] as well as the growth rate of nearby islands.

Several cases of diffusion-mediated island–island interaction are illustrated in Fig. 3.27, which shows a typical dendritic-skeletal island landscape after deposition of 0.2 ML Pt at 640 K:

A. The islands indexed by A are located in areas that are relatively free of other islands. Of particular importance is the fact that there are no closely located islands or step edges in the directions of the corners of the triangular shape. The island tips are surrounded by large diffusion fields and can develop undisturbed.
B. In contrast, the situation of the islands indexed by B is typical for islands that interact via their diffusion field. The tip of island 2 pointing towards island 1 is screened by the vicinity of island 1 and cannot grow as rapidly as the other two tips of island 2 (the latter clearly exhibits the features of instability-II). Thus, the existence of a triangular island shape is not sufficient for the onset of instability-II. In addition, a sufficiently large diffusion field has to be present which provides the enhanced adatom supply to the corners.
C. The islands indexed by C are even closer, and interact more strongly. They display a surprising aspect of the diffusion-mediated interaction: The islands have a bias not to coalesce. As soon as their edges come close to each other, the islands seem to be mutually repelling. The narrower the space between islands, the smaller the number of atoms deposited in this space. The growth of an island towards a neighboring one slows down compared to the growth in the open directions, where other islands are further away. This is a first indication of the *Zeno effect*, which leads to the formation of deep crevices in multilayer growth (see Sect. 4.2).

[27] See Sect. 2.3.1.

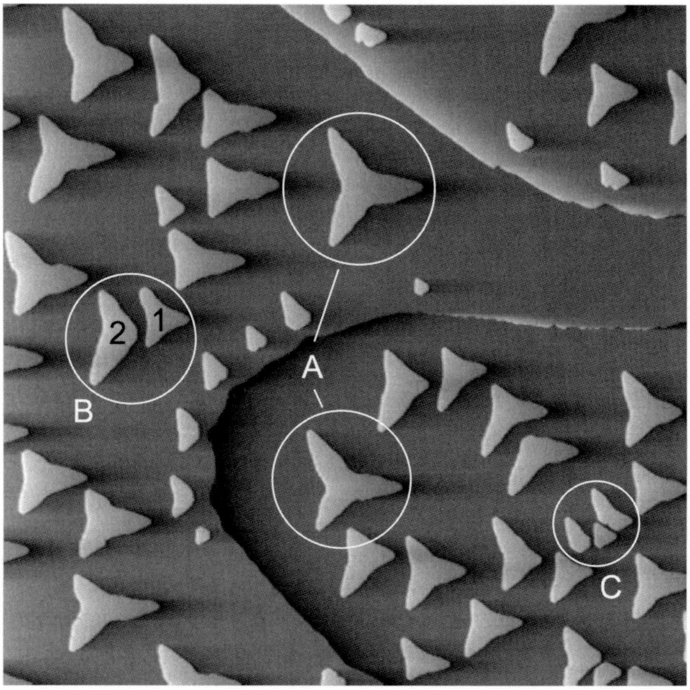

Fig. 3.27. Influence of the diffusion field on the growth of individual islands. Image taken after deposition of 0.2 ML Pt on Pt(111) at 640 K with a flux of 7×10^{-3} ML/s. The shapes of the islands in the areas indexed by A, B, and C are discussed in the text

3.7 Supplementary Topics

3.7.1 Migration Along Rough Island Edges

Mass transport along island edges plays an important role in most parts of this chapter: It allows the islands to reach their equilibrium shapes at high temperatures; it triggers the transition from fractal to compact growth shapes at low temperatures; and it provides the dominant smoothening mechanism which counteracts diffusional shape instabilities. To some extent the motion of a step adatom along an island edge can be viewed as the one-dimensional analog of the surface diffusion process treated in Sect. 2.2, but the analogy is not complete, because island edges generally contain a large number of kinks, which act as sinks and (at sufficiently high temperatures) sources of adatoms. Since the failure to appreciate this point may lead to an erroneous interpretation of experimental data, the pertinent conceptual issues are summarized here. We consider the high temperature regime, where detachment of step adatoms from kinks is active and the step edge relaxes to its equilibrium shape in the absence of a deposition flux.

The most dramatic consequence of the trapping of step adatoms at kinks is that the one-dimensional tracer diffusion coefficient, defined through the analog of (2.1), *does not exist*: A detailed analysis of a model of an isolated step shows that the mean square displacement of a marked step adatom grows sublinearly, as $\langle (r(t) - r(0))^2 \rangle \sim t^{7/8}$ [310]. This reflects the fact that a trapped adatom runs a considerable risk of being overgrown by a large step fluctuation, and hence the probability distribution of trapping times has a very broad tail. The relevant quantity describing the mass transport along the edge in a continuum picture is therefore not the diffusion coefficient, but rather the collective step atom mobility σ_{st}. It is defined as the factor of proportionality between the mass current j_{st} along the step and the step chemical potential gradient driving it,

$$j_{\mathrm{st}} = -\sigma_{\mathrm{st}} \frac{\partial \mu_{\mathrm{st}}}{\partial s}. \tag{3.38}$$

Here s denotes the arclength measured along the (generally curved) step. According to the Gibbs–Thomson relation (3.9), the step chemical potential μ_{st} is enhanced at step protrusions, and hence the current (3.38) smoothens the step.

The mobility σ_{st} can be calculated exactly for a model of an isolated, linear step in which overhangs are neglected (Solid–On–Solid constraint) and the energy barrier for the hopping of an atom along the step is proportional to the coordination number at the initial site [311]. One finds that the rate limiting process is the detachment of step adatoms from kinks (process 8 in Table 3.1), in the sense that

$$\sigma_{\mathrm{st}} = \frac{a\nu_0}{2k_{\mathrm{B}}T} \mathrm{e}^{-E_{\mathrm{det}}/k_{\mathrm{B}}T}. \tag{3.39}$$

In this expression E_{det} is the detachment barrier and ν_0 is the attempt frequency. A similar result has been derived within a linear fluctuation theory [301].

To illustrate the application of (3.39), consider a typical STM step fluctuation experiment, in which the step position is monitored as a function of time [172, 255]. When migration along the step edge is the dominant smoothening process, a Langevin fluctuation theory[28] based on (3.38) predicts that the mean square displacement of the step grows as

$$\langle (\zeta(y,t) - \zeta(y,0))^2 \rangle \approx a_{\perp}^2 (t/\tau_{\mathrm{st}})^{1/4}, \tag{3.40}$$

where the characteristic time scale τ_{st} (the time required for the step to fluctuate by one atomic row) is given by

$$\tau_{\mathrm{st}} \approx \frac{\Omega^2 \, \tilde{\delta}^3}{(k_{\mathrm{B}}T)^4 \, \sigma_{\mathrm{st}}}. \tag{3.41}$$

[28] Comprehensive accounts of this theory can be found in [255, 312, 313].

Recalling from Sect. 3.2.4 that the step stiffness $\tilde{\delta}$ is inversely proportional to the step diffusivity b^2, and hence proportional to the kink spacing at low temperatures, we conclude that the activation energy for the prefactor of the temporal step fluctuation (3.40) is given by

$$E_{\mathrm{a}} = \frac{1}{4}(E_{\mathrm{det}} + 3\epsilon), \tag{3.42}$$

where ϵ denotes the kink energy. For the closed packed steps on Pt(111), the measurement of the step fluctuations [215] together with the experimentally determined value $\epsilon = 0.167\,\mathrm{eV}$ for the kink energy yields the estimate $E_{\mathrm{det}} = 1.50 \pm 0.16\,\mathrm{eV}$ for the detachment barrier quoted in Table 3.1 (we shall see at the end of this subsection why this is only an upper bound).

In the literature [171,172,215,314] the interpretation of experimental step fluctuation data is often based on an Einstein relation

$$\sigma_{\mathrm{st}} = \frac{n_{\mathrm{st}}^{\mathrm{eq}} D_{\mathrm{st}}}{k_{\mathrm{B}}T} \tag{3.43}$$

for the mobility, where $n_{\mathrm{st}}^{\mathrm{eq}}$ is the (one-dimensional) equilibrium concentration of step adatoms and D_{st} denotes the tracer diffusion coefficient for an adatom migrating along a kinked step. The latter is estimated by considering the motion of an adatom in a model potential where kink sites are represented as traps of depth E_{det} spaced at the mean kink distance. The resulting activation energy for D_{st} is $E_{\mathrm{det}} - \epsilon$. Since a step adatom can be viewed as a double kink (see Sect. 3.2.3), the concentration of step adatoms is $n_{\mathrm{st}}^{\mathrm{eq}} \sim \mathrm{e}^{-2\epsilon/k_{\mathrm{B}}T}$, and hence the activation energy of σ_{st} is predicted by (3.43) to be $E_{\mathrm{det}} + \epsilon$, in disagreement with the exact result (3.39).

With hindsight, the failure of the Einstein relation (3.43) should be no surprise. Since adatoms are continually absorbed and emitted at kinks, they do not constitute a conserved species, and it is difficult to consistently define D_{st} and n_{st} [33]. In fact, as we have mentioned above, strictly speaking $D_{\mathrm{st}} \equiv 0$. Representing the migration of an adatom along a kinked step by the diffusion of a particle in an external potential neglects both the possibility of long term trapping due to step fluctuations, and the fact that a kink site is not actually "filled" when an adatom attaches to it – it is merely shifted. Nevertheless a relation of the form (3.43) does hold, if D_{st} is replaced by the diffusion coefficient D_{e} for the migration along a *straight* close packed step, which was introduced in Sect. 3.4. Then (3.43) and (3.39) simply express the balance between the detachment and attachment of step adatoms at the kinks.

To lend further credibility to the fundamental expression (3.39), we provide here a simple, general derivation based on an estimate of the characteristic time scale τ_{st}. This will also indicate how (3.39) has to be modified in the presence of an additional *kink rounding barrier*[29] (process 7+3a in Table 3.1) [316].

The elementary process contributing to the motion of the step is the shift of a kink by one lattice spacing. This requires, first, that a step adatom detaches from a kink, and second, that it reattaches to a another kink *before* returning to the kink where it originated. The detachment rate is $\nu_{\rm det} = \nu_0 {\rm e}^{-E_{\rm det}/k_{\rm B}T}$. The probability $p_{\rm att}$ for the attachment of the emitted step adatom at the distant kink can be obtained from a one-dimensional random walk calculation, which yields an expression of the form [317]

$$p_{\rm att} \approx \frac{1}{l_{\rm k}/a + 1/p_{\rm kr}}. \tag{3.44}$$

Here $l_{\rm k} = (a/2){\rm e}^{\epsilon/k_{\rm B}T}$ is the mean kink spacing, and $p_{\rm kr}$ is the probability that the step adatom actually attaches to the distant kink once it has reached it. This is of order unity for a forward kink, but it can be much smaller, of order $p_{\rm kr} \approx {\rm e}^{-E_{\rm kr}/k_{\rm B}T}$, for a backward kink, which can be reached only through a kink rounding process associated with an additional energy barrier $E_{\rm kr}$. It can be seen from (3.44) that such a kink rounding barrier is relevant if it is larger than the kink energy ϵ, since then $l_{\rm k}/a \ll 1/p_{\rm kr}$, and irrelevant otherwise[30].

It follows that the effective hopping rate of a kink is $p_{\rm att}\nu_{\rm det}$. For the step to move by one atomic row, the kink has to diffuse over a distance comparable to the kink spacing $l_{\rm k}$, which requires a time of the order of

$$\tau_{\rm st} \approx \frac{(l_{\rm k}/a)^2}{p_{\rm att}\,\nu_{\rm det}}. \tag{3.45}$$

In the absence of strong kink rounding barriers $p_{\rm att} \approx a/l_{\rm k}$, hence $\tau_{\rm st} \approx (l_{\rm k}/a)^3/\nu_{\rm det}$, which agrees with (3.41) provided the step adatom mobility $\sigma_{\rm st}$ is set proportional to the detachment rate, in accordance with (3.39). On the other hand, in the presence of kink rounding barriers with $E_{\rm kr} > \epsilon$, we have $p_{\rm att} \approx p_{\rm kr}$, and the comparison of (3.45) with (3.41) shows that the activation energy for $\sigma_{\rm st}$ is now $E_{\rm det} - \epsilon + E_{\rm kr} > E_{\rm det}$. For this reason the measurement of $\sigma_{\rm st}$ through observations of temporal step fluctuations yields, in general, only an upper bound of the detachment barrier.

3.7.2 The Transition from Fractal-Dendritic to Compact Islands: Theory Versus Experiment

For Pt islands on Pt(111) the transition from fractal-dendritic to compact islands takes place between 200 K and 400 K (compare Fig. 3.1a and Fig. 3.1c).

[29] Further discussion of the effects of kink and corner rounding barriers can be found in Sects. 3.7.3, 4.7.3 and 4.9.7. Direct experimental evidence for such a barrier has been reported in a FIM study of Rh(100) [315].

[30] This is analogous to the distinction between strong and weak step edge barriers in Chap. 4.

We have seen above in Sect. 3.3 that the first step in this transition is the onset of step adatom migration (process 4 in Fig. 3.2), as indicated by the increase of the in-layer coordination number above 275 K in Fig. 3.10. According to the data collected in Table 3.1, and the theoretical potential energy diagram of Fig. 3.17a, step adatom migration sets in first along A-steps, and the onset at B-steps occurs at somewhat higher temperatures. The diffusion of adatoms along dense packed steps leads to a thickening of the island branches, but the transition to compact island shapes is completed only when adatom transport around corners (process 7 +3b) becomes possible. Once this kinetic channel is opened, the step adatoms are able to react to the binding energy difference between the two types of island edges and the triangular shapes described in Sect. 3.4 form.

The fractal-dendritic to compact shape transition for Al islands on Al(111) follows the same general scenario. It is nevertheless worthwhile to discuss this case in somewhat more detail, because for the Al/Al(111) system DFT calculations related to homoepitaxial growth [233, 265, 318] were performed *prior* to the corresponding experiments [137, 186, 202, 232]. The results of the DFT calculations were used as input for KMC simulations [266, 275–277], which predicted the temperature dependence of the island shape without knowledge of the experimental facts. Comparison of these KMC simulations with the experimentally observed Al island shapes on Al(111) may therefore yield some insight into the question as to whether such a theoretical approach has true predictive power.

In Fig. 3.28 the experimental results shown in the first row are compared to two sets of DFT-based KMC simulations. The simulations displayed in the second row (set I), taken from [275], reveal a severe disagreement with the experiment. The simulated low temperature islands in Fig. 3.28d are isotropic and miss the experimentally observed dendritic aspect, the simulated intermediate temperature islands in Fig. 3.28e are compact with an A-step envelope instead of the experimentally observed B-step envelope, and finally the simulated high temperature islands in Fig. 3.28f exhibit concave edges all the way to the triangle corners in contrast to the rounded triangular shape observed in the experiment[31]. The comparison with the simulations depicted in the third row (set II), taken from [266], is more favorable. The sequence of island shapes in Fig. 3.28g, h and i is based on the DFT calculations in the GGA represented by the potential energy diagram in Fig. 3.17b. This KMC simulation reproduces the low temperature fractal-dendritic islands and the high temperature triangular, compact islands in a satisfactory manner. A marked difference between simulation and experiment remains only for the transition shape between these regimes, as can be seen by comparing Fig. 3.28b and h: In the experiment the islands exhibit three fat, well oriented branches,

[31] The simulated concave island edges also strongly differ from the appearance of the edges of dendritic-skeletal islands. These typically exhibit a concave central region and convex parts towards the corners (compare to Figs. 3.25 and 3.27).

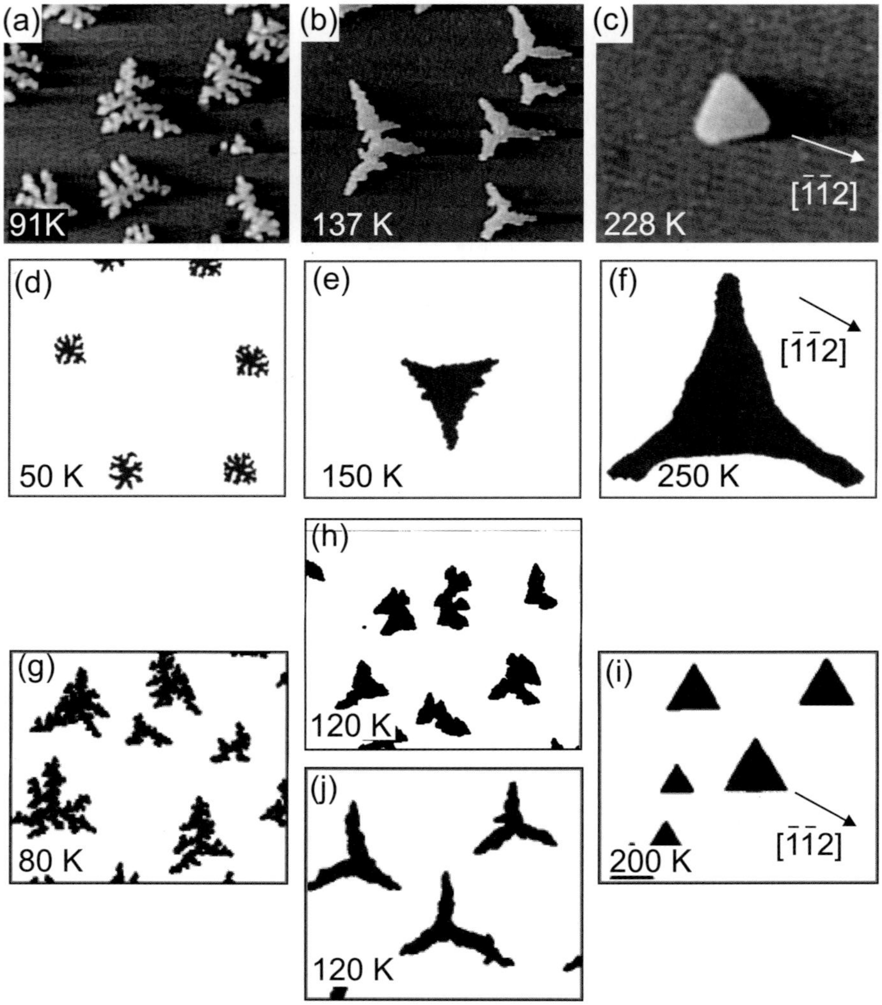

Fig. 3.28. Comparison of Al island shapes on Al(111) from experiment (first row) and KMC simulations based on DFT calculations (second and third row). The temperatures used in experiments and simulations are indicated. The field of view is 570 Å × 680 Å in all cases. The experiments (**a**) to (**c**) were performed with a deposition rate of 7.2×10^{-3} ML/s and the coverage is 0.1 ML. In (**d**)–(**f**) the deposition rate is 8×10^{-2} ML/s and the coverage is 0.04 ML. In (**g**)–(**j**) the deposition rate would be 1.8×10^{-3} ML/s at 160 K and is adjusted to keep the ratio of deposition rate and monomer diffusion coefficient constant over the temperature range investigated. The coverage is 0.1 ML. For details see text ((**d**)-(**f**) based on Fig. 18 of [275] reprinted with permission from Elsevier Science, (**g**)-(**j**) based on Figs. 3 and 4 of [266])

while in the simulation they are more irregularly branched and less oriented. Full agreement between the observed and simulated shapes (Fig. 3.28j) is obtained by reversing the edge diffusion anisotropy with respect to the DFT results shown in Fig. 3.17b; instead of using the lower activation energy for the motion of step adatoms along the B-steps as found in DFT, in Fig. 3.28j the activation energy for motion along A-steps is assumed to be lower than along B-steps.

It is interesting to note that the DFT/LDA results also published in [265], which were not used as a basis for the KMC simulations [266], yield a much larger activation energy for step adatom motion along the B-step (0.36 eV instead of 0.26 eV in the GGA), thereby indeed reversing the edge diffusion anisotropy[32]. After the remark in Sect. 3.2.3 concerning step edge energies, we find here another example that, for the low coordinated surface configurations of interest, the DFT in the GGA is not necessarily more reliable than results from the LDA. This example shows how a detailed comparison between experiments and DFT-based KMC simulations may help to improve the reliability and predictive power of the ab initio methods.

Two issues remain to be addressed. The first question is why the transition shapes in Fig. 3.28h and Fig. 3.28j react so strongly to the reversal of the step edge diffusion ansiotropy. In the absence of efficient diffusion around corners (which would allow for the formation of compact islands), two anisotropies compete in the island shape formation: the corner asymmetry effect discussed in Sect. 3.3, and the anisotropy of step edge diffusion. The corner asmmetry implies preferential incorporation of atoms into A-steps and favors the formation of B-steps, while the faster edge diffusion along B-steps predicted by the DFT/GGA calculations [265] (see Fig. 3.17) leads to preferential formation of A-steps, as discussed at the end of Sect. 3.4. Thus in the KMC simulations based on the DFT/GGA energy barriers the two anisotropies counteract each other, giving rise to the branched heterogeneous appearance of the islands in Fig. 3.28h. In contrast, if both effects prefer the B-step orientation as in Fig. 3.28j, a well-defined island shape consisting of three fat branches is formed.

Second, one may ask why the simulations of set I, presented in the second row of Fig. 3.28, fail so drastically to reproduce the experimental island shapes. This is even more surprising as the DFT energy barriers underlying set I and II agree with each other to within a few hundreths of an eV. A first part of the answer may be the fact that the DFT/LDA values used as a basis for set I [233,318] were certainly less complete than those underlying set II [265]. In particular, the corner energetics was missing and had to be introduced ad hoc into the KMC simulations [276]. As a consequence, the construction of a potential energy diagram for step adatom motion, as in

[32] For most other processes the predictions of LDA and GGA are quantitatively very similar. Both LDA and GGA predict that the motion of a step adatom along the B-step occurs through exchange of the step adatom with an adatom in the step.

Fig. 3.28b, is impossible without further assumptions, and uncertainties regarding the binding energies of the step adatoms at the two different types of step edges may have been introduced. As discussed in Sect. 3.4, this binding energy difference is decisive for the orientation of the triangular compact islands.

An additional factor which could contribute to the failure of the set I simulations is a peculiar choice of the attempt frequencies ν_0 for step adatom migration. Based on an estimate of the theoretical potential energy surface [193] as well as on experimental data for other systems, attempt frequencies of $\nu_{0,\mathrm{A}} = 2.5 \times 10^{12}\mathrm{s}^{-1}$ and $\nu_{0,\mathrm{B}} = 2.5 \times 10^{14}\mathrm{s}^{-1}$ for step adatom motion along the A- and B-step were assumed. The large difference in prefactors leads to a reversal in the ordering of the edge diffusion coefficients as a function of temperature: Below 250 K step adatom migration is faster along the A-steps, while above this temperature it is more facile along the B-steps. In [275] the change in the orientation of the triangular islands from Fig. 3.28e to f was attributed to this reversal. However, as we explained in Sect. 3.4, the effect of the edge diffusion anisotropy on the growth shape should in fact be *opposite* to what is observed! Again, this suggests that uncontrolled binding energy differences may be involved.

In conclusion, the discussion in the present section shows that a sound basis for KMC simulations requires a full, DFT based, potential energy diagram for the motion of an atom around the island, exploring all relevant sites. The results of ab initio calculations themselves, though not perfect, are less likely to be a source of error in a combined DFT/KMC approach to growth prediction than the choice of relevant processes for the KMC simulation. In particular, it is of utmost importance to ensure that the calculated energy barriers for the individual processes included in the simulation indeed add up to a consistent binding energy landscape (see also Sect. 6.5.3).

3.7.3 Stability Criteria for Compact Islands

In this section we develop some simple estimates for the critical island size L_c at which the transition from compact to dendritic shapes occurs. In contrast to the linear stability analyses described in Sect. 3.6, the present derivation does not explicitly involve the adatom diffusion field. Rather, we ask under what conditions the redistribution of mass along the edge of a compact island is sufficiently fast to counteract the destabilizing tendencies arising from the inhomogeneous supply of adatoms, due to the point effect of diffusion, as well as from randomly generated protrusions. The resulting stability criteria should apply in particular to the transition from facetted, compact islands to dendritic-skeletal shapes (instability-II of Sect. 3.6).

We consider a compact island with L edge sites. The total flux impinging onto the island edge is F/N, where N denotes the island number density. The one-dimensional flux per unit edge length is then

$$F_{1\mathrm{d}} = \frac{F}{NLa}, \tag{3.46}$$

and the time between the subsequent arrivals of adatoms at the island edge is

$$\Delta t_\mathrm{e} = N/F. \tag{3.47}$$

For the island to remain in its compact, facetted shape, it is surely sufficient that each newly arrived adatom can find a kink site before the next atom arrives. If there are only a few kinks along the island perimeter, then the time required to reach a kink is of the order of the diffusion time

$$\tau_\mathrm{e} \approx a^2 L^2 / D_\mathrm{e}. \tag{3.48}$$

Setting $\tau_\mathrm{e} \sim \Delta t_\mathrm{e}$ we obtain the estimate [303, 319]

$$L_c^{(1)} \approx \left(\frac{D_\mathrm{e} N}{a^2 F} \right)^{1/2}. \tag{3.49}$$

At this point we need to check the consistency of the assumption that the number of kinks along the island edge is small. According to the nucleation theory of Sect. 2.4.3, the kink spacing along a one-dimensional edge subject to a flux $F_{1\mathrm{d}}$ is of the order of $l_{1\mathrm{d}} \approx (aD_\mathrm{e}/F_{1\mathrm{d}})^{1/4}$ [see (3.27)]. Setting $La \approx l_{1\mathrm{d}}$ defines a second critical island size [301]

$$L_c^{(2)} \approx \left(\frac{D_\mathrm{e} N}{a^2 F} \right)^{1/3}. \tag{3.50}$$

Islands larger than $L_c^{(2)}$ have edges with an extensive number of kinks.

Since $L_c^{(1)} > L_c^{(2)}$, we see that the assumption leading to (3.49) was in fact inconsistent. The expression (3.49) can nevertheless be justified in the context of a theory of layer–by–layer growth, which will be developed in Sect. 5.6. To apply it here, we regard the island edge as a one-dimensional surface of length L, which is closed to a ring ("periodic boundary conditions" in the jargon of statistical physics). It can be shown that such a surface grows in a layer–by–layer fashion, i.e. with a bounded (small) number of exposed layers, as long as L is smaller than the *coherence length* $\tilde{l}$, which is much larger than the kink spacing; in one dimension $\tilde{l}/a \approx (l_{1\mathrm{d}}/a)^2$ [320]. Demanding that $L \approx \tilde{l}/a$ leads precisely to the critical island size (3.49). Thus islands with sizes between $L_c^{(2)}$ and $L_c^{(1)}$ have kinked edges which nevertheless remain straight, in the sense that the island grows row by row. Without explicit consideration of the diffusion field, it is difficult to say whether the edge roughness which is present in this intermediate regime is sufficient to trigger the diffusional shape instability. KMC simulations [300, 319] support the stability criterion (3.49).

In general, the stability of the island depends also on the possible existence of additional energy barriers for the rounding of kinks (process 7+3a in Table 3.1) and corners (process 7+3b) of the kind briefly discussed in

Sect. 3.7.1. For example, if the rounding of kinks is completely prohibited, then the step edge undergoes the one-dimensional version of the mounding instability described in Sect. 4.2 (see Sect. 4.7.3). A stability criterion that includes corner barriers can be derived by replacing the diffusion time (3.48) by the *residence time* $\tau_{1\mathrm{d}}$ which a step adatom spends on a straight edge of the island boundary before it escapes to a neighboring edge via a corner crossing process [321]. This is the one-dimensional analog of the residence time of an adatom atop an island, which will be computed in Sect. 4.3.2. Similar to the expression (4.39) for the two-dimensional case, the residence time at an island edge of length L is given by

$$\tau_{1\mathrm{d}} = \frac{L^2 a^2}{12 D_\mathrm{e}} + \frac{L}{2\nu_\mathrm{cr}}, \tag{3.51}$$

where ν_cr denotes the rate of corner rounding.

As in Sect. 4.3.2, two regimes can be distinguished according to which of the two terms on the right hand side of (3.51) dominates the residence time. When the first term dominates, the residence time is of the same order as the diffusion time (3.48), and the stability considerations developed previously apply; in this case the corner barriers are irrelevant. On the other hand, when the corner rounding rate is sufficiently small, so that the second term dominates in (3.51), the residence time becomes $\tau_{1\mathrm{d}} \approx L/2\nu_\mathrm{cr} \gg \tau_\mathrm{e}$. Setting this equal to the time interval Δt_e between subsequent arrivals at the island edge then defines yet another critical radius,

$$L_c^{(3)} \approx \frac{\nu_\mathrm{cr} N}{F}, \tag{3.52}$$

which governs the shape stability under conditions of dominant corner barriers. The stability criterion based on (3.51) appears to be in good qualitative agreement with KMC simulations and available experimental results [321]. In particular, it accounts for the observed absence of fractal–dendritic growth on lattices of square symmetry [264].

Using the expressions for the island number density N derived in Sect. 2.3.1, it is easy to see that the critical island size (3.49) increases with increasing temperature and decreasing flux, as would be expected from the general stability considerations in Sect. 3.6. From a practical point of view, it is often more relevant to consider the *critical coverage* Θ_c at which the islands become unstable, rather than the critical island size [303]. This can be estimated as

$$\Theta_c \approx N(aL_c^{(1)})^2. \tag{3.53}$$

Inserting the scaling law (2.19) for the island number density, we find that the critical coverage scales with flux as $F^{2\chi-1}$, where $\chi = i^*/(i^*+2)$ is the island density scaling exponent. As the flux is lowered, the critical nucleus size i^* increases. When it passes the value $i^* = 2$, so that χ become larger than 1/2, the flux dependence is reversed, and the islands (observed at fixed coverage) become more dendritic with *decreasing* flux. This type of *reentrant* morphological instability has been observed in KMC simulations [303].

We noted above in Sect. 3.1 that there is some evidence for such a reentrant shape evolution as a function of *temperature*, in the sense of a re-destabilization of the compact shapes with increasing T, for adatom islands on Pt(111). To see whether this can be explained along the lines of [303], we combine (3.49) and (3.53) with the expression (2.18) of the island density N to deduce that the temperature dependence of the critical coverage is of the form $\Theta_c \sim \mathrm{e}^{-E_\mathrm{a}/k_\mathrm{B}T}$ with the activation energy

$$E_\mathrm{a} = E_\mathrm{e} - \frac{2}{i^* + 2}(i^* E_D + E_{i^*}), \tag{3.54}$$

where E_e is the activation energy for diffusion along a straight step edge. At low temperatures the critical nucleus size $i^* = 1$, hence $E_\mathrm{a} = E_\mathrm{e} - (2/3)E_D$, which is at least 0.34 eV for Pt(111) (see Table 3.1). A positive activation energy E_a implies "normal" temperature dependence in the sense that the islands become more compact at higher temperatures. As was shown in Sect. 2.3.4, a transition to $i^* = 2$ occurs around 430 K. Evaluating (3.54) for $i^* = 2$ using the dimer binding energy derived in Sect. 2.3.4 and the smallest among the estimates for the step edge diffusion barrier given in Table 3.1 ($E_\mathrm{e} = 0.60$eV) we find indeed a slightly negative value $E_\mathrm{a} = -0.03$eV, which would indicate a decrease of Θ_c with increasing temperature. Thus, though not quite conclusive because of a lack of accurate estimates for E_e, the data are not inconsistent with the scenario for a reentrant morphological instability suggested in [303].

4. Pattern Formation in Multilayer Growth

Was die Schönheit für Täler gehalten hat, waren in Wirklichkeit Berge.[1]
Elfriede Jelinek, Der Tod und das Mädchen

While the growth of islands discussed in the last chapter could be considered as an essentially two-dimensional problem, the morphology of a multilayer film is determined not only by the transport of atoms within an atomic layer (*intralayer transport*), but in addition by the transport of atoms between different atomic layers (*interlayer transport*). In the same way as the evolution of island shapes with temperature can be understood in terms of the activation of atomic processes along the step edge, the sequence of multilayer growth regimes that unfold as a function of temperature is governed by the activation of atomic processes which enable the exchange of atoms between different atomic layers.

4.1 The Temperature Dependence of Multilayer Growth

In this introductory section an overview over the temperature dependence of multilayer growth morphologies in Pt(111) homoepitaxy will be given. Before presenting the experimental examples, it may be useful to schematically distinguish four idealized growth modes (Fig. 4.1).

• *Step flow growth*: Even the most careful preparation of a low Miller index single crystal surface leaves a certain number of atomic steps on the surface. Steps originate from the mosaic spread of the single crystal (i. e. the presence of low angle grain boundaries) or from limitations in the perfection of crystal orientation prior to and during the processes of crystal cutting and polishing. For metal and semiconductor surfaces the average step spacing l may be as large as 5000 Å and the spread of step spacings around their average value may allow even terraces of several μm width to be present on the surface. Nevertheless, at high temperature the adatom diffusion coefficient D and the size i^* of the critical nucleus will be large. According to (2.18), this implies that the island number density N is small, and hence the average island separation

[1] "What Beauty took for valleys were in fact mountains."

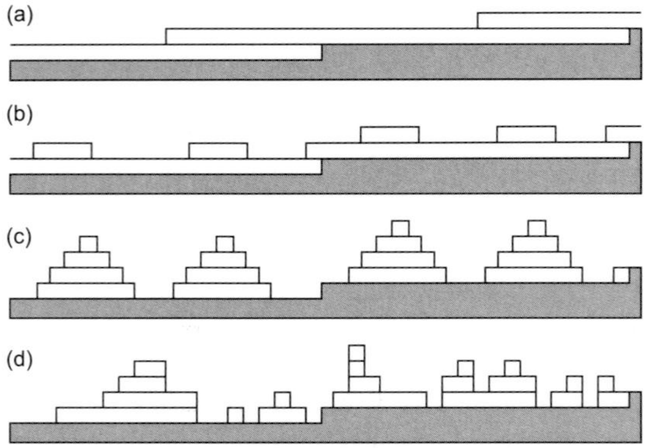

Fig. 4.1. Schematic representation of idealized growth regimes in homoepitaxy. (**a**) At sufficiently high temperature (large adatom diffusion coefficient) or small step spacing step flow takes place. At lower temperatures, adatom islands nucleate on the terraces, giving rise to either (**b**) layer–by–layer growth (perfect interlayer transport) or (**c**) mound formation (no interlayer transport). (**d**) At very low temperatures adatoms are immobile and a random, self-affine morphology without a characteristic length-scale forms

$$l_D = N^{-1/2} \tag{4.1}$$

may well exceed the terrace width l, $l \ll l_D$. Consequently, all adatoms are captured by preexisting steps and no nucleation takes place. As shown in the one dimensional sketch of Fig. 4.1a, under these conditions no new steps are created on the surface during growth and the steps move laterally with an average speed imposed by the deposition rate F and the step spacing l.

Besides this unavoidable high temperature step flow growth, it is possible to induce step flow growth at any temperature – provided adatom diffusion is still possible – by deposition on an intentionally miscut surface. The miscut with respect to a low index orientation has to be chosen such that the resulting l is significantly smaller than the expected island spacing l_D on the corresponding perfect low index surface.

- *Layer–by–layer (LBL) growth*: Decreasing the temperature leads to the onset of island nucleation on the terraces when $l_D < l$. In the case of perfect interlayer transport all atoms arriving on the islands are incorporated in the island step. The islands coalesce and nucleation of the second layer takes place only when the first layer is completely filled. The conditions for nucleation of the second layer are then (almost) identical to those on the original substrate, and a periodic oscillation in the surface morphology with a periodicity of $1/(F\Omega)$ is observed.
- *Mound formation*: Nucleation of islands on the original surface together with inhibited interlayer transport leads to mound formation. Once the islands are formed, atoms arriving on top of them will form new second layer nuclei, on top of which the third layer nucleates, and so forth. Mound formation leads to a steadily growing number of exposed atomic layers and thus to

an increase in roughness at the growth front. Note that both in LBL growth and in mound formation the average island separation in the first layer defines the initial lateral length scale.

• *Self-affine growth*: At sufficiently low temperatures adatom diffusion is effectively turned off, when the hopping rate ν becomes smaller than $F\Omega$. In this regime of diffusionless growth the atoms remain at the lattice site where they are deposited. As in mound formation, the roughness increases during growth, but this is not associated with the selection of a characteristic lateral length scale. Instead, the morphology has a random appearance, with fluctuations in the film thickness occurring over a wide range of length scales. These surfaces can be described as self-affine, which implies that the morphology is statistically self-similar under magnification, provided different magnification factors are chosen in the lateral and the vertical directions. Both in mound growth and in self-affine growth *dynamic scaling* is often observed, in the sense that the morphology at a later stage of growth is similar to that at an earlier stage if lateral and vertical dimensions are scaled by appropriate, time-dependent factors.

***

As we shall see in the following, these growth modes are generally not realized in their idealized form in nature, and in any case they appear only in a limited range of film thicknesses. Step flow growth and associated growth instabilities are analyzed in Sect. 4.7 and LBL growth is discussed in Chap. 5. In metal–on–metal homoepitaxy, self-affine growth is observed only at the very lowest temperatures, and experiments under well controlled growth conditions are rare [322]. Theoretically this topic has been treated extensively [32, 36, 106]. A brief summary of the relevant concepts will be given in Sect. 4.8.

Mound formation is the central theme of the present chapter, and will be investigated in Sects. 4.2–4.6 in its different facets. A brief guide to the reader seems in order. Section 4.2 elaborates on the oversimplified but highly instructive limit of zero interlayer transport, which already explains several features of mound formation and mound shapes on Pt(111). A detailed exposition of the atomistic processes involved in interlayer transport, including the means to experimentally determine interlayer transport rates, is provided in Sect. 4.3. In Sect. 4.4 it is shown how the notion of an effective interlayer transport rate which depends on the atomic structure of steps leads to a unified view of the complex mound morphologies observed on Pt(111) under varying growth conditions. Finally, Sects.4.5 and 4.6 introduce and illustrate concepts which are necessary to understand some global aspects of mound growth, such as the delayed emergence of the mounds under conditions of weakly suppressed interlayer transport, the selection of the mound slope, and the coarsening of the mound size with increasing film thickness.

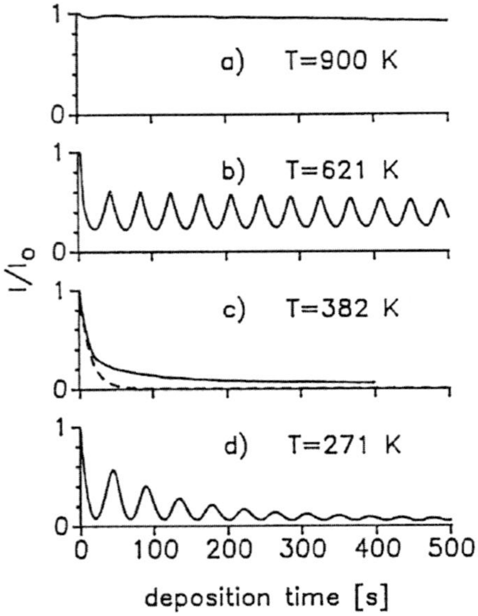

Fig. 4.2. Intensity of the specular reflected thermal energy He atom beam under anti-phase scattering conditions during homoepitaxy on Pt(111) for different deposition temperatures. The deposition rate is in the range of $1.7 - 2.5 \times 10^{-2}$ ML/s and all other deposition parameters are identical (Fig. 1 of [323])

Sections 4.3, 4.5 and 4.6 emphasize the generality of the developed concepts, while the specificities of the Pt/Pt(111) system are in the focus of Sect. 4.4. To set the stage, the following two subsections describe the main experimental findings which underlie our present understanding of multilayer growth of Pt on Pt(111).

4.1.1 Growth Modes on Pt(111): TEAS

The unique temperature dependence of multilayer growth of Pt on Pt(111) was discovered by thermal energy atom scattering (TEAS) in 1990 by Kunkel et al. [324]. The underlying physical picture was uncovered gradually on the basis of TEAS experiments and STM investigations[2]. The development was by no means linear and included bold conjectures as well as subsequent refutations. Nevertheless, even today Pt/Pt(111) is likely to be the most intensely studied, and best understood growth system.

Figure 4.2 exhibits the intensity of the specular reflected thermal energy He beam under anti-phase scattering conditions during Pt deposition at four different temperatures. In the anti-phase scattering condition, destructive interference of reflected contributions from neighboring terraces and terraces

[2] Key references are [323–329] for the TEAS work, and [130,131,206,257,280,330–334] for the STM work.

separated by an odd number of steps takes places, while terraces separated by an even number of layers interfere constructively (see Sect. 6.3).

At the highest temperature of 900 K, the reflected intensity exhibits hardly any change during deposition (Fig. 4.2a). This implies a nearly time-independent terrace distribution and suggests step flow growth. Lowering the temperature below 800 K leads to the onset of pronounced intensity oscillations in the scattering signal, indicating an oscillatory terrace distribution at the growth front. At 621 K in Fig. 4.2b these oscillations are of almost constant amplitude. Accordingly, the signal is interpreted in terms of nearly perfect LBL growth. Below 450 K the behavior of the specular reflected intensity changes qualitatively. The intensity now drops rapidly and continuously with increasing film thickness. This behavior is attributed to inhibited interlayer transport which results in mound formation. However, interlayer transport is not completely inhibited. The dashed line in Fig. 4.2c, with an even faster intensity drop than that observed experimentally, corresponds to a situation where interlayer transport is completely absent (statistical growth, see Sect. 4.2).

Surprisingly, at room temperature and down to 80 K, the specular He intensity again shows oscillations, as illustrated in Fig. 4.2d for 271 K. The amplitude of the oscillations decreases rapidly with film thickness – the oscillations are *damped* – but nevertheless more than ten oscillation periods are discernible. The signal is again attributed to LBL growth, however this growth mode is apparently not perfectly realized, since it includes a gradual build-up of surface roughness. The reappearance of LBL growth at low temperatures was highly unexpected when it was discovered [324]. It has since become known as the phenomenon of *reentrant layer–by–layer growth.*

The original interpretation [324] of the observed sequence of growth mode transitions can be summarized as follows. As was mentioned above, the transition from step flow to LBL growth is caused by an increase in nucleation probability with decreasing temperature, leading to the formation of islands on the terraces between preexisting steps. The explanation for the transition from LBL growth to mound formation invoked as a key element an additional barrier for the hop of an adatom sitting at a descending step into a step position one layer below. This *step edge barrier* will be discussed in detail in Sect. 4.3 (see also Fig. 4.11). It was assumed that the barrier is easily overcome at high temperature, leading to LBL growth. Lowering the temperature makes the barrier more efficient, thus enhancing the nucleation probability on top of islands and initiating mound formation. The reentrance of LBL growth was attributed to a low temperature breakdown of the step edge barrier. This breakdown was conjectured to be caused either by an increase of step roughness, or by a decrease in island size with decreasing temperature.

It will be seen below that the step edge barrier does indeed play a decisive role in determining the growth regimes, but that additional elements are necessary to understand the temperature dependence of growth on Pt(111).

4.1.2 Growth Modes on Pt(111): STM

The real space view of the temperature dependent growth modes on Pt(111) is shown in Fig. 4.3. Each column of the figure contains topographs obtained after deposition at one specific temperature, with the deposited film thickness increasing from 0.15 ML at the top to 1.0 ML in the middle and 5 ML in the bottom row. The first row exemplifies the decrease of the island number density with temperature as expected from nucleation theory and discussed in Sect. 2.3, as well as the temperature dependent shape evolution as analyzed in Chap. 3. The second row shows nucleation and adatom island formation on top of the first coalesced and nearly closed layer. The third row exhibits the largely developed, distinct multilayer growth morphologies at the different temperatures.

We start the analysis with the reentrant LBL regime, represented by the columns for 200 K and 300 K. After deposition of 1.0 ML the first layer is almost perfectly closed and only occasionally narrow trenches are visible (see the insets in Fig. 4.3). The adatom islands formed on top of the almost closed atomic layer are, in density and shape, nearly identical to the first layer nuclei after deposition of 0.15 ML. Thus the terrace distribution and surface morphology appears to reproduce itself within the period of one monolayer, consistent with the periodic intensity variations in TEAS. After deposition of 5 ML, the dominant part of the surface area is still distributed over two uncovered atomic levels, but the morphology is no longer representative of perfect LBL growth. Island formation proceeds on flat mesas, which are separated by two to three layer deep trenches. Correspondingly, we refer to this peculiar morphology as the *mesa mound morphology*. A periodic TEAS signal is still possible, since the trenches are narrow and deep and do not contribute significantly to the scattering signal, which is thus dominated by the two top terrace levels, the mesa and the island terrace level. It is interesting to note that even after deposition of 5 ML the top layer islands are fractal-dendritic, almost indistinguishable in shape from the first layer islands. In contrast, the mesa terraces observed after 5 ML deposition are compact, but irregular. The mesa mound morphology is investigated in more detail in Sects. 4.4 and 4.5.

The mound formation regime is represented by the STM topographs of the 400 K column. Although at 400 K after deposition of 1.0 ML the first layer islands have coalesced, it is also apparent that this coalescence led to the least perfect closure of the layer among the temperatures shown. After deposition of 5 ML a clear mound morphology emerges, which is composed of pyramids consisting of triangular adatom islands. The number of visible layers has increased from five at 300 K to eight at 400 K. Moreover, while at 300 K the mesa level could be identified as the dominant surface level, a more uniform distribution of the areas of the uncovered terrace levels is present at 400 K. The investigation of mound formation and mound growth for large film thicknesses of up to several hundred atomic layers is at the heart of the Sects. 4.2 and 4.3.

At 500 K, the TEAS data lead one to expect the reappearance of LBL growth [324]. In contrast, the STM exhibits a rather heterogeneous morphology, which is actually the result of a growth bifurcation, as will be further discussed in Sect. 4.4. While after deposition of 1.0 ML the first layer closes almost perfectly – suggesting the possibility of LBL growth –, after 5 ML adatom island pyramids (in the right top and bottom corners of the topograph in Fig. 4.3) are present together with large flat areas and coalesced islands. The morphology is obviously not representative of LBL growth. It is interesting to note that the adatom islands close to the top of the pyramids are precisely triangular, while lower level islands are dendritic-skeletal.

Another surprise comes with the STM topographs of LBL growth at 600 K in Fig. 4.3. Between the initial nucleation on the original surface at 0.15 ML and the LBL morphology present after deposition of 5 ML, the lateral length scale has *decreased*! Quantitatively, the island number density has increased by about a factor of 20. The island number density after deposition of 5 ML at 600 K is even higher than the mesa mound density after deposition of 5 ML at 300 K. A characteristic change of the surface structure is already visible at 1.0 ML, where a faint mesh or network can be perceived on the dominant terrace level. This is a deposition-induced surface dislocation network, which forms upon incorporation of a few percent of Pt atoms into the surface layer. The network flushes large terraces with monolayer period, giving rise to a decrease in the island separation and thereby indirectly to LBL growth. The formation of the network and the associated induced LBL growth are analyzed in detail in Sect. 5.4.

4.1.3 Roughness Measures

At this point it is useful to introduce some quantitative measures of surface roughness evolution. The vertical film structure is conveniently described by the layer coverages θ_n, $0 \leq \theta_n \leq 1$, where $n = 1, 2, 3...$counts the layer and $\theta_0 = 1$ represents the original substrate. The total deposited coverage is given by

$$\Theta = \sum_{n=1}^{\infty} \theta_n = \Omega F t \,, \tag{4.2}$$

and the *exposed* coverage of layer n is

$$\varphi_n = \theta_n - \theta_{n+1} \geq 0 \,. \tag{4.3}$$

Since φ_n is the probability that an arbitrary point on the surface resides on layer n, it can also be viewed as the probability distribution of the local film height, measured in units of the layer thickness d. The mean height is Θ, and the standard deviation of the φ_n defines the *surface width* W, a common measure of film roughness, through

$$W^2 = \sum_{n=0}^{\infty} (n - \Theta)^2 \varphi_n \ . \tag{4.4}$$

For perfect LBL growth, only the top layer at level n_{top} is incomplete at any given time. Then $\varphi_{n_{\text{top}}} = \theta_{n_{\text{top}}}$ and $\varphi_{n_{\text{top}}-1} = 1 - \theta_{n_{\text{top}}}$, and the evaluation of (4.4) yields

$$W^2_{\text{LBL}} = \theta_{n_{\text{top}}}(1 - \theta_{n_{\text{top}}}) = (\Theta - [\Theta])(1 - \Theta + [\Theta]) \ . \tag{4.5}$$

Here $[x]$ denotes the integer part of the real number x, so that $\theta_{n_{\text{top}}} = \Theta - [\Theta]$. The surface width vanishes for integer values of Θ (completed layers) and reaches its maximum value of 1/2 at half-integer coverages.

The second roughness measure, the growth number g, compares the layer distribution of an actual morphology to the two possible extreme cases of surface roughness, perfect LBL growth and *statistical growth*, which is the limiting case of growth without any interlayer transport (see Sect. 4.2). The growth number is defined by

$$g = \frac{\sum_{n=1}^{\infty} |\theta_{n,\text{exp}} - \theta_{n,\text{LBL}}|}{\sum_{n=1}^{\infty} |\theta_{n,\text{stat}} - \theta_{n,\text{LBL}}|} \ . \tag{4.6}$$

Here $\theta_{n,\text{exp}}$, $\theta_{n,\text{LBL}}$ and $\theta_{n,\text{stat}}$ are the layer coverages determined from experiment, according to ideal LBL growth, and according to statistical growth for the same total coverage Θ, respectively. The layer coverages for statistical growth are given in (4.10). By definition, $g = 1$ in the absence of interlayer transport, and $g = 0$ for ideal LBL growth.

A third useful characteristic parameter of the layer distribution is the anti-phase Bragg intensity defined by

$$I_{\text{anti}} = \left| \sum_{n=0}^{\infty} (-1)^n \varphi_n \right|^2 , \tag{4.7}$$

see Sect. 6.3 for a derivation and further discussion. In the case of perfect LBL growth

$$I_{\text{anti,LBL}} = (1 - 2(\Theta - [\Theta]))^2 \ , \tag{4.8}$$

which is a periodic sequence of parabolae with peaks at integer coverages and zeros at half-integer coverages.

Figure 4.4 shows the surface width and the growth number as a function of temperature for the 5 ML deposits. Both roughness measures exhibit a similar dependence on temperature. From the values of the growth number g it is particularly obvious that at 600 K LBL growth is nearly perfect with $g = 0.03$, while at 400 K interlayer transport is severely impaired with $g = 0.65$. It is also apparent that the reentrant LBL growth is far from perfect with $g \approx 0.35$.

Comparison of Figs. 4.2 and 4.4 shows that the presence of TEAS oscillations can be related to a reduced roughness and a layer distribution with $g < 0.5$, while a strongly damped TEAS signal without oscillations is charac-

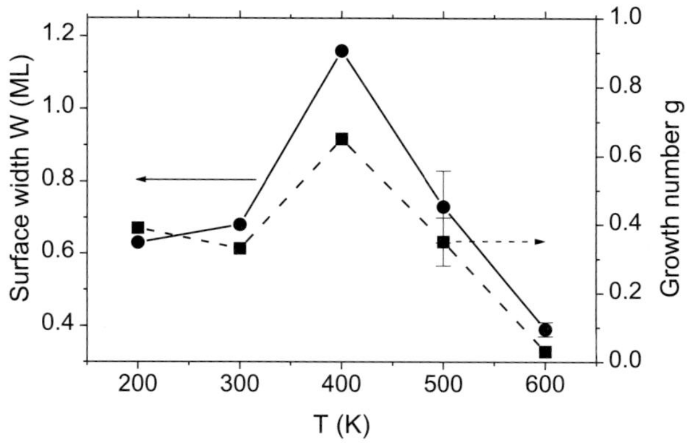

Fig. 4.4. Evolution of surface width W (*full circles*) and growth number g (*full squares*) with temperature for a deposited amount of 5 ML Pt on Pt(111). Lines to guide the eye. In the absence of interlayer transport (the statistical growth limit) the surface width would have the value $W = \sqrt{5} \approx 2.24$ at this coverage (see Sect. 4.2)

teristic of a large roughness and a layer distribution with $g > 0.5$. The long lived TEAS oscillations observed in the high temperature regime were thus correctly interpreted in terms of nearly perfect LBL growth. However, the microscopic origin of the LBL growth – the surface reconstruction mentioned above (see Sect. 5.4) – could not be inferred from TEAS measurements, but required the real space STM information [329, 332].

Similarly, the damped TEAS oscillations associated with the reentrant LBL growth are an indicator for reduction of the roughness at low temperatures. However, the STM images show that the initial LBL growth turns into a mesa mound morphology already after deposition of 5 ML. As will be shown in the next section, mound formation does not generally preclude transient TEAS oscillations: For an oscillatory behavior of the anti-phase Bragg intensity it is sufficient that a significant fraction of the surface is dominated by one or two uncovered atomic layers, which is clearly the case in the mesa mound regime. In Sects. 4.5 and 4.9.1 other growth systems will be described in which damped transient oscillations of a scattering signal are followed by mound growth; in fact this seems to be a rather general scenario.

The rapid overview of growth behavior which is available through scattering methods thus does not relieve one of the slow and cumbersome process of obtaining real space STM images at different temperatures and coverages, in order to develop an adequate physical picture of the growth mechanisms.

4.2 Statistical Growth and the Shape of Wedding Cakes

We have already seen that the most important new kinetic process which enters the scene in multilayer growth is the migration of adatoms between different atomic layers of the surface. By necessity this involves crossing a step edge, which is often associated with an additional energy barrier in excess of the surface diffusion barrier within a layer. We postpone a detailed dicussion of these step edge barriers to Sect. 4.3. Here we give a first demonstration of the relevance of interlayer transport for the morphology of multilayer films by assuming, for the moment, that it is completely suppressed.

In the absence of interlayer transport, each adatom remains in the layer in which it was first deposited, and is incorporated into that layer at an ascending step edge. This implies a simple evolution of the layer coverages: The rate at which layer n grows is proportional to the exposed coverage φ_{n-1} of the layer $n-1$ below, and therefore the θ_n satisfy [335]

$$\frac{\mathrm{d}\theta_n}{\mathrm{d}t} = \Omega F(\theta_{n-1} - \theta_n) \tag{4.9}$$

with the initial conditions $\theta_{n \geq 1}(0) = 0$ and the boundary condition $\theta_0 = 1$. It is straightforward to check that the solution reads [336]

$$\theta_{n,\mathrm{stat}} = 1 - \mathrm{e}^{-\Theta} \sum_{k=0}^{n-1} \frac{\Theta^k}{k!} \,. \tag{4.10}$$

Correspondingly the exposed coverages follow a Poisson distribution with parameter Θ,

$$\varphi_n = \frac{\mathrm{e}^{-\Theta}\Theta^n}{n!} \,. \tag{4.11}$$

For the Poisson distribution, the variance is equal to the mean, and therefore the surface width is given by

$$W_{\mathrm{stat}} = \sqrt{\Theta} \tag{4.12}$$

for growth without interlayer transport. Remarkably, the expression (4.12) is independent of the in-layer diffusion rate. This is because the surface width depends only on the amount, but not on the lateral distribution of the deposited material within a given layer. As far as W is concerned, growth without interlayer transport is therefore equivalent to *random deposition*, where atoms are stacked irreversibly on top of each other without any lateral or vertical mobility [32, 337]. Equation (4.12) represents the maximum surface roughness that can be generated by the randomness in the deposition flux, and is referred to as the *statistical growth* limit. Interlayer transport is solely responsible for reducing the roughness below this limit.

In the same spirit, the statistical growth model puts a limit on how rapidly the anti-phase intensity measured with a scattering probe can decay to zero due to the roughening of the growth front. Evaluating the expression (4.7) for the layer distribution (4.10) yields [335]

$$I_{\mathrm{anti,stat}} = \mathrm{e}^{-4\Theta} \,, \tag{4.13}$$

which is included as a dashed line in Fig. 4.2c.

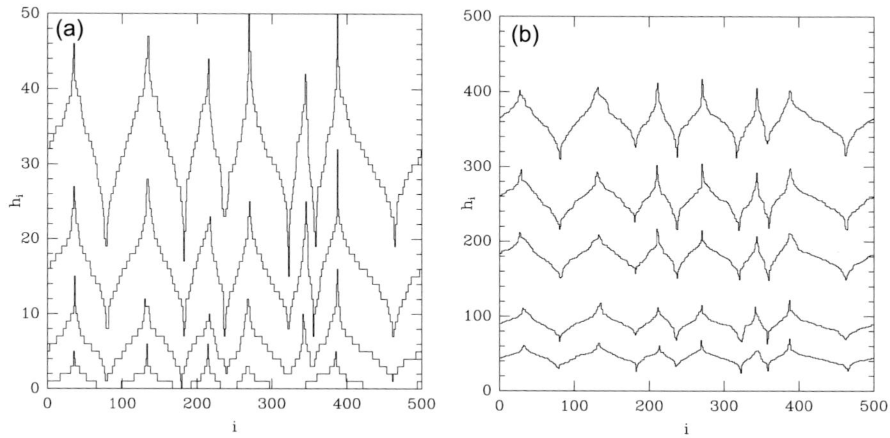

Fig. 4.5. Surface morphology in a one-dimensional growth model without interlayer transport in a system of 500 sites. (**a**) Morphology after deposition of 1, 5.6, 16 and 32 monolayers. (**b**) Morphology after deposition of 45.25, 90.5, 181, 256 and 362 ML. The ratio of diffusion to deposition rate is $\nu/(\Omega F) = 5 \times 10^6$ (Fig. 1 of [338])

While in-layer diffusion does not affect the *vertical* surface morphology, as encoded in the layer coverages θ_n, it is certainly reflected in the *lateral* mass distribution along the surface. This is illustrated in Fig. 4.5 by a one-dimensional Monte Carlo simulation. It shows the emergence of a fairly regular pattern of mound-like surface features with a characteristic pointed shape. Each mound consists of a tapering stack of islands upon islands, reminiscent of a *wedding cake*. In the following we argue that (i) many properties of this pattern follow immediately from the expression (4.10) for the layer coverages [338], and (ii) the wedding cake model provides a reasonable zeroth order description of Pt growth on Pt(111) at 440 K [333].

Figure 4.5 clearly demonstrates that the lateral positions and sizes of the mounds originate in the growth of the first layer: The wedding cakes grow on the templates of the first layer islands, and their spacing is simply determined by the density N of first layer nuclei, which in one dimension is proportional to $(F/D)^{1/4}$ (see Sect. 2.4.3). The persistence of the first layer pattern throughout the deposition of hundreds and thousands of layers requires, first, that no additional mounds nucleate during the later stages of growth, and, second, that neighboring mounds do not merge. The first requirement is met because the step spacing on the sides of the mounds decreases with increasing coverage (the mounds steepen), and therefore nucleation on the vicinal terraces making up the sides becomes highly unlikely[3]. The merging of mounds is suppressed because the lateral positions of the maxima and minima of the

[3] The nucleation rate on a one-dimensional vicinal terrace of width l is of the order of $F^2 l^4/D$, see (4.96) [339].

surface profile are strongly correlated from one layer to the next. For the maxima this reflects the fact that nucleation occurs only in the top layer of each mound, while for the minima it can be attributed to the *Zeno effect*, the appearance of deep crevices between neighboring mounds [339]. Such a crevice develops, because fewer and fewer deposited atoms find their way to its bottom terrace, and the step edge barrier prevents it to be filled by surface diffusion. In particular, according to (4.10) the exposed fraction of the substrate $\varphi_0 = 1 - \theta_1 = \mathrm{e}^{-\Theta}$ approaches zero only asymptotically, but does not vanish at any finite time.

The resulting picture of the pattern forming process is readily generalized to growth on real, two-dimensional substrates. The nucleation of the first layer islands partitions the substrate into capture zones (compare to Sect. 2.3.2). Each zone supports a single mound, which is fed by the atoms deposited into the zone. This implies that the typical shape of the mounds can be read off from the layer distribution (4.10): The area A_n of the n'th layer of a mound will be equal to $\theta_n A_0$, where A_0 is the area of the corresponding capture zone. A more transparent form of (4.10) is obtained in the limit of thick layers, $\Theta \gg 1$, when the Poisson distribution (4.11) can be replaced by a Gaussian of width $\sqrt{\Theta}$,

$$\varphi_n \approx \frac{1}{\sqrt{2\pi\Theta}} \mathrm{e}^{-\frac{(n-\Theta)^2}{2\Theta}} . \tag{4.14}$$

The layer distribution then follows by integration,

$$\theta_n = \Phi((n - \Theta)/\sqrt{\Theta}) \tag{4.15}$$

where

$$\Phi(s) = 1 - \frac{1}{\sqrt{2\pi}} \int_{-\infty}^{s} dy \, \mathrm{e}^{-y^2/2} = \frac{1}{2}[1 - \mathrm{erf}(s/\sqrt{2})] \tag{4.16}$$

and $\mathrm{erf}(s)$ denotes the error function. Equation (4.15) shows that the mounds attain a time-independent limiting shape when rescaled vertically by $W = \sqrt{\Theta}$. Suppose for simplicity that the mounds are composed of circular islands. Then the mound radius at height n is $\sqrt{A_n/\pi}$, and by inverting the relation (4.15) the height h of the mound surface at radius r is obtained in the form

$$h(r)/d = \Theta + \sqrt{\Theta}\mathcal{H}(r/r_0) , \tag{4.17}$$

where $r_0 = \sqrt{A_0/\pi}$ and the shape function $\mathcal{H}$ is the inverse of $\sqrt{\Phi}$.

The main features of the wedding cake model agree well with deposition experiments[4] on Pt(111) in the mound formation regime, carried out at 440 K and for film thicknesses ranging from 0.3 to 300 ML [333]. In Fig. 4.6 the evolution of the surface morphology from submonolayer islands at 0.35 ML to well developed mounds at 90 ML is presented. The onset of mound formation

[4] Quantitative evidence for statistical growth has also been obtained, among other growth systems, for Au and Ag on Ag(111) [340–342] and for Fe on Fe(110) [343].

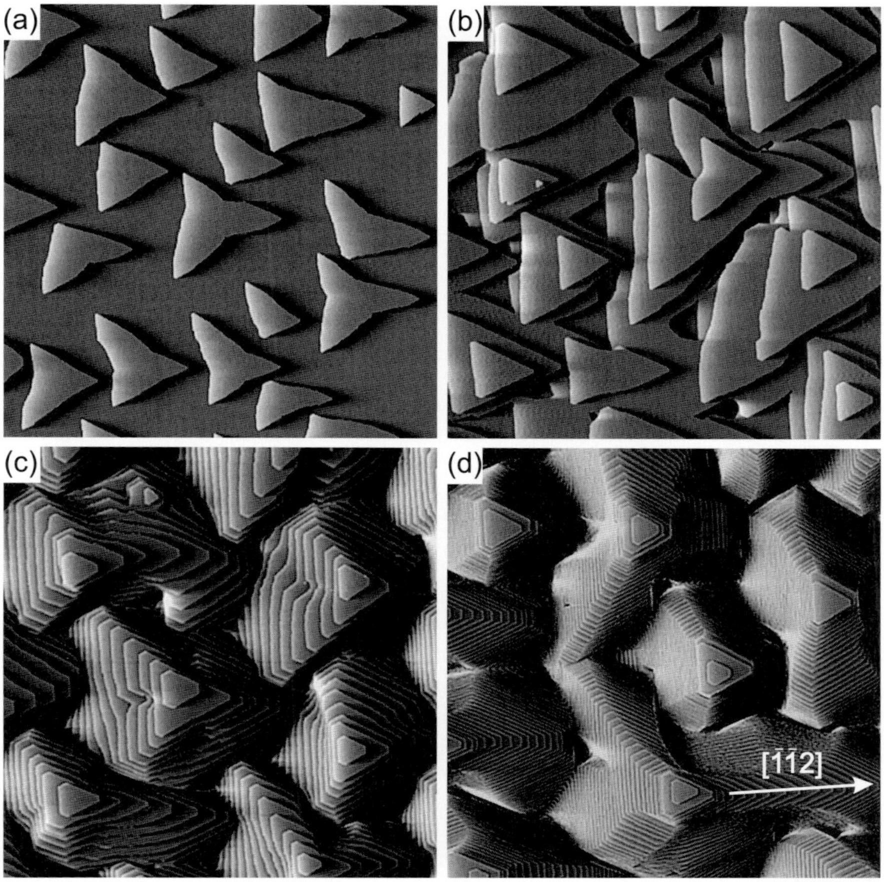

Fig. 4.6. STM topographs of Pt on Pt(111) after growth at 440 K with $F = 7 \times 10^{-3}$ ML/s. (**a**) 0.35 ML, (**b**) 3 ML, (**c**) 12 ML, (**d**) 90 ML. Topographs do not show the same sample location. Topograph size is 3450 Å × 3450 Å (based on Fig. 1 of [333] reprinted with permission from Elsevier Science)

in Figs. 4.6a and b is quite similar to the situation at 400 K shown in Fig. 4.3. A number of qualitative features are obvious from the topographs in Fig. 4.6: (i) From Fig. 4.6a to d the surface roughness increases continuously. (ii) The density of mounds and hence their average separation λ varies only slowly and remains almost constant from Fig. 4.6c to d. (iii) As the average terrace width decreases from Fig. 4.6a to d, no specific slope is selected on the hillsides of the mounds. (iv) In Fig. 4.6c and d, deep grooves separating the mounds appear. (v) In Fig. 4.6d, there is an abrupt transition from the mound slopes to the flat mound tops. (vi) The appearance of the steps bounding the top terrace islands gradually changes with increasing Θ. With increasing film thickness, they align more and more closely with the close packed $\langle 110 \rangle$ orientations.

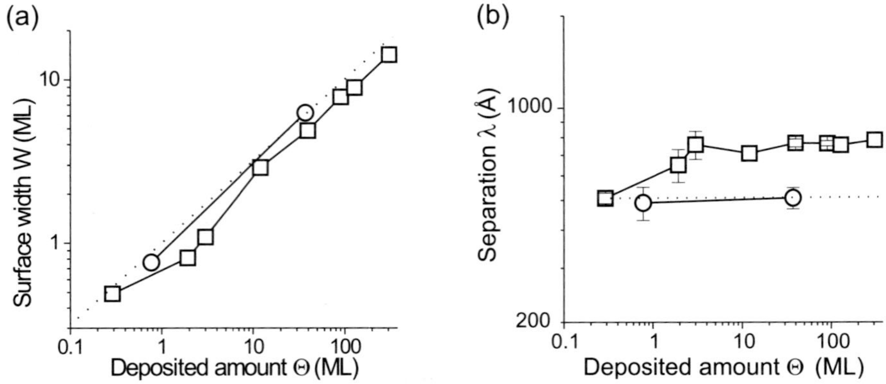

Fig. 4.7. (**a**) Surface width W and (**b**) mound separation λ versus the deposited amount Θ. *Squares*: clean growth at 440 K. *Circles*: growth at 440 K in a CO partial pressure of $p_{CO} = 2 \times 10^{-9}$ mbar (see Sect. 4.4.2 for a discussion of the CO data). The dotted lines indicate the predictions of the wedding cake model $W = \sqrt{\Theta}$ and $\lambda =$ const.. The error bars in (**a**) are smaller than the symbol size (based on Fig. 2 of [333] reprinted with permission from Elsevier Science)

Indeed, prior to nucleation of a new island on top, the top terrace step edges in Fig. 4.6a have a kink density of ≈ 0.2, i. e. there is one kink per 5 step atoms, whereas in 4.6d the kink density has decreased to $\approx 10^{-2}$.

The first observations (i)-(iii) are substantiated by a quantitative evaluation of the surface width W and the average mound separation λ as a function of film thickness shown in Fig. 4.7. The mound separation was determined from the height–height correlation function

$$G(r) = \langle (h(\boldsymbol{r}' + \boldsymbol{r}) - \bar{h})(h(\boldsymbol{r}') - \bar{h}) \rangle \, , \tag{4.18}$$

with $r = |\boldsymbol{r}|$. Here $h(\boldsymbol{r})$ is the height of the surface above the substrate point $\boldsymbol{r}$, $\bar{h}$ is the mean height and the angular brackets refer to a spatial average. Common measures of the mound separation are twice the position of the first zero of $G(\boldsymbol{r})$, and four times its full width at half maximum. The data for λ shown in Fig. 4.7 represent the mean of these two quantities. The surface width (4.4) can also be expressed in terms of the correlation function (4.18) as

$$W = d^{-1} \sqrt{G(0)} \, . \tag{4.19}$$

After a transient regime lasting up to about 10 ML, the surface width follows closely the predicted square root law (4.12), and the mound separation λ stays constant, indicating that the lateral distribution of the mounds does not change.

The mound shape at 130 ML, obtained by averaging the layer coverages over 10 mounds, is displayed in Fig. 4.8. Anisotropies in the mound shape have been removed by transforming the layer coverages into circular discs. The predicted shape function (4.16) matches the observed height profile both

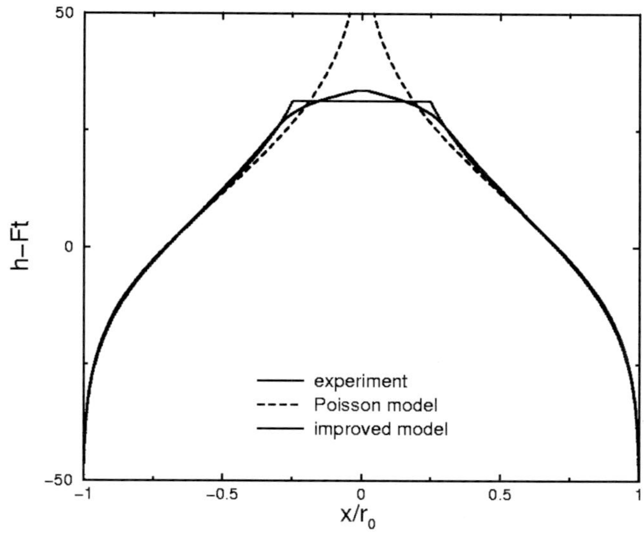

Fig. 4.8. Full line shows the experimentally determined average mound shape from [333], the dashed line the prediction of the wedding cake model, and the dotted line shows the best fit to the shape function (4.20). The critical coverage is $\theta_c = 0.122$

in the valleys [observation (iv)] and on the slopes of the mounds, but differs significantly near the tops: Instead of the pointed peaks seen in Fig. 4.5, the real mounds terminate in flat terraces [(observation (v)] of a characteristic lateral size (see Fig. 4.6d).

This discrepancy should be no surprise, since the model assumption of zero interlayer transport becomes extremely unrealistic for an adatom that has been deposited onto the freshly nucleated, small top island of a mound. In the absence of ascending step edges, such an atom has no choice but to interrogate the island edge many times and eventually cross it, even if the crossing probability in each attempt is very small. As the island grows, the residence time of the adatom increases and the time between the successive arrivals of adatoms decreases, until the probability for two atoms to be present simultaneously on the island becomes appreciable, and a new nucleus is formed on top of the island. We are thus lead to the notion of a *critical radius* for nucleation on top of an island, which will be explored in depth below in Sect. 4.3.3. Within the minimal growth model defined by (4.9), this effect can be incorporated[5] by allowing a layer n to start to grow only when the preceding layer $n-1$ has reached a critical coverage θ_c. The resulting layer distribution still approaches the scaling form (4.15) for large Θ, but

[5] Such a model was first introduced in [345] for the description of erosion pits. A more realistic nucleation criterion based on the time-integrated nucleation probability [compare to (4.49)] was considered in [344], and leads to the same asymptotic shape.

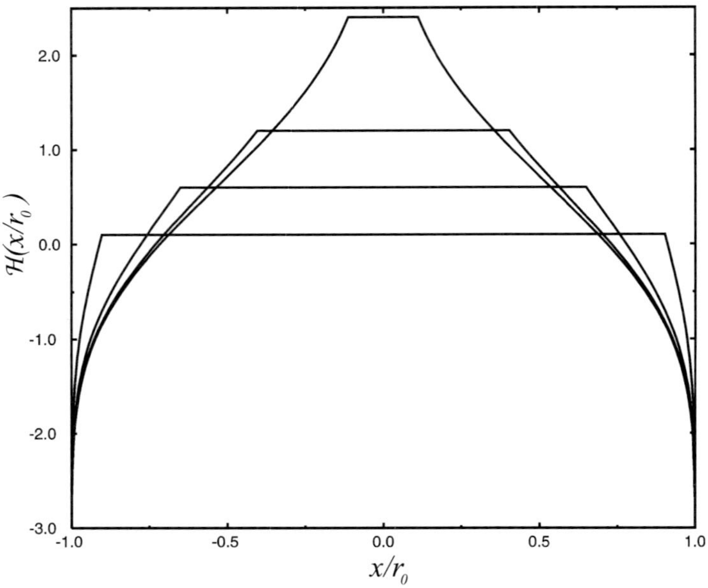

Fig. 4.9. Asymptotic mound shapes according to (4.20). From top to bottom, the critical coverage is $\theta_c = 0.01$, 0.15, 0.4 and 0.8 (Fig. 8 of [344] reprinted with kind permission of Kluwer Academic Publishers)

with a modified shape function given by [344]

$$\Phi(s) = \begin{cases} 1 - C(\theta_\mathrm{c}) \int_{-\infty}^{s} dy\, \mathrm{e}^{-y^2/2} & : \quad s \leq s_{\max}(\theta_\mathrm{c}) \\ 0 & : \quad s > s_{\max}(\theta_\mathrm{c}) \,. \end{cases} \tag{4.20}$$

The constants C and $s_{\max}$ are fixed by the relation $\Phi(s_{\max}) = \theta_\mathrm{c}$ and the normalization condition (4.2). Figure 4.9 illustrates how the shape varies with θ_c. The height of the mound top, relative to the mean film thickness, is equal to $s_{\max}\sqrt{\Theta}$. The surface width still increases as $\sqrt{\Theta}$ but with a reduced coefficient; the explicit calculation shows that the statistical roughening law (4.12) is replaced by

$$W = \sqrt{(1 - \theta_c)\Theta} \,. \tag{4.21}$$

This provides a natural interpretation for experiments where a linear increase of the height variance W^2 is observed, but the prefactor is smaller than the statistical growth limit (4.12), see e.g. [342].

Despite the rapid increase of the surface roughness, after some initial damping the model displays persistent oscillations of the anti-phase Bragg intensity (4.7), with an amplitude of order θ_c^2 (Fig. 4.10). The oscillations reflect the periodic nucleation and growth of the top terrace. With increasing roughness the contribution of the mound slopes to the scattering intensity vanishes, because the corresponding exposed coverages shrink to zero, but

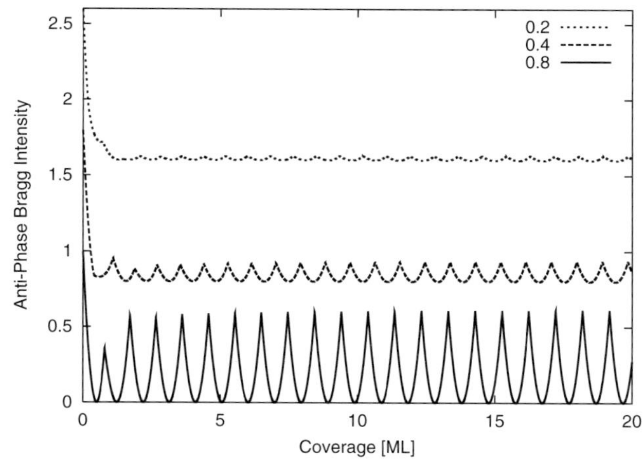

Fig. 4.10. Anti-phase scattering intensity for the wedding cake growth model at different values of the critical coverage θ_c for second layer nucleation. The behavior shown corresponds to the growth of a single mound, or an array of mounds growing perfectly in phase. The curves have been shifted vertically for greater clarity

the top terrace gives a contribution of order θ_c to the sum in (4.7), and hence a contribution of order θ_c^2 to I_{anti}, at all times. This example shows that mound formation is not incompatible with an oscillatory scattering signal. Eventually the persistent oscillations will be damped due to the fluctuations in the number of atoms deposited onto different mounds, but this effect sets in at much later times (see Sect. 5.6).

A fit of the modified shape function (4.20) to the experimental data is shown in Fig. 4.8. The match is excellent apart from the region very near the top, where the experimental data show some rounding, due to the averaging over several mounds with top terraces of different size. The fit yields the value $\theta_c = 0.122$ for the critical coverage. It will be explained in Sect. 4.3.3 how this number can be used to estimate the rate of interlayer transport.

So far we have not addressed the transient growth regime which is seen in Fig. 4.7 for $\Theta < 10$. In this regime the surface width grows more slowly than the statistical growth limit (4.12) and at the same time the mound separation increases by about a factor of two. In comparison with the statistical growth limit, both features imply a significant amount of interlayer transport. We postpone the discussion of this growth regime to Sect. 4.4.2, where it will be shown in detail how it is related to the gradual change of the atomic structure of the top terrace island boundaries during growth, i.e. to observation (vi) noted above.

4.3 Interlayer Transport

Although it is customary to assign a single kinetic rate of standard Arrhenius form to interlayer diffusion [see (4.22)], this is a radical simplifaction, which may be unwarranted in many cases. In contrast to the hopping (or exchange) of an adatom between different adsorption sites within a crystal plane, the move between different layers is not an elementary process, because it always entails the crossing of a step, which is a spatially extended object with considerable internal structure. This implies at least two types of complications: First, the actual crossing rate depends on the step morphology, i. e. on the step orientation as well as the presence of kinks and corners. We will return to this point in Sect. 4.4. Second, the step may induce long range distortions of the surface lattice, which affect the adatom motion long before it reaches the step (see Sect. 2.4.1 for an example).

Despite its somewhat fictional character, however, the notion of a single overall interlayer transport rate is highly useful in systematizing the different regimes of multilayer growth. We will therefore adhere to, and elaborate on this concept for the most part of this chapter. A more accurate view of the relationship between step structure and step crossing, and its consequences for multilayer growth, will be sketched in Sect. 4.4.

4.3.1 The Step Edge Barrier and Some of Its Consequences

In their pioneering FIM study of surface diffusion of tungsten on tungsten, Ehrlich and Hudda [16] observed that W adatoms are often reflected by the boundaries of the W(110), W(211) and W(321) lattice planes. They estimated that an energy barrier for step crossing that exceeds the diffusion barrier E_{D} by about 0.065 eV would be sufficient to account for this observation, and pointed out that the existence of such an additional barrier is consistent with a simple pair-bond picture, in which the potential energy of the adatom is proportional to its coordination number[6]. In a subsequent paper on the stability of growing stepped surfaces (see Sect. 4.7.1), Schwoebel and Shipsey [17] introduced the schematic potential energy landscape near a step, that has since become the signature of what is often referred to as the *Ehrlich–Schwoebel–effect* (Fig. 4.11). It contains the minimal modifications of the adatom potential energy that have to be expected from coordination considerations: (i) An increased binding energy on the lower terrace next to the step, and (ii) an increased energy barrier for crossing the step from the upper terrace. The additional binding energy at the step implies that the rate for detachment from the step is smaller than the in-layer hopping rate

[6] Later, more detailed investigations yielded a much larger estimate of about 0.20 eV for the excess barrier for an W adatom to descend from a W(110) lattice plane [346].

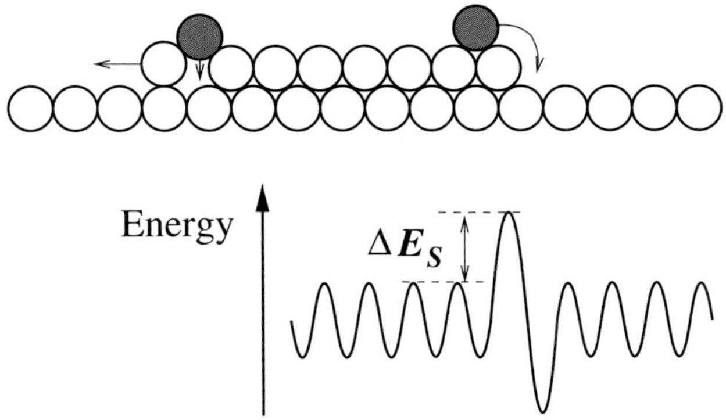

Fig. 4.11. Upper part of the figure shows the descent of adatoms (shaded) from an island by hopping (right hand side) and exchange (left hand side). The lower part illustrates the energy landscape for hopping and the definition of ΔE_S

ν, while the rate for step attachment from the lower terrace (in the idealization of Fig. 4.11) is equal to ν. In contrast, step attachment from the upper terrace requires descent of the adatom and is governed by a separate rate

$$\nu' = \nu_0' \, e^{-E_S/k_B T} \tag{4.22}$$

with its own energy barrier E_S and attempt frequency ν_0'. The *additional* step edge barrier is then given by[7]

$$\Delta E_S = E_S - E_D \, . \tag{4.23}$$

To be precise, we define ν' as the rate at which an adatom poised at an edge site descends from the upper terrace. In contrast to ν, which counts the adatom jumps per unit time irrespective of their direction, ν' is the rate for a *directed* move. We do not specify at this point if the descent occurs by hopping or exchange, though the latter appears to be the typical situation (see Sect. 4.3.4). The possibility of interlayer mass transport by an exchange process was in fact suggested already by Schwoebel and Shipsey in their 1966 paper [17]. Experimentally, exchange at a descending step edge has been observed by FIM for W atoms on Ir(111) islands [164].

To explore the consequences of the step edge barrier in some simple geometries, we adopt a continuum approximation in which the step is treated as a sharp, one-dimensional boundary separating structureless two-dimensional terraces[8]. This can be justified for terraces that extend over many atomic distances. Adatom diffusion on the terraces is described by the diffusion equation

[7] Alternatively, the additional barrier could be defined as the difference in energies between the *transition states* for adatom descent and for diffusion on the terrace [347]. For the simple situation shown in Fig. 4.11, the two definitions are equivalent.

[8] This approach goes back to the classic work of Burton, Cabrera and Frank [22]. For contemporary reviews see [35, 348].

(2.3), which has to be supplemented by the deposition flux F and by boundary conditions representing attachment and detachment at the step. In general, the adatom fluxes cause the steps to move, which in turn changes the adatom concentration profile $n(\boldsymbol{r},t)$. The problem simplifies if the step motion is slow on the time scale of adatom diffusion, because then the adatom density can be computed *quasistatically* from the stationary diffusion equation

$$D\nabla^2 n + F = 0 \,, \tag{4.24}$$

which is to be solved for a fixed configuration of steps.

The following estimate shows that the quasistatic approximation suffices for most purposes. For steps spaced at a typical distance l, the time scale for adatom diffusion is $\tau_D \sim l^2/D$. The time required for a step to propagate across a terrace is of the order of the monolayer growth time $\tau_{\mathrm{ML}} = (\Omega F)^{-1}$. The ratio of the two time scales defines the *Péclet number* [349]

$$\mathrm{Pe} = \tau_D/\tau_{\mathrm{ML}} = \Omega F l^2/D \,. \tag{4.25}$$

The stationary approximation (4.24) is applicable if $\mathrm{Pe} \ll 1$, or

$$l \ll \left(\frac{D}{\Omega F}\right)^{1/2} \,. \tag{4.26}$$

According to the prediction (2.18) of atomistic nucleation theory, the distance between nuclei is of the order of

$$l_D = N^{-1/2} \approx \Omega^{1/2}(D/F\Omega^2)^{\chi/2}\mathrm{e}^{-E_{i^*}/2(i^*+2)k_{\mathrm{B}}T} \tag{4.27}$$

with $\chi < 1$ and $E_{i^*} > 0$. Therefore terraces large enough to violate (4.26) never appear on a growing surface: If they did, new steps would immediately be created by island nucleation.

The attachment–detachment boundary conditions for $n(\boldsymbol{r},t)$ take the form [349, 350]

$$D\boldsymbol{n}\cdot\nabla n|_+ = a\nu(n - n^{\mathrm{eq}})|_+ \tag{4.28}$$

$$-D\boldsymbol{n}\cdot\nabla n|_- = a\nu'(n - n^{\mathrm{eq}})|_- \,. \tag{4.29}$$

Here $\boldsymbol{n}$ denotes a unit vector normal to the step which by convention points towards the lower terrace, and the expressions are evaluated approaching the step from below (+) and above (–), respectively. For growth far from equilibrium, the equilibrium adatom concentration n^{eq} can be neglected compared to the concentration of deposited adatoms. Both sides of the relations (4.28,4.29) express the net adatom flux attaching to the step from below and above. The left hand sides represent the diffusion flux arriving from the terrace in the vicinity of the step, while the right hand side is obtained by balancing the attachment/detachment hops between the step edge and the sites next to the

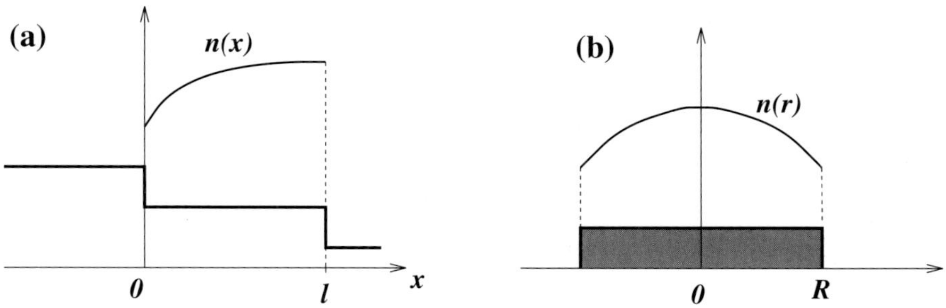

Fig. 4.12. Schematic adatom density profiles on (**a**) a vicinal terrace between two straight steps, and (**b**) a circular island. In (**a**) it is assumed that $\nu' = 0$, so that no adatom flux crosses the descending step, whereas (**b**) corresponds to a finite interlayer transport rate

step[9]. Setting the two equal implies continuity of the adatom flux between the large scale diffusion field on the terrace, and the microsopic events at the step. A derivation of (4.29) from a more microscopic picture is given in Sect. 4.9.3.

The adatom concentration is typically discontinuous across the step. In fact three different values of the adatom concentration at the step have to be distinguished: The limiting concentrations $n|_+$ and $n|_-$ attained when approaching the step from above or below, and the concentration n_s of step adatoms attached to the step. The boundary conditions (4.28,4.29) assume that n_s takes on its equilibrium value at all times. A generalized formalism taking into account the detachment/attachment of step adatoms at kinks has been developed in [351, 352].

In the following we present solutions to (4.24,4.28,4.29) for two special cases (Fig. 4.12).

• *Vicinal terrace between straight steps.* The x–axis is oriented perpendicular to the steps, with the ascending step located at $x = 0$ and the descending step at $x = l$. The density profile is parabolic,

$$n(x) = n(0) + 4(n(0) - n^{\text{eq}})(x/a) - \frac{F}{2D}x^2 , \tag{4.30}$$

with the adatom density at the ascending step given by

$$n(0) - n^{\text{eq}} = \left(\frac{Fl}{a\nu}\right) \frac{1 + 2(l/a)(\nu'/\nu)}{1 + \nu'/\nu + 4(l/a)(\nu'/\nu)} . \tag{4.31}$$

[9] In (4.28) the attachment rate to the ascending step has been set equal to the jump rate ν on the terrace. This corresponds approximately to the assumption that the potential energy landscape for adatoms approaching the step from below is not distorted by the step, as implied in Fig. 4.11. A more accurate, microscopic treatment should include geometry-dependent corrections accounting for the fact that not all atoms that reside near the step actually jump towards it, see Sect. 4.9.3.

The right hand side varies only mildly, from $Fl/2a\nu$ to $Fl/a\nu$, as the interlayer transport rate is decreased from $\nu' = \nu$ to $\nu' = 0$. The obstruction of interlayer diffusion does not strongly enhance the adatom density on a vicinal terrace, because the atoms retain the option of attaching to the ascending step.

The fraction p_+ of atoms captured at the ascending step is computed from the diffusion fluxes to the steps as

$$p_+ = \frac{Dn'(0)}{Dn'(0) - Dn'(l)} = \frac{2l + l_{\mathrm{ES}} + a}{4l + l_{\mathrm{ES}} + 2a} , \tag{4.32}$$

where $n' = \mathrm{d}n/\mathrm{d}x$. In the last step the *Ehrlich–Schwoebel length*[10] [339, 353]

$$l_{\mathrm{ES}} = (\nu/\nu' - 1)a \tag{4.33}$$

has been introduced. It is a measure for the strength of the step edge barrier, normalized to vanish when $\nu = \nu'$; for $\nu/\nu' \gg 1$ the approximation $l_{\mathrm{ES}} \approx (\nu/\nu')a$ can be used. Equation (4.32) shows that the attachment asymmetry depends on the ratio of the terrace length l to l_{ES}: For $l \gg l_{\mathrm{ES}}$ nearly half of the atoms attach to the descending step ($p_+ \approx 1/2$), while for $l \ll l_{\mathrm{ES}}$ almost all atoms attach to the ascending step ($p_+ \approx 1$). The important, general lesson is that the strength of the step edge barrier is not determined simply by the ratio ν/ν', but can be gauged only by comparing the Ehrlich–Schwoebel length to some other appropriate lateral scale of the surface morphology. This observation will be a recurring theme in what follows.

• *Circular island.* Solving the stationary diffusion equation (4.24) in polar coordinates, the adatom density at distance r from the center of a circular adatom island of radius R is found to be

$$n(r) = n(0) - \frac{F}{4D} r^2 , \tag{4.34}$$

where

$$n(0) - n^{\mathrm{eq}} = \frac{FR}{a\nu} [(R/a) + (\nu/2\nu')] . \tag{4.35}$$

Because the island is bounded entirely by a descending step, here the confinement of the atoms by the step edge barrier can lead to a manyfold increase of the adatom density compared to the value for perfect interlayer transport, provided that $l_{\mathrm{ES}} \gg R$. The resulting enhancement of nucleation on top of islands will be analyzed next.

4.3.2 Theory of Second Layer Nucleation

The process that most clearly displays the influence of interlayer transport on the growth morphology is the onset of nucleation in the second atomic

---
[10] The reader should be cautioned that different definitions for l_{ES} are encountered in the literature.

layer. The way in which it determines the growth mode of a multilayer film was described on a qualitative level in Sect. 4.1: If second layer nucleation begins only after the first layer islands have started to coalesce, then the first layer is completed before much growth in higher layers has occurred, and the exposed surface at monolayer coverage $\Theta = 1$ is almost perfectly flat[11]; this is the regime of layer–by–layer growth, which will be described in detail in Chap. 5. If, on the other hand, second layer nucleation sets in while the first layer islands are still small, repeated nucleation in the top layer takes place prior to first layer coalescence, and a three-dimensional wedding cake morphology results (see Sect. 4.2).

It is therefore natural to discuss the conditions for layer–by–layer growth, including strategies for its manipulation, in terms of second layer nucleation [264, 324]. In addition, studies of second layer nucleation have become an important method for the experimental determination of step edge barriers (Sect. 4.3.3). Here we provide a simple theoretical analysis for the case of irreversible aggregation ($i^* = 1$), which is based on a comparison of the various time scales involved in the process [334]. Some remarks concerning the extension to $i^* > 1$ can be found in Sect. 4.9.5. An earlier approach based on rate equations [354] and the reason for its failure is described in Sect. 4.9.4.

We consider a circular island of fixed radius R, onto which atoms are deposited at flux F. There are three relevant time scales in the process: The mean time interval between subsequent arrivals of atoms on the island

$$\Delta t = \frac{1}{F\pi R^2} \tag{4.36}$$

the *diffusion time* required for an atom to diffuse across the island,

$$\tau_{\mathrm{D}} \sim R^2/D \tag{4.37}$$

and the mean *residence time* τ that an atom spends on the island, if no nucleation occurs. Under steady state conditions the residence time is related to the spatially averaged adatom density $\bar{n}$ on the island through the mass balance relation

$$\tau = \bar{n}/F \,. \tag{4.38}$$

Computing $\bar{n}$ from (4.34,4.35) yields[12]

$$\tau = \frac{R^2}{8D} + \frac{R}{2a\nu'} \,. \tag{4.39}$$

The first term dominates when $R \gg l_{\mathrm{ES}}$ (*weak barrier regime*). The residence time is then comparable to the diffusion time (4.37), because the atom leaves the island after a few excursions to the edge. Conversely, in the *strong barrier*

[11] Note, however, that other roughening mechanisms may lead to a breakdown of layer–by–layer growth at a much later stage of film evolution (Sect. 4.5.2 and Sect. 5.6).

[12] The full distribution of residence times is calculated in [355].

regime $R \ll l_{\mathrm{ES}}$ the residence time is dominated by the second term, and becomes independent of D. In this limit the adatom density profile (4.34) becomes flat, and $\tau^{-1} = 2a\nu'/R$ is simply equal to the interlayer hopping rate multiplied by the probability of finding the atom at the edge. In this case $\tau \gg \tau_{\mathrm{D}}$. We will moreover assume that

$$\tau \ll \Delta t \tag{4.40}$$

which is true under typical growth conditions. If (4.40) is violated, no steady state is established because a nucleus forms as soon as the second atom is deposited on the island[13].

The formation of a second layer nucleus requires, first, that (at least) two atoms are present on the island simultaneously, and, second, that the two meet before one of them escapes from the island. In the case of strong step edge barriers the two atoms will almost certainly meet, because their residence time τ is much larger than the encounter time, which can be roughly identified with the diffusion time τ_{D}. The probability for a freshly deposited atom to form a nucleus is therefore equal to the probability that another atom is present on the island at its deposition.

To compute this probability, suppose the first adatom has arrived on the island at time $t = 0$, and denote by t_1 and t_2 the departure time of the first adatom and the arrival time of the second one. Nucleation occurs if $t_1 > t_2$. Since deposition is a random (Poisson) process, t_2 is exponentially distributed with mean Δt. Denoting the distribution of residence times by P_{res}, we have

$$\mathrm{Prob}[t_1 > t_2] = \frac{1}{\Delta t} \int_0^\infty dt_1 \; P_{\mathrm{res}}(t_1) \int_0^{t_1} dt_2 \; \mathrm{e}^{-t_2/\Delta t} \; . \tag{4.41}$$

Under the condition (4.40) the integration over t_1 is effectively limited to $t_1 \ll \Delta t$. Then $t_2 < t_1 \ll \Delta t$ also, and we may set $\mathrm{e}^{-t_2/\Delta t} \approx 1$ in the second integral. The first integral then simply yields the mean residence time, and it follows that $\mathrm{Prob}[t_1 > t_2] \approx \tau/\Delta t$. Multiplying this with the number $(\Delta t)^{-1}$ of atoms deposited on the island per unit time, we obtain an expression for the rate of nucleation on the island,

$$\omega = \frac{\tau}{(\Delta t)^2} = \frac{\pi^2 F^2 R^5}{2a\nu'} \tag{4.42}$$

which is exact under the stated conditions: $\tau \ll \Delta t$ and $R \ll l_{\mathrm{ES}}$.

It is instructive to try to visualize the events involved in nucleation on top of an island bounded by a step with a large step edge barrier. Let us assume an island has grown to the size where nucleation typically takes places. An atom arriving on the island diffuses around rapidly, being reflected many times at the island edge but eventually leaving the island after a mean time τ. Then, for a long time nothing happens (see (4.40)). For a typical case with large edge barriers, like the systems discussed in the next section, the island

[13] This *nonstationary* nucleation regime is discussed in [356,357].

indeed remains empty for a mean time interval $\Delta t - \tau$ which exceeds the lifetime τ by a factor of 100. The ratio $\tau/\Delta t$ has a simple physical meaning [334]: It is the time-averaged mean number of adatoms present on the island. Under the condition (4.40), this number is much smaller than one.

One may ask how the *lonely adatoms*, separated by times far longer than their lifetime, ever manage to find each other for nucleation. Naively one might expect that nucleation takes place only once the island area has grown so large that Δt and τ become comparable, i.e. once the adatoms are no longer lonely. The key to understanding why nucleation takes place much earlier is the fact that the time intervals τ and Δt have broad distributions. The distributions are biased towards time intervals shorter than the mean, which are compensated by a few time intervals much longer than the mean. Thus, nucleation typically takes place when two atoms are successively deposited within a time interval much shorter than Δt, and the lifetime of the first deposited atom is much longer than the mean lifetime τ. Nucleation events require time intervals far from the average. Nucleation therefore relies crucially on *large fluctuations* of the adatom density. This suggests that a calculation of the nucleation rate on top of an island based on the average adatom density may not be appropriate[14] [334, 356, 360–362], as will be explained in more detail in Sect. 4.9.4. The fluctuation-dominated nucleation mode breaks down for large values of the critical nucleus size i^*, see Sect. 4.9.5.

For intermediate and weak barriers the probability for an encounter between the two atoms is no longer close to unity, and the simple expression (4.42) does not apply. A detailed analysis of second layer nucleation in this general case has been presented in [360–363]. Here we will use an approximation for the encounter probability that can be derived from the stationary diffusion equation (4.24) applied to a circular terrace with a second adatom (represented as a circle of radius a) placed at the center. The encounter probability is then given by the fraction of the total adatom flux which attaches to the central adatom, rather than to the island edge. For $a \ll R$ this yields the expression[15]

$$p_{\mathrm{enc}} = \frac{1 + C_1 a\nu' R/D}{1 + C_2 (a\nu' R/D)\log(R/a)}, \tag{4.43}$$

with constants $C_1 = 1/2$ and $C_2 = 1$. It is clear that placing one of the adatoms in the center of the terrace overestimates the encounter probability. To partly compensate for this effect, we treat C_1 and C_2 as free parameters which are adjusted such that (4.43) optimally matches the exact encounter probability for a circular island. The latter can be calculated numerically[16] as described in [362]. The comparison shows that a good fit is obtained for

---

[14] Interestingly, a similar effect has been noted in the context of chemical reactions taking place on small dust grains in interstellar space [358, 359].

[15] A systematic derivation of a similar approximation formula can be found in [363].

[16] The data were kindly provided by Claudio Castellano. The calculation was carried out on a square lattice.

$C_1/C_2 \approx 1/4$ and $C_1 \approx 2/3$. The nucleation rate is then given by $\omega = p_{\mathrm{enc}}\tau/(\Delta t)^2$, which reduces to

$$\omega = \frac{\pi^2 F^2 R^6}{32 D \log(R/a)} \tag{4.44}$$

in the limit of weak barriers, $l_{\mathrm{ES}} \ll R$. The logarithmic factor in the denominator is related to the marginal character of diffusion in two dimensions, see Sect. 2.4.3.

It is easy to generalize (4.42) to islands of arbitrary (compact) shapes [334]. Since the adatom density profile becomes completely flat in the limit of strong step edge barriers, the residence time τ can be evaluated from a simple mass balance argument without invoking the diffusion equation (4.24). We describe the island shape by a parameter ζ, which relates the number of edge sites (L) to the number of adsorption sites on the island (A) through

$$A = \zeta L^2 . \tag{4.45}$$

For example, $\zeta = 1/12$ for a regular hexagon, $\zeta = 1/18$ for a triangle, and $\zeta = \pi/(24\sqrt{3}) \approx 1/13.2$ for a circular island on an fcc(111) surface. In the strong barrier regime the rate for descent from the island, which equals the inverse residence time τ^{-1}, is simply given by the product of ν' and the probability L/A for the atom to reside at the edge. Hence $\tau = \zeta L/\nu'$, and the nucleation rate becomes

$$\omega = \frac{\zeta^3 (F\Omega)^2 L^5}{\nu'} . \tag{4.46}$$

An approximate formula for ω which is valid also in the weak barrier regime can be obtained by replacing the island radius in (4.39) and (4.43), which were derived for circular islands, by an effective radius $R_{\mathrm{eff}} = \sqrt{\Omega\zeta/\pi}L$ in such a way that the exact expression (4.46) is recovered for strong barriers. This results in the expression

$$\omega = \frac{\zeta^3 (F\Omega)^2 L^5}{\nu'} \frac{(1 + R_{\mathrm{eff}}\nu'/a\nu)(1 + 4C_1 R_{\mathrm{eff}}\nu'/a\nu)}{1 + 4C_2 (R_{\mathrm{eff}}\nu'/a\nu)\log(R_{\mathrm{eff}}/a)} , \tag{4.47}$$

which will be used for the analysis of experimental data in the following section.

4.3.3 Experimental Determination of the Step Edge Barrier

Compared to the methods for the experimental determination of surface diffusion rates within an atomic plane, which were described in Sect. 2.2, any measurement of interlayer transport rates faces the additional difficulty that an adatom typically crosses a step only once, since it is then incorporated into the step. The analogue of the mean square displacement (2.1), which averages over many in-layer hops, is therefore not available for interlayer diffusion. Instead, one has to trace single step crossing events and accumulate their statistics, or resort to indirect methods based e.g. on second layer nucleation. The present subsection describes approaches that have been applied to the Pt/Pt(111) system. Some other methods are collected in Sect. 4.9.6.

Field Ion Microscopy. The most straightforward approach to measuring interlayer diffusion is the observation of the escape of individual atoms from an island using the FIM [52, 346]. Figure 4.13 (a) shows the result of such a measurement for Pt on Pt(111) [364]. The adatom escape rate $1/\tau$ is seen to follow a clean Arrhenius law, with an activation energy of $E_a = 0.32 \pm 0.01$ eV and a prefactor $1/\tau_0 = 7.3(\times 2.2^{\pm 1}) \times 10^{10}\mathrm{s}^{-1}$. Assuming that the experiment operates in the strong barrier regime, comparison with (4.39) suggests to identify E_a with E_S in (4.22), and yields the estimate $\nu_0' \approx 10/\tau_0 = 7.3(\times 2.2^{\pm 1}) \times 10^{11}\mathrm{s}^{-1}$ for the preexponential factor and $\Delta E_S = 0.06 \pm 0.01$ eV for the additional step edge barrier. Using the experimental value for the in-layer diffusion rate ν obtained in Section 2.2, one finds that $\nu/\nu' \approx 1000 \gg R/a$ at 140 K, so the strong barrier assumption is self-consistent.

The detailed analysis of the spatial distribution of adatoms on the island indicates, however, that contrary to the model assumptions in Sect. 4.3.1 and 4.3.2, the energy landscape on the island has a fairly complicated structure (Fig. 4.13). The interior of the island is separated from the edge by an "empty zone" of weaker binding, and furthermore the sites next to the edge form an "attractive ring" in which the binding is *stronger* than in the island interior. The latter feature has also been seen in calculations using semiempirical

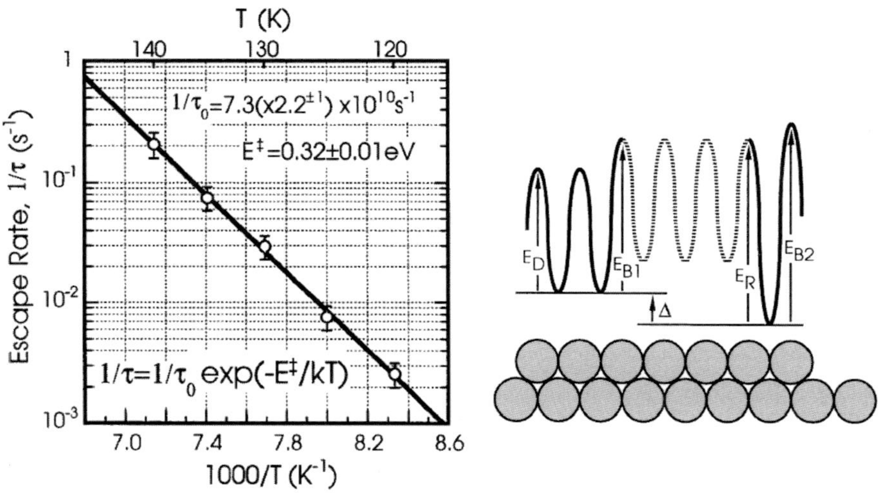

Fig. 4.13. Left panel: Arrhenius plot of the Pt adatom escape rate from a Pt(111) island of radius $20a$ (Fig. 3 of [364]). Right panel: Schematic energy landscape on a Pt(111) island near a step edge. The island cross section is sketched below the energy diagram; the island extends to the left, the terrace to the right (Fig. 4 of [364])

potentials [90,365]; it can be attributed to the lower coordination of the edge atoms, which implies stronger bonding [139]. The effective energy barrier for the escape of an adatom from the island interior is therefore given by $E_{\rm a} = E_{\rm B2} - \Delta$, where $E_{\rm B2}$ is the barrier for descending from the attractive edge ring to the lower terrace, and Δ is the difference in binding energy between the attractive ring and the island interior; in the experiment, it is estimated that $\Delta \approx 0.04$ eV. The nonuniform energy landscape on the island implies a nonuniform adatom distribution, and therefore the expression (4.39) for the residence time cannot be used to reliably estimate the preexponential factor ν_0'. For example, if the adatoms are completely localized at the island edge, the residence time in the strong barrier limit ($\tau \gg \tau_D$) becomes simply $\tau = 1/\nu'$ independent of the island size. Nevertheless it is clear that the energy barrier $E_{\rm a}$ obtained from Fig. 4.13 (a) is the relevant *effective* barrier for interlayer transport, in the sense described at the beginning of this section.

Unfortunately, FIM of Pt(111) apex planes does not allow to image the step edge structure in atomic resolution. Therefore it is not clear what kind of island edge structure the estimate for $\Delta E_{\rm S}$ actually refers to. For Ir(111), a chemically similar noble metal for which the atomic step edge structure could be imaged, apex planes of similar size as used for the Pt(111) experiments usually contain not only straight A– and B–steps, but also a considerable number of kinks [164].

Second Layer Nucleation Experiments. The direct measurement of interlayer transport by FIM is subject to the same restrictions as in the case of in-layer diffusion (Sect. 2.2.3). A more versatile approach is based on observing the onset of second layer nucleation during the deposition of the first layer [354]. For this purpose the theory of the second layer nucleation rate developed in Sect. 4.3.2 has to be supplemented by a model for the growth of the first layer islands. In its most simple form, the model assumes an array of circular islands of areal density N and identical radius R, which nucleate simultaneously at the beginning of the deposition process. The total coverage is[17] $\Theta = \pi R^2 N$, and hence the island radius grows according to

$$R = \sqrt{\Theta/\pi N} = \sqrt{F\Omega t/\pi N} \; . \tag{4.48}$$

In the quasistatic approximation described in Sect. 4.3.1, the instantaneous nucleation rate ω is given by the expressions derived in Sect. 4.3.2 for fixed size islands, evaluated at the radius (4.48). The quantity of interest in a growth experiment is the fraction $f_2(t)$ of first layer islands covered by a second layer nucleus at time t, or, equivalently, the probability for second layer nucleation to have occurred on a single island up to time t. Since the probability for nucleation in an infinitesimal time interval dt is $\omega(1 - f_2)dt$, f_2 is obtained by integration as

[17] This relation holds only as long as the second layer coverage θ_2 is negligible compared to Θ.

$$f_2(t) = 1 - \exp\left[- \int_0^t ds\, \omega(R(s)) \right] . \qquad (4.49)$$

Using the expression (4.42) derived for $i^* = 1$ in the strong barrier limit, this becomes

$$f_2(t) = 1 - \exp[-(R(t)/R_c)^7] , \qquad (4.50)$$

where R_c denotes a *critical radius* defined by[18]

$$R_c = \left(\frac{7}{\pi^3} \frac{a \Omega \nu'}{FN} \right)^{1/7} . \qquad (4.51)$$

Similar, if less precise, expressions can be derived for $i^* > 1$ using the appropriate estimates for the nucleation rate (see Sect. 4.9.5).

Rather than following the fraction f_2 as a function of coverage, the critical island radius can also be determined from a single experiment carried out at a coverage at which islands with and without second layer nuclei are present in comparable numbers. By averaging the sizes of the largest islands without, and the smallest islands with a second layer nucleus for many different STM images, one obtains an estimate of the radius $R_{1/2}$ defined by $f_2(R_{1/2}) = 1/2$. Figure 4.14 contains the results of such a measurement for Pt growth on Pt(111) at 400 K, in the presence of controlled CO exposure [131]. The figure shows the number of edge sites (island edge length in units of nn–distances) $L_{1/2}$, at which nucleation has taken place with probability 1/2, as a function of the applied CO partial pressure. The probability for second layer nucleation increases with increasing CO exposure. This effect is due to the decoration of the step edge atoms by CO molecules, as was discussed in Sect. 3.5; presumably the CO molecules prevent Pt adatoms from descending into the step edge.

From $L_{1/2}$ the effective critical island radius $R_{\mathrm{eff},1/2}$ was determined, taking into account the appropriate geometry factor ζ as defined in (4.45). These data were converted into estimates of ν' using the expression (4.47) for the nucleation rate to evaluate (4.49) numerically[19]. The nucleation experiments in Sect. 2.3.4 show that the Pt dimer is stable at 400 K, hence the assumption $i^* = 1$ underlying (4.51) is justified. The resulting values for ν' can be transformed into estimates for the step edge barrier ΔE_S by choosing a value for the preexponential factor ν'_0 in (4.22). In the absence of specific information on the value of the prefactor, it is customary to set[20] $\nu'_0 = \nu_0$. In the present

[18] An alternative definition of the critical radius has been based on the assumption that second layer nucleation occurs when the adatom density reaches the "critical" value at which the first layer islands nucleated [366,367]. This assumption is generally incorrect because of the different boundary conditions for the adatom density on an island [334,368].

[19] An independent analysis of the same data based on a more accurate expression for the nucleation rate can be found in [363]. The resulting values for ΔE_S are very similar to those presented here.

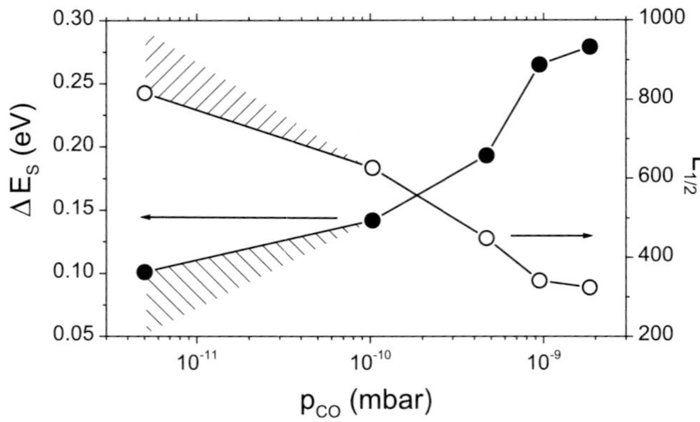

Fig. 4.14. Critical island step edge length $L_{1/2}$ for nucleation of a second layer island on Pt(111) at 400 K in units of nn–distances (right y–axis) and additional step edge barrier ΔE_S (left y–axis) as a function of applied CO partial pressure. The CO exposure was started 100 s prior to, and maintained during deposition. As indicated by the hatching, the value $L_{1/2}$ for the lowest possible CO partial pressure under nominally clean growth conditions is only a lower bound, as nearly all islands coalesce prior to nucleation in the second layer. Accordingly, the resulting value of ΔE_S for clean growth is only an upper bound

case, the value $\nu_0' \approx 7.3(\times 2.2^{\pm 1}) \times 10^{11} \mathrm{s}^{-1}$ obtained above on the basis of field ion microscopy experiments for the descent of Pt adatoms from Pt(111) islands in the absence of adsorbates will be used. The inherent uncertainty in the absolute height of the excess barrier caused by the uncertainty in ν_0' does not affect the variation of ΔE_S with the CO–pressure or other external parameters.

The analysis is most reliable for the largest CO pressure, where the strong barrier criterion $R_c \ll l_{ES}$ is satisfied, and the expression (4.47) for the nucleation rate is exact. One finds $\Delta E_S = 0.28 \pm 0.01\,\mathrm{eV}$, which is close to the estimate that will be derived below from multilayer wedding cake morphologies[21]. For nominally clean growth at the lowest possible CO partial pressure of about 5×10^{-12} mbar the values given are only lower and upper bounds on $L_{1/2}$ and ΔE_S, respectively. The reason is simply that most islands have coalesced prior to nucleation of second layer islands. The upper bound value $\Delta E_S \leq 0.10\,\mathrm{eV}$ for clean growth is compatible with the FIM based estimate $\Delta E_S = 0.06\,\mathrm{eV}$.

The large power of R/R_c in the argument of the exponential in (4.50) implies that the onset of second layer nucleation is quite abrupt[22]. This is

[20] For the same reasons discussed in connection with the boundary condition (4.28), it can be argued that a more reasonable choice is $\nu_0' = (z_{\mathrm{down}}/z_{\mathrm{all}})\nu_0$, where z_{all} is the total number of diffusion channels out of the edge site and z_{down} is the number of channels leading to an interlayer move, see Sect. 4.9.3.

[21] The error on the additional step edge barrier given here and in the following does not include the uncertainty in the FIM estimate of ν_0'.

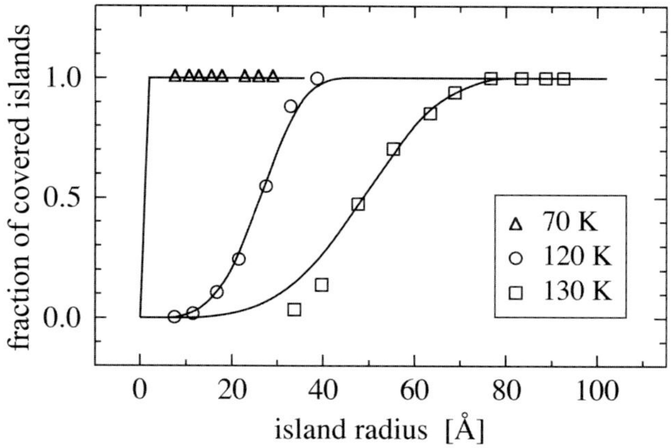

Fig. 4.15. Fraction f_2 of first layer islands with a second layer nucleus as a function of island radius, for predeposited Ag islands on Ag(111). The full lines are fits to the theoretical prediction derived from the rate equation theory of second layer nucleation (Fig. 2a of [369])

illustrated in Fig. 4.15 for Ag growth on Ag(111) [369]. In these experiments, ensembles of islands of uniform radius R_0 were prepared by deposition of a coverage Θ_0 followed by annealing at high temperature, and the fraction of second layer nuclei that had appeared after deposition of an additional coverage $\Delta\Theta$ was monitored as a function of R_0. The second dose was deposited under conditions of irreversible growth ($i^* = 1$). During the second dose the island radius grows according to $R = R_0\sqrt{1 + F\Omega t/\Theta_0}$, which replaces (4.48) in the analysis above. This yields $f_2 = 1 - \exp[-(R_0/\tilde{R}_c)^5]$, with a critical radius now given by

$$\tilde{R}_c = \left(\frac{7}{\pi^2[(1 + \Delta\theta/\theta_0)^{7/2} - 1]} \frac{\nu'}{F\theta_0} \right)^{1/5} . \tag{4.52}$$

The data shown in Fig. 4.15 were originally analyzed in the framework of the rate equation theory of second layer nucleation [354], which yielded the estimates $\Delta E_S = 0.120 \pm 0.015\,\text{eV}$ and $\nu'_0 = 1 \times 10^{13\pm1}\text{s}^{-1}$ for the interlayer diffusion parameters.

A reanalysis of the data for 120 K and 130 K using the correct expression (4.52) for the critical radius under the assumption of a constant, temperature independent value for the step edge barrier produces physically unreasonable values for ΔE_S and ν'_0 [334, 363, 370]. Only allowing for a temperature dependence of ΔE_S leads to a physically meaningful picture. As an example, assuming the value $\nu'_0 = 1 \times 10^{13}\,\text{s}^{-1}$ as in the analysis above results in

[22] This becomes even more pronounced when $i^* > 1$. For $i^* \geq 3$ one finds $f_2 = 1 - \exp[-(R/R_c)^{i^*+5}]$ [168, 354, 357].

$\Delta E_S = 0.16\,\text{eV}$ at 120 K and $\Delta E_S = 0.14\,\text{eV}$ at 130 K. The necessity to allow for a variation of ΔE_S with temperature indicates that slight temperature dependent changes in the island shape and step atomic structure[23] may influence the effective step edge barrier; we will return to this issue in Sect. 4.4. The most direct experimental estimate of the interlayer transport rate for Ag(111), obtained from the decay of vacancy islands at 300 K [239] (see Sect. 4.9.6), gives $\Delta E_S = 0.13 \pm 0.04\text{eV}$ and $\nu_0'/\nu_0 = 10^{-0.6\pm0.5}$, consistent with the original analysis of [369]. Semiempirical potential calculations for Ag(111) yield numbers for ΔE_S ranging from 0.03 eV [93] to 0.15 eV [371], with preferred descent by exchange at the B–step (see Sect. 4.3.4).

The Top Terrace Size of a Wedding Cake. The interlayer transport rate can also be estimated from the size of the flat top terraces terminating the multilayer mounds of Sect. 4.2 (see Fig. 4.6). The basic mechanism through which they form was described already: A new island appears on a mound when the top terrace has reached a size at which the nucleation probability is sufficiently large [372]. The scaling form of the coverage profile (4.15,4.20) implies that, up to a correction of the order of $1/\sqrt{\Theta}$, the mound top grows at the same speed as the mean film thickness. This requires on average one nucleation event to occur on each mound during the growth time of one monolayer, and consequently the size R_{top} of the top terrace is given, in order of magnitude, by the relation

$$\omega(R_{\text{top}})/(F\Omega) \approx 1 \,. \tag{4.53}$$

A more precise estimate is obtained from a simple model, which assumes that the top terrace grows on a base terrace of fixed size, and suppresses any downward transport from the base terrace [334]. The distribution of nucleation times can then be derived from the expression (4.42) for the nucleation rate, and demanding that the mean time interval between nucleation events equals the monolayer growth time one finds for the case $i^* = 1$

$$R_{\text{top}}/a = 0.60 \cdot (\nu'/\Omega F)^{1/5} \,. \tag{4.54}$$

This expression is generalized to mounds with non-circular cross sections by using (4.46) for the derivation of the nucleation times. For the analysis of experimental data it is useful to describe the top terrace size in terms of the number Λ of edge sites of the base terrace, which has a rather narrow distribution (see Fig. 4.16 and Fig. 4.17). This is related to the growth parameters by

$$\Lambda = 1.193 \cdot \zeta^{-3/5} (\nu'/\Omega F)^{1/5} \,. \tag{4.55}$$

[23] Note, however, that the quantitative analysis indicates that an increase of the kink concentration by a factor of two or three is necessary to explain the observed decrease of the effective step edge barrier between 120 K and 130 K [363]. Such a dramatic change in step structure seems hard to rationalize.

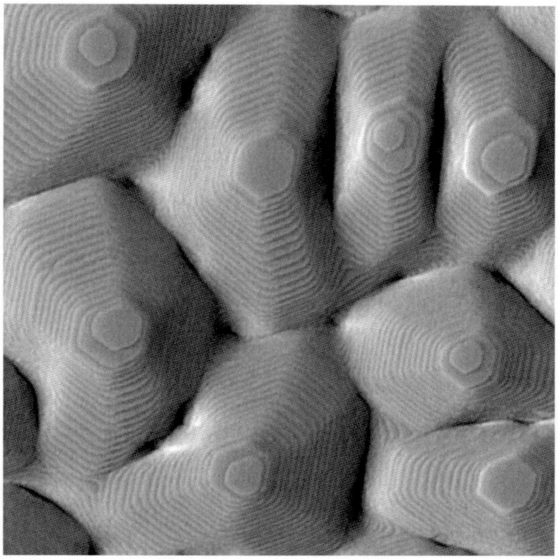

Fig. 4.16. STM topograph of mounds after the deposition of 37.1 ML Pt on Pt(111) in a CO partial pressure of 1.9×10^{-9} mbar and with $F = 1.3 \times 10^{-2}$ ML/s. Topograph size 1660 Å× 1660 Å (Fig. 3 of [334])

The distribution of the number of edge sites of the top terraces is then predicted to be of the form

$$P(L) = \frac{2L}{\Lambda} \exp[-C(\Lambda/L)^7] , \tag{4.56}$$

where the constant C is fixed by the normalization of $P(L)$.

The model has been applied to multilayer Pt films grown on Pt(111) at 440 K in the presence of CO (Fig. 4.16). As was discussed above, the adsorbates increase the step edge barrier. The mounds in Fig. 4.16 also show a roughly equal presence of A– and B–steps, in contrast to growth under nominally clean conditions at the same temperature, where B–steps dominate (compare to Fig. 4.6). As explained in Sect. 3.5, the presence of A–steps is indicative for CO adsorption. At 440 K the assumption of a critical cluster size $i^* = 1$ for nucleation on the top terrace holds according to the criterion (4.119), and the dimer binding energy obtained in Sect. 2.3.4. From images such as Fig. 4.16, the sizes of 145 top terraces and base terraces were evaluated. In Fig. 4.17 the histogram for the top terrace sizes L is compared to the prediction (4.56). In view of the limited statistics, the agreement is satisfactory. The size distribution of the base terraces is seen to be quite narrow, which shows that the model assumption of a base terrace of fixed size is reasonable.

Using (4.54) and choosing the geometrically appropriate value $\zeta = 1/12$, the estimate $\nu' = 1.13 \times 10^6$ s^{-1} for the interlayer transport rate was obtained from the mean of the top terrace size distribution $\langle L \rangle = 212$ [334]. At 440 K,

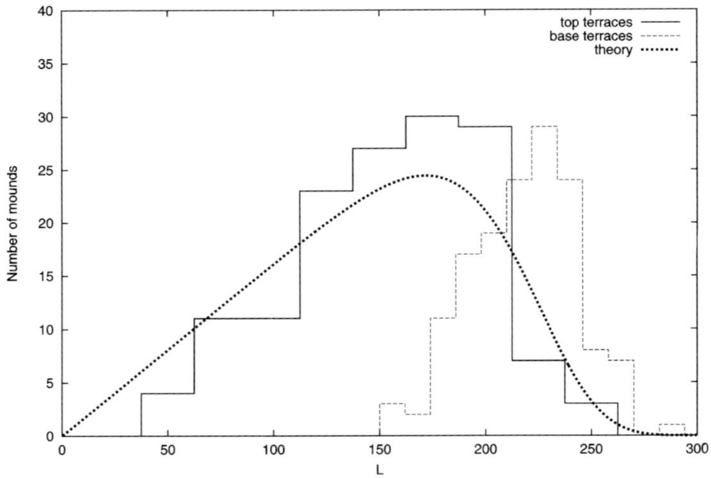

Fig. 4.17. The experimental size distribution of top terraces, represented by the terrace perimeters L, is shown as full line. The thin dashed line shows the (narrow) size distribution of the base terrace perimeters Λ, and the dotted line is the prediction (4.56) (Fig. 2 of [334])

this implies $\nu/\nu' \approx 5000 \gg R_{\rm top}/a \approx 23$, which confirms that the experiment is conducted in the strong barrier regime. Assuming again the preexponential factor for interlayer transport to be $\nu_0' = 7.3 \times 10^{11}\,{\rm s}^{-1}$, the value of ν' translates into the additional step edge barrier $\Delta E_{\rm S} = 0.25 \pm 0.01\,{\rm eV}$. This value is similar to $\Delta E_{\rm S} = 0.28 \pm 0.01\,{\rm eV}$ obtained from the second layer nucleation experiments of Sect. 4.3.3 at the same CO partical pressure but at 400 K, rather than 440 K.

The same method has also been applied to mounds grown under nominally clean conditions at 440 K. After deposition of 90 ML we find $\nu' = 7.47 \times 10^6\,{\rm s}^{-1}$ and after 306 ML $\nu' = 6.53 \times 10^6\,{\rm s}^{-1}$. With the precise experimental deposition rates and the usual choice for ν_0' these values for ν' translate into the additional step edge barrier $\Delta E_{\rm S} = 0.18 \pm 0.01\,{\rm eV}$ at both coverages. This result differs significantly from both the FIM value $\Delta E_{\rm S} = 0.06\,{\rm eV}$ and the estimate $\Delta E_{\rm S} \leq 0.10\,{\rm eV}$ based on the second layer nucleation experiment. The discrepancy will be resolved in Sect. 4.4.

A closely related approach uses the fit of the mound shape to the modified wedding cake model described in Sect. 4.2. The critical coverage θ_c at which the next top terrace nucleates is related to the perimeter Λ of the base terrace by $\theta_c = (\Lambda/\Lambda_0)^2$, where $\Lambda_0 = \sqrt{A_0/\Omega\zeta}$ and A_0 is the area of the capture zone of the mound. In this way the value $\theta_c = 0.122$ obtained from the fit in Fig. 4.8 translates into the estimate $\nu' = 1.96 \times 10^7\,{\rm s}^{-1}$ for the interlayer jump rate at 440 K, corresponding to an additional step edge barrier of $\Delta E_{\rm S} = 0.14\,{\rm eV}$. This method differs from that based only on the size of the top terrace in that the full visible layer distribution is included in the fit. On the other hand, since the deviation of the layer distribution from the statistical growth limit is small in the regime of interest here (compare the statistical growth shape with the modified shape in Fig. 4.8), it is probably less sensitive to the precise value of ν'. An additional uncertainty is introduced because the mean mound shape constitutes an average over mounds of different sizes. This may well explain the discrepancy between the estimates obtained from the two approaches.

4.3.4 Theoretical Estimates of Step Crossing Rates on Pt(111)

Microscopic calculations of interlayer transport processes for Pt(111) have been carried out using a variety of semiempirical potentials [90, 93, 95, 213, 347, 365]. While the numbers for the energy barriers scatter considerably (see Table 4.1), all these investigations agree on the following three points: (i) Step crossing occurs by *concerted substitution*, an exchange process of the kind discussed in Sect. 2.4.2 and illustrated in Figs. 4.11 and 4.18, rather than by simple hopping. (ii) Descent at the {111} microfacetted B–step is favored compared to descent at the {100} microfacetted A–step. (iii) Descent is facilitated near kinks. The position of minimum additional step edge barrier is always a kink in a B–step. The values of these minimal additional barriers $\Delta E_{\mathrm{S}}^{\mathrm{B,kink}}$ range from 0.17 eV to –0.036 eV. A negative value implies that descent is more facile than a jump on the terrace.

Table 4.1. Calculated additional interlayer diffusion barriers for Pt(111). $\Delta E_{\mathrm{S}}^{(\mathrm{A})}$ and $\Delta E_{\mathrm{S}}^{(\mathrm{B})}$ are the additional barriers for descent from straight steps of type A and B. $\Delta E_{\mathrm{S}}^{(\mathrm{A,kink})}$ and $\Delta E_{\mathrm{S}}^{(\mathrm{B,kink})}$ are the calculated additional barriers for descent in the vicinity of a kink or step adatom. The barriers refer to exchange processes; when available, the corresponding barriers for hopping are given in italics. Adding the in-layer hopping barrier E_{D} displayed for reference in the second column gives the respective total interlayer barriers. All energy barriers are given in eV. The acronyms in the first column refer to various semiempirical potentials [embedded atom method (EAM), effective medium theory (EMT), corrected effective medium theory (CEM) and the tight-binding based Rosato-Guillope-Legrand potential (RGL)], and ab initio density functional theory (DFT); see Sect. 6.4 for further discussion of these methods.

Method	E_{D}	$\Delta E_{\mathrm{S}}^{(\mathrm{A})}$	$\Delta E_{\mathrm{S}}^{(\mathrm{B})}$	$\Delta E_{\mathrm{S}}^{(\mathrm{A,kink})}$	$\Delta E_{\mathrm{S}}^{(\mathrm{B,kink})}$
EAM [90]	0.08	0.22	0.10	0.00	-0.02
EMT [213]	0.16	*0.25*	0.21	–	0.10
CEM [93]	0.05	0.25 (*0.83*)	0.00 (*0.84*)	–	-0.04
EMT [365]	0.13	0.43 (*0.33*)	0.18 (*0.33*)	0.35	0.17
RGL [95]	0.17	0.33 (*0.36*)	0.23 (*0.35*)	0.25	0.09
EAM [347]	0.05	0.24 (*0.79*)	0.12 (*0.76*)	0.11	0.06
DFT [87]	0.29	0.02 (*0.24*)	0.35 (*0.51*)	–	–

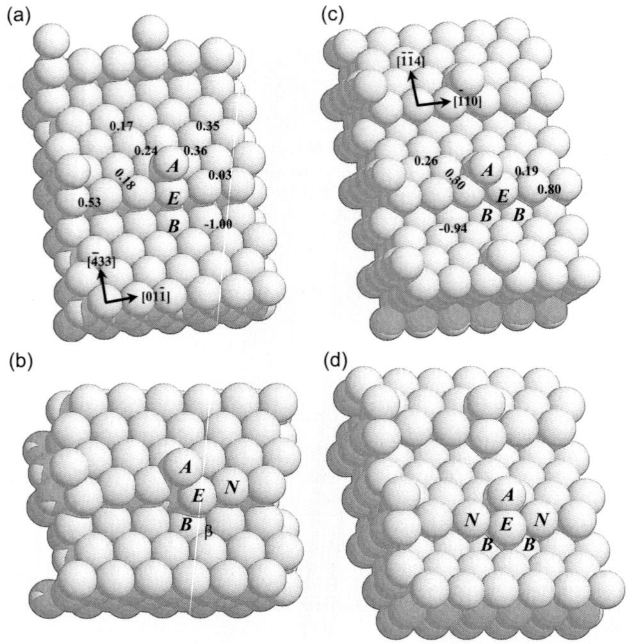

Fig. 4.18. Geometry of step crossing on Pt(111), according to DFT calculations. Left panels show (**a**) initial state and (**b**) transition state for adatom descent at an A–step, right panels [(**c**) and (**d**)] for descent at a B–step. The descending adatom is labeled with A, the emerging edge atom with E. The numbers in the upper panels denote the energies (in eV) of the adatom at the corresponding sites (Figs. 1 and 2 of [87])

The preference for descent at B–edges observed in the semiempirical calculations has been rationalized in terms of the geometry of the local atomic configurations (Fig. 4.18). At the B–step, the step atom involved in the exchange can move out normal to the edge, passing through a bridge site between the two atoms in the layer below. This is not possible at the A–step, because the edge atom is blocked by an atom in the layer below which sits directly in its way. Rather than passing on top of the blocking atom, the exchanging edge atom initially moves sideways, which requires considerable displacement of the adjacent edge atoms [90].

In view of the failure of semiempirical work to establish the activation energy for adatom diffusion E_{D}, as illustrated previously in Table 2.1, it was perhaps not a major surprise when an ab initio DFT study [87] arrived at a picture that was qualitatively different from the one developed on the basis of the semiempirical calculations. While the exchange mechanisms identified earlier were confirmed, the additional energy barrier at the straight B–step was found to be 0.35 eV, more than an order of magnitude larger than the minute barrier of 0.02 eV found at the A–step. Ironically, it is precisely the geometric feature that was argued previously to impede descent at the A–

step that is now suggested to be responsible for the low A–step energy barrier in the ab initio calculations: Because the emerging step atom moves initially along, rather than perpendicular to the step, it maintains a higher coordination at the transition state than the normally displaced atom at the B–step (see Fig. 4.18).

The preference for descent at the A–step is not universal for fcc(111) surfaces. Both FIM experiments on Ir(111) [164] and ab–initio calculations for Al(111) indicate a slightly lower barrier at the B–step; in the latter case $\Delta E_S^{(A)} = 0.04$eV and $\Delta E_S^{(B)} = 0.02$eV [233]. The failure of the semiempirical calculations to even qualitatively account for the energetics of interlayer transport on Pt(111) seems to be a common feature, which may be related to the general difficulty of correctly describing exchange processes with these methods [373]. For example, for the Ag(100) surface, ab initio calculations show that step crossing occurs via exchange without any additional barrier ($E_S = E_D$) [374], while EMT favors descent by hopping with $\Delta E_S = 0.114$ eV [375].

The large anisotropy of the Pt(111) step edge barrier predicted by DFT implies a strong dependence of the residence time of an adatom on top of an island on the island shape, specifically on the fraction of A– and B–type step edges along the island boundary. The semiempirical estimates shown in Table 4.1 suggest in addition an important effect of kinks, in particular at the B–step which has a large step edge barrier[24]. A straightforward quantitative comparison of the calculated step edge barriers with the residence time measurements of Kyuno and Ehrlich (Sect. 4.3.3) is therefore not possible, since the edge morphology of the islands used in the experiments is not known. The small sample size used in the DFT calculation (a six layer slab in a $3 \times 2\sqrt{3}$ supercell, see Fig. 4.18) further adds to the uncertainty, because it prevents the complex energy landscape observed experimentally on the islands (see Fig. 4.13) to fully unfold.

Summarizing this section, it has to be concluded that our grasp of interlayer transport kinetics still lags considerably behind the state of the art in in-layer surface diffusion, as it was presented in Sect. 2.2. This is particularly true if the attention is turned towards the preexponential factor ν_0' in (4.22). We are aware of a single MD study for Au(111) using many–body tight binding potentials, which yields conventional prefactors of order 10^{12}s^{-1} but also indicates a significant lowering of the step edge barrier by 25–30 % compared to its zero temperature value [376].

[24] The existence of easy descent paths near kinks at the B–step has been confirmed by DFT calculations [273].

4.4 Step Atomic Structure and Growth Modes on Pt(111)

4.4.1 A Model for the Effective Step Edge Barrier

In the preceding section we have encountered a bewildering variety of values for ΔE_{S} on Pt(111) obtained from FIM and STM experiments, as well as from ab initio and semiempirical calculations, which span the range between –0.04 eV [93] and 0.35 eV [87]. This arbitrariness challenges the very meaning of the concept of an additional step edge barrier. The barrier seems to be a moving target and no simple way to remove the discrepancies between the different measurements and calculations appears to be in sight.

Here a possibility to make sense of the maze of ΔE_{S} values will be presented. We will argue that the observed diversity of numbers reflects the fact that the effective additional step edge barrier $\Delta E_{\mathrm{S}}^{\mathrm{eff}}$ of an island as measured in experiments is determined by the atomic step edge structure of the island. It constitutes an average over step edge barriers associated with different local atomic step configurations, which are present in the island edge in concentrations specific to the growth conditions and the environment of the island. As $\Delta E_{\mathrm{S}}^{\mathrm{eff}}$ is a function of the island shape, it is clear that, according to the findings of Chap. 3, it varies with temperature and also with coverage. Even for a fixed temperature and coverage, island shapes differ according to their time of nucleation and the surrounding diffusion field (see Sect. 3.6). In short: The notion of a unique additional step edge barrier ΔE_{S} is a fiction, which is useful for systematic reasons, but may obscure an understanding of growth in real systems.

The following set of three rules for the additional step edge barrier associated with specific step configurations is sufficient to rationalize the observed morphological evolution on the Pt(111) surface:
(i) Straight B–steps have a large barrier $\Delta E_{\mathrm{S}}^{\mathrm{B}}$.
(ii) Kink positions have a low barrier $\Delta E_{\mathrm{S}}^{\mathrm{k}}$.
(iii) If present, straight A–steps have a large barrier $\Delta E_{\mathrm{S}}^{\mathrm{A}}$ which is comparable to $\Delta E_{\mathrm{S}}^{\mathrm{B}}$.
The first two rules are expected on the basis of the calculations discussed in Sect. 4.3.4, while the third rule apparently contradicts the ab initio calculations, which find a negligible barrier for the A–step [87]. Its precise meaning will be explained below in Sect. 4.4.5.

In order to get a crude, but transparent model for the effective step crossing rate $\nu'^{\,\mathrm{eff}}$, which allows to define the effective barrier $\Delta E_{\mathrm{S}}^{\mathrm{eff}}$, we assume that the island boundary (of length L) consists of only two types of inequivalent sites, with populations L_1 and L_2 satisfying $L_1 + L_2 = L$. The step crossing frequencies at the two kinds of sites are ν'_1 and ν'_2 with associated step edge barriers $E_{\mathrm{S}}^1 = E_{\mathrm{D}} + \Delta E_{\mathrm{S}}^1$ and $E_{\mathrm{S}}^2 = E_{\mathrm{D}} + \Delta E_{\mathrm{S}}^2$. Assuming that the occupation probability of the edges sites is uniform along the boundary[25], the averaged step crossing rate is given by

$$\nu'^{\,\mathrm{eff}} = \frac{L_1}{L}\nu_1' + \frac{L_2}{L}\nu_2' \,. \tag{4.57}$$

The effective additional step edge barrier is then defined by

$$\Delta E_{\mathrm{S}}^{\mathrm{eff}} = k_{\mathrm{B}}T\ln(\nu_0'/\nu'^{\,\mathrm{eff}}) - E_{\mathrm{D}} \,. \tag{4.58}$$

It is evident from (4.58) that $\Delta E_{\mathrm{S}}^{\mathrm{eff}}$ will be largely determined by the fraction L_i/L of sites with the lower barrier $\Delta E_{\mathrm{S}}^{i}$, as long as the two barriers are significantly different. In the following we will show that for $\nu_0' = 7.3\times10^{11}\,\mathrm{s}^{-1}$ the values $\Delta E_{\mathrm{S}}^{\mathrm{k}} = 0.00\,\mathrm{eV}$ for kink positions and $\Delta E_{\mathrm{S}}^{\mathrm{A,B}} = 0.33\,\mathrm{eV}$ for the dense packed steps A– and B–steps adequately reproduce the measured effective additional barrier $\Delta E_{\mathrm{S}}^{\mathrm{eff}}$ in a variety of situations.

4.4.2 Film Thickness Dependence of $\Delta E_{\mathrm{S}}^{\mathrm{eff}}$ at 440 K

In Sect. 4.2 we noticed considerable deviations from the wedding cake model for the initial evolution of the roughness W and the mound separation λ (see Fig. 4.7). Similar to the situation at 400 K and 500 K depicted in Fig. 4.3, at 440 K the first layer islands coalesce prior to nucleation of new second layer islands. An analysis for growth at 440 K identical to that performed in Sect. 4.3.3 for 400 K results in $\Delta E_{\mathrm{S}}^{\mathrm{eff}} \le 0.09\,\mathrm{eV}$ for the first deposited layer. On the other hand $\Delta E_{\mathrm{S}}^{\mathrm{eff}} = 0.18\pm0.01\,\mathrm{eV}$ for the top terraces after deposition of 90 ML and 306 ML (see Sect. 4.3.3).

This apparent discrepancy can be explained by the changes in the island edge structure during growth. Figure 4.19 displays the change in island morphology with coverage Θ. The average kink concentration decreases from 2×10^{-1} in the submonolayer regime after deposition of 0.35 ML in Fig. 4.19a to 9×10^{-3} in the top terrace steps after deposition of 306 ML in 4.19c. Application of (4.57) using $\Delta E_{\mathrm{S}}^{\mathrm{k}} = 0.00\,\mathrm{eV}$ and $\Delta E_{\mathrm{S}}^{\mathrm{A,B}} = 0.33\,\mathrm{eV}$ as defined above results in $\Delta E_{\mathrm{S}}^{\mathrm{eff}} = 0.06\,\mathrm{eV}$ for $\Theta = 0.35\,\mathrm{ML}$ and $\Delta E_{\mathrm{S}}^{\mathrm{eff}} = 0.18\,\mathrm{eV}$ for $\Theta = 306\,\mathrm{ML}$. The negligible step edge barrier at kink positions and the change in kink concentration thus provide a straightforward explanation for the change in $\Delta E_{\mathrm{S}}^{\mathrm{eff}}$ with Θ. Note that in the present application the value for $\Delta E_{\mathrm{S}}^{\mathrm{eff}}$ is only marginally dependent on the precise value of $\Delta E_{\mathrm{S}}^{\mathrm{A,B}}$, as long as it is large in magnitude. However, $\Delta E_{\mathrm{S}}^{\mathrm{eff}}$ is very sensitive to the kink concentration.

The low initial value of $\Delta E_{\mathrm{S}}^{\mathrm{eff}}$ causes significant interlayer transport during the early phase of growth, which explains the initial slow increase of W as compared to the wedding cake model (Fig. 4.7a). It also explains the initial

[25] This assumption is adopted here mainly for simplicity. It is strictly valid only if both barriers $\Delta E_{\mathrm{S}}^{1}$ and $\Delta E_{\mathrm{S}}^{2}$ are strong in the sense of Sect. 4.3, so that the adatom density is uniform on the island [363].

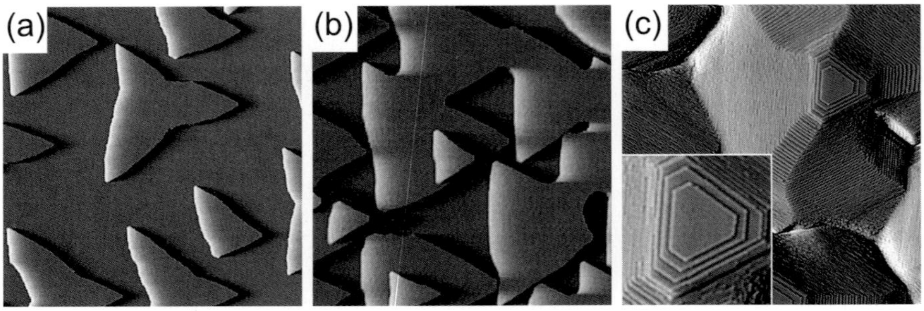

Fig. 4.19. STM topographs after deposition of (**a**) 0.35 ML, (**b**) 0.8 ML and (**c**) 306 ML Pt on Pt(111) at 440 K with a deposition rate $F = 7 \times 10^{-3}$ ML/s^{-1}. Inset in (**c**) shows top terrace in double magnification. The kink concentration of the highest layer islands decreases from 2×10^{-1} in (**a**) to 9×10^{-3} in (**c**). Topograph size is 1650 Å × 1650 Å

increase of the characteristic mound separation λ as seen in Fig. 4.7b in the transient regime below 10 ML film thickness: Islands coalesce and the second layer nuclei on the coalesced terraces deplete areas larger than the first layer capture area $1/N$. This mechanism of *coalescence coarsening* is visible in Fig. 4.19b after deposition of 0.8 ML. Coalescence coarsening ceases with the increase of $\Delta E_{\mathrm{S}}^{\mathrm{eff}}$ and the development of mounds.

The direct link between the initial coarsening and the change in the step edge barrier can be demonstrated by growth in a partial pressure of CO, which due to the step edge decoration with CO gives rise to a very high step edge barrier right from the beginning of growth (see Sect. 4.3.3). For growth in a CO partial pressure of 1.9×10^{-9} mbar the density of mounds after deposition of 37 ML (Fig. 4.16) and the density of islands in the submonolayer regime are identical. The corresponding data points are found in Fig. 4.7b as open symbols and apparently follow perfectly the wedding cake model predictions. In the CO experiments also the initial delay in roughness evolution is absent as seen in Fig. 4.7a.

It remains to be understood why the kink concentration in the edges of the top terraces changes so drastically during growth. Two mechanisms can be identified. First, the origin of the large kink concentration in the boundaries of the submonolayer islands is the onset of dendritic-skeletal growth beyond the critical edge length L_{c}, as discussed in Sects. 3.6 and 3.7.3. While an island on the initially flat surface grows from an area of $\approx 1/N$, an island on top of a mound grows only from the small area of the base terrace. Thus the maximum linear dimension of the top terrace on a mound is much smaller than that of an island on the initial surface [for a quantitative expression see (4.73)]. Therefore the top terrace of a mound may never reach L_{c}, although under the same growth conditions an island on the initial surface rapidly exceeds L_{c} and develops rough steps.

Second, the value of L_c and, more generally, the probability for the formation of growth-induced kinks in the island edge depends itself on the one-dimensional flux F_{1d} of adatoms to the step edge, i.e. the number of arriving atoms per unit time and edge length (see Sect. 3.7.3). The one-dimensional fluxes onto a submonolayer island and a top terrace differ drastically due to the difference in the respective capture areas, which are, as noted above, given by the inverse submonolayer island number density and the size of the base terrace, respectively. More specifically, for the top terrace the capture area relevant for the formation of growth induced kinks is the *difference* between the base and top terrace areas: The adatoms descending from the top terrace into its step form no kinks, since they only descend at kink positions. Therefore as the top terrace grows, its capture area decreases substantially. Thus kink formation is largely suppressed for the slowly growing top terraces, and L_c is considerably larger for a top terrace on a mound compared to an island on the initial surface.

In summary, due to the growing step concentration during deposition F_{1d} decreases, thus allowing the step adatoms to find existing kinks rather than to nucleate new ones and inhibiting the onset of dendritic-skeletal growth. The development of surface roughness during growth thereby reduces the kink concentration in the steps and increases the effective additional step edge barrier ΔE_S^{eff}. This implies a feedback mechanisms for roughness build-up, which is somewhat reminiscent of the feedback mechanism for dendritic-skeletal growth caused by the point effect of diffusion (Sect. 3.6).

4.4.3 Low Temperature Growth on Pt(111)

The growth morphology changes significantly upon lowering the temperature below 300 K. The origin of the morphological difference between the mound formation regime discussed so far and growth at $T \leq 300\,\text{K}$ is the fact that in the latter regime the step edge barrier is low initially, and remains low during subsequent growth. This is because step edge diffusion is largely absent at these temperature, leaving the step edges rough in all layers, with a kink concentration of about 50%.

Both the first layer islands and all newly nucleated islands in higher layers now have a fractal-dendritic appearance. The islands are ramified with a large fraction of lattice sites in the step edge, corresponding to a small value of the shape parameter ζ introduced in (4.45). Further growth fattens the branches of the freshly nucleated islands and eventually irregular, nearly compact islands result. These islands begin to coalesce when the coverage approaches one monolayer. Subsequently, but *before* the first layer is perfectly closed, second layer nucleation takes place.

The remaining holes in the first layer – vacancy islands – are difficult to fill for reasons related to the step edge barrier. Although the step edge barrier is too small to induce second layer nucleation prior to the onset of island coalescence, for atoms arriving in the second layer the barrier implies a

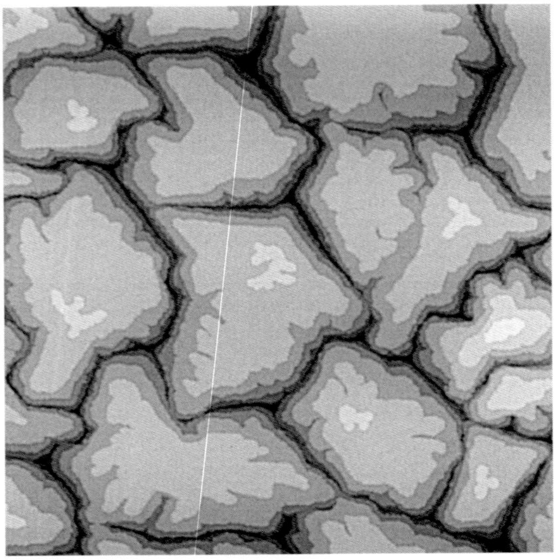

Fig. 4.20. Grey scale STM topograph after deposition of 10 ML Pt on Pt(111) at 300 K with $F = 1 \times 10^{-2}$ ML/s^{-1}. Topograph size is 1400 Å × 1400 Å

strong preference to incorporate into the second layer island steps, rather than descending to the first layer. Thus the vacancy islands in the first layer shrink only due to the adatoms which are deposited directly into them, a process that slows down indefinitely as the area of the vacancy islands decreases; this is the Zeno effect already mentioned in Sect. 4.2. The persistence of the vacancy islands in the lower layers makes island coalescence in higher layers more and more difficult, such that after the deposition of 5 ML the situation is already reversed (see Fig. 4.3): Instead of continuous layers we see a connected network of grooves that largely inhibits further coalescence. As shown in Fig. 4.20, for larger Θ nucleation of new islands takes place on mesa–shaped mounds separated by deep crevices, and coalescence of top layer islands does not take place.

While mound formation at 440 K is driven by enhanced nucleation on the top terraces, it appears that at 300 K mound formation results from the difficulty to close the crevices separating the mesas. This reflects the importance of the local step density in determining the efficiency of the step edge barrier. At 300 K its effect on nucleation becomes negligible because the Ehrlich–Schwoebel length falls below the top terrace size. Nevertheless the step edge barrier continues to influence the growth dynamics on the sides of the crevices, where l_{ES} is still larger than or comparable to the local terrace width. We will return to this growth regime in Sect. 4.5.

An analysis of the average base terrace size after deposition of 10 ML at 300 K, assuming (4.54) to hold, yields indeed a small barrier[26] of $\Delta E_{\mathrm{S}}^{\mathrm{eff}} \approx 0.01\,\mathrm{eV}$. Application of (4.57) with $\Delta E_{\mathrm{S}}^{\mathrm{k}} = 0.00\,\mathrm{eV}$, $\Delta E_{\mathrm{S}}^{\mathrm{A,B}} = 0.33\,\mathrm{eV}$ and a kink concentration of 50% results in $\Delta E_{\mathrm{S}}^{\mathrm{eff}} = 0.02\,\mathrm{eV}$ in decent agreement with the experimental estimate. This lends further credibility to our simple kink concentration model for the step edge barrier.

4.4.4 Growth Bifurcation Around 500 K

The heterogeneous growth morphology which appears in the temperature range from 480 K–520 K was already mentioned in Sect. 4.1.2. After deposition of 5 ML at 500 K, the surface represented by Fig. 4.3 exhibits steep pyramids with small top terraces bounded by straight B–steps. In addition to the pyramids also flat areas with large top terraces are visible. While the step edges of these large terraces roughly follow the dense packed orientations, they also exhibit a significant concentration of kinks. A similar behavior is observed after deposition of 20 ML at 520 K, as shown in Fig. 4.21. Straight, triangular top terraces free of kinks are located on pyramids and coexist with large top terraces bounded by curved, kinked step edges, e.g. in the middle of the left side of Fig. 4.21.

The existence of top terraces several times larger than the average base terrace size apparently contradicts the model developed in Sect. 4.3.3 and points to a new mechanism for the selection of the top terrace size. Indeed, the top terrace sizes fall into two distinct classes: Top terraces with $L \approx 250$ and top terraces with $L \approx 1500$. The small top terraces have straight B–steps, and form the top layer islands of the pyramids, whereas the large top terraces always possess rough steps and typically share their base terrace with other islands. Estimating the barrier height for the pyramids on the basis of (4.54) results in a large effective step edge barrier of $\Delta E_{\mathrm{S}}^{\mathrm{eff}} \approx 0.29\,\mathrm{eV}$. Calculations for the large top terraces are difficult to perform, because their growth behavior is not known. Nevertheless, it is clear that their step edge barrier must be negligible, because otherwise the nucleation of an additional layer would have taken place. With a kink concentration between 0 and 3×10^{-3} (zero or one kink in the island edge) for the pyramid top terraces the two site model yields $\Delta E_{\mathrm{S}}^{\mathrm{eff}}$ varying between 0.33 eV (no kink in island edge) and 0.25 eV (one kink in the island edge), in reasonable agreement with the experimental estimate. With a kink concentration of about 2×10^{-1} for the large terraces in the flat areas, one arrives at $\Delta E_{\mathrm{S}}^{\mathrm{eff}} = 0.07\,\mathrm{eV}$, again in agreement with expectation. It may thus be concluded that the large variations in kink concentration of the island step edges are decisive for the heterogeneous surface morphology.

---

[26] This value presumably underestimates the true barrier, since the encounter probability p_{enc} for such small barriers is no longer close to unity.

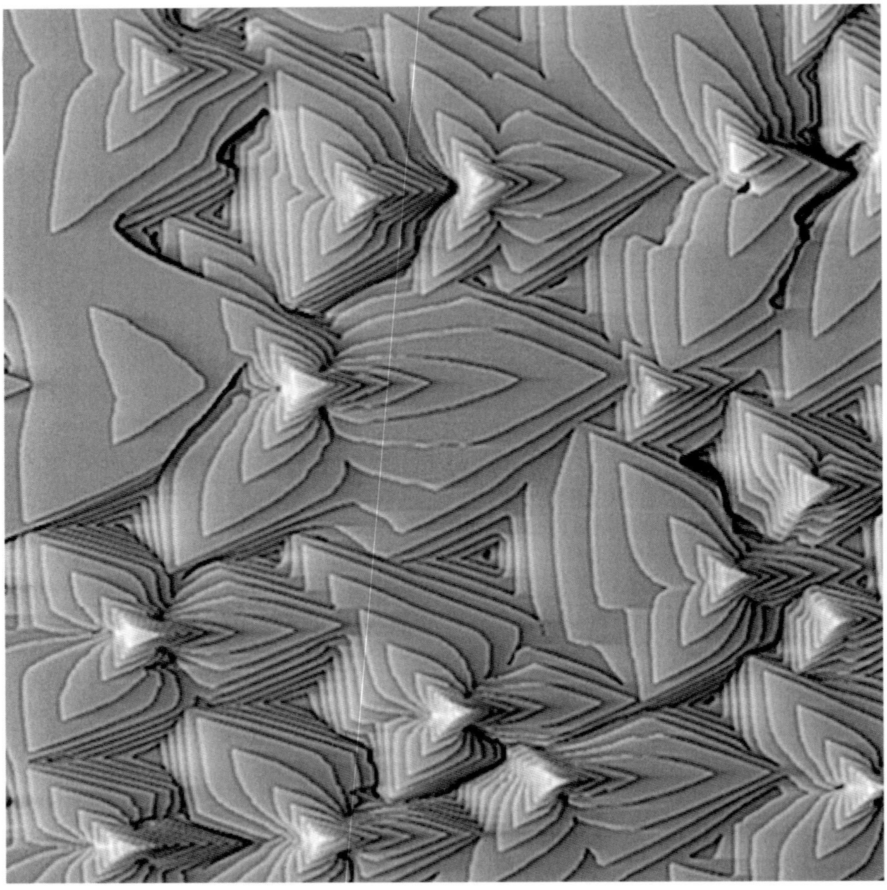

Fig. 4.21. STM topograph after deposition of 20 ML Pt on Pt(111) at 520 K with $F = 1 \times 10^{-2}$ ML/s^{-1}. Topograph size is 7900 Å × 7900 Å

The sharp, needle-like pyramids in Fig. 4.21 indicate that the same destabilizing feedback mechanism mentioned in Sect. 4.4.2, which couples the roughness build-up to the increase in the effective step edge barrier, is operative also at 520 K. However, in the high temperature regime the onset of the feedback mechanism appears to be delayed. At a coverage of 20 ML the surface is separated into areas where the feedback mechanism has produced pyramids, and areas which have not yet entered the feedback regime. In contrast, at 440 K the entire surface is covered with pyramids already after deposition of 12 ML (see Fig. 4.6c). It has to be expected that during growth larger and larger parts of the surface switch from flat areas to the pyramid growth regime until finally the flat areas go extinct. At 480 K this transition takes place between 5 ML and 43 ML, while at 520 K the coexistence regime extends at least up to the deposition of 20 ML, see Fig. 4.21. In this sense the coexistence regime can be viewed as a continuation of the transient growth regime observed at 440 K (see Sect. 4.2 and 4.4.2) to larger film thicknesses[27].

We can only speculate about the origin of the difficulty to initiate the feedback mechanism at 520 K. It appears likely that the probability for nucleation on top of an island is a nonmonotonic function of the island size in this growth regime: If nucleation takes place, it has to occur before the island reaches the critical size L_c for the onset of dendritic-skeletal growth, because beyond that point the kink density increases dramatically and the effective step edge barrier drops almost to zero. Islands which succeed in forming a second layer nucleus enter the feedback regime, whereas the other islands coalesce and new large fast growing islands appear on top of them after coalescence. In this way a bimodal distribution of top terrace sizes may result.

The following three factors could contribute to lowering the probability for early second layer nucleation at 520 K, thus extending the regime in which the two morphologies coexist. First, interlayer transport is trivially enhanced at higher temperatures. Second, the apparent shift of the onset of dendritic-skeletal growth to lower coverages when going from 400 K to 500 K (see Sects.3.1 and 3.7.3) would further reduce the "window of opportunity" for second layer nucleation. Third, the dependence of the size of the critical nucleus on top of an island on the effective step edge barrier may play a role. According to (4.119), at about 500 K the critical nucleus size fluctuates between $i^* = 1$ and $i^* = 2$ depending on the actual barrier height. This would certainly lead to large fluctuations in the nucleation probability for top terraces, enhancing the tendency towards a bimodal size distribution.

4.4.5 CO Adsorption and Mound Growth on Pt(111)

The adsorption of CO on Pt(111) and more specifically the preferential adsorption of CO at step edges has tremendous effects on island shapes (Sect. 3.5) and increases the effective step edge barrier, as shown in Sect. 4.3.3. In earlier measurements with non-negligible CO partial pressure, mound formation was significantly influenced by the action of CO [131]. Even for the mound formation experiments discussed above, which were performed in an extremely low total background pressure below 2×10^{-11} mbar with the CO partial pressure below 5×10^{-12} mbar, CO adsorption may be of importance. In mound formation experiments the surface is exposed to a small partial pressure of CO for times up to several hours, making it more difficult to exclude the influence of CO than in submonolayer experiments.

We have shown in Sect. 3.5 that the presence of CO at island edges is signaled by the presence of straight portions of A–steps. The sensitivity of

[27] The way in which the mound morphology invades the flat surface is somewhat similar to the breakdown of step flow in one-dimensional growth models, which has been associated with the decay of metastable thermodynamic states [377] (see Sect. 4.7.4).

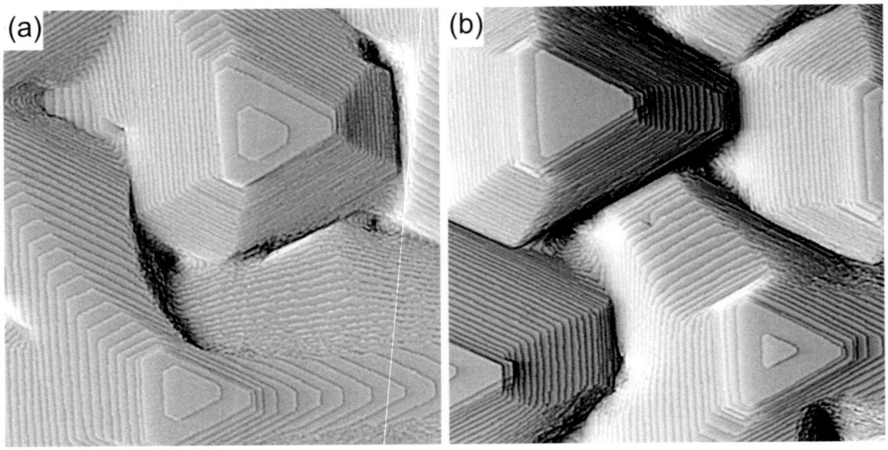

Fig. 4.22. (**a**) STM topograph after deposition of 90 ML Pt on Pt(111) at 440 K with $F = 7 \times 10^{-3}$ ML/s^{-1}. Note the presence of short, straight A–step segments in the top and base terraces, which are otherwise dominated by B–steps. (**b**) After annealing for 30 s to 560 K and additional deposition of 3 ML the presence of A–step segments is significantly reduced in the top and base terraces. This implies the presence of decorating CO molecules at the A–step segments in (**a**) (see text). Topograph sizes are 1400 Å × 1400 Å

mound formation to CO adsorption will be greatest for a temperature that is low enough to retain the CO at the step for a significant length of time, but still high enough to keep the step density sufficiently small, so that a considerable amount of CO accumulates at each step. While at 440 K the equilibrium coverage of CO on a Pt(111) terrace site is only 3×10^{-6} [294] and thus marginal, it is certainly larger for a step site, where the CO molecule binding is considerably stronger. An experiment which directly probes the presence of CO during mound formation is shown in Fig. 4.22, and was performed in the following way. After deposition of 90 ML (compare Fig. 4.22a), the sample was annealed for 30 s at 560 K, thereby safely removing all CO which was possibly adsorbed to steps. Deposition of another 3 ML at 440 K leads to top terraces, which have a larger aspect ratio $R_{\mathrm{B/A}}$ of B–step length L_{B} to A–step length L_{A} (compare Fig. 4.22b): $R_{\mathrm{B/A}}$ increases from 3.8 ± 1.6 to 8.8 ± 3.2. Similarly, for the base terraces $R_{\mathrm{B/A}}$ increases from 6.5 ± 1.8 to 15.7 ± 5.9. Consistent with these findings, growth at 440 K in a larger background pressure leads to mounds with significantly extended A–step portions of the top layer islands (see Fig. 4.16), which in extreme cases may even dominate the mound shape [281].

Therefore, if A–step segments are visible in mound growth below 500 K, it has to be assumed that they are decorated by CO molecules, which implies a significant additional step edge barrier at these steps[28]. This resolves the apparent discrepancy between growth rule (iii) in Sect. 4.4.1, stating that A–steps have a step edge barrier comparable to the large B–step barrier, and the ab initio calculations yielding only a small A–step edge barrier [87]. Granting the presence of CO at step edges during mound growth, one might be tempted to attribute the large value of $\Delta E_{\mathrm{S}}^{\mathrm{eff}}$ estimated above for mound growth above 400 K exclusively to the action of CO. However, the clear correlation between CO background pressure and A–step length implies that for the top and base terraces consisting predominantly of B–steps most of the step sites are *not* decorated by CO. The dominant step site species, the position in a straight B–step, then must have a large $\Delta E_{\mathrm{S}} \geq 0.18\,\mathrm{eV}$. This statement is underlined by the mound growth experiment performed at 520 K. There only B–steps are present in the boundaries of the top terraces, signifying the absence of CO molecules at the straight steps, and nevertheless a large step edge barrier was estimated in Sect. 4.4.4[29].

4.4.6 A Criterion for Mound Growth

To summarize the content of this section, we have shown that mound growth on Pt(111) can be explained qualitatively in a two site model for the effective step edge barrier, in which the concentration of the low barrier sites, the kink sites, determines the magnitude of $\Delta E_{\mathrm{S}}^{\mathrm{eff}}$. The experiments imply that $\Delta E_{\mathrm{S}} \geq 0.18\,\mathrm{eV}$ for straight B–steps.

As the actual microscopic structure of a step edge is a complicated function of the growth temperature, of the deposited coverage, and even of the position and local environment of the step, the same is true for the effective step edge barrier. Nevertheless, we are able to identify two distinct scenarios for mound evolution, which correspond to two different values of the step edge barrier. For small $\Delta E_{\mathrm{S}}^{\mathrm{eff}}$, like in the case of deposition at $T = 300\,\mathrm{K}$, the initial layer–by–layer growth changes to mesa mound formation, because the step edge barrier prevents crevices from closing up. In the second scenario, when $\Delta E_{\mathrm{S}}^{\mathrm{eff}}$ is large, the step edge barrier enhances the nucleation on the top terraces and wedding cakes form, as was described in Sect. 4.2. This regime is realized in growth at 440 K, but only after the deposition of several layers during which $\Delta E_{\mathrm{S}}^{\mathrm{eff}}$ continuously increases.

Neglecting this induction phase, a simple criterion for mound growth can be formulated in terms of the onset of second layer nucleation: If this occurs

---

[28] On the other hand, the experiments supply no evidence for the ab initio prediction that the clean, straight A–step has a negligible ΔE_{S} [87].

[29] At 520 K, CO molecules may still occasionally decorate their strongest binding sites on Pt(111), the kink sites [283]. As only few kinks are present in the boundaries of the top terraces at this temperature, the occasional adsorption of a CO molecule could possibly enhance the fluctuations in the top terrace nucleation probability.

prior to coalescence of the first layer islands, mounds form, while in the opposite case the initial growth proceeds layer–by–layer. Since the island size near coalescence is comparable to the distance between islands, a quantitative criterion is obtained by comparing the island spacing $l_D = N^{-1/2}$ to the critical island radius R_c for second layer nucleation [354]. Assuming a square lattice array of islands, second layer nucleation takes place prior to coalescence if

$$l_D > 2R_c \ . \tag{4.59}$$

For the critical nucleus size $i^* = 1$, the expressions (4.51) for R_c and (2.26) for N can be used to derive a criterion which directly relates mound growth to the fundamental activation energies $\Delta E_{\mathrm{S}}^{\mathrm{eff}}$ and E_{D}. For Pt(111) growth at a typical deposition rate $F = 5\times10^{-3}$ ML/s and with the attempt frequencies $\nu_0 = 5 \times 10^{12}\,\mathrm{s}^{-1}$ and $\nu_0' = 7.3 \times 10^{11}\,\mathrm{s}^{-1}$, one obtains

$$\Delta E_{\mathrm{S}}^{\mathrm{eff}} + \frac{1}{6}E_{\mathrm{D}} > \left[1 + \frac{1}{6}\ln\left(\frac{\nu_0}{\Omega F}\right) + \ln\left(\frac{\nu_0'}{\nu_0}\right)\right] k_{\mathrm{B}}T \approx 4.8\,k_{\mathrm{B}}T \ . \tag{4.60}$$

The left hand side of (4.60) points to the decisive role of the additional step edge barrier for mound formation. Using $E_{\mathrm{D}} = 0.26$ for Pt(111), mound growth at 440 K is seen to require $\Delta E_{\mathrm{S}}^{\mathrm{eff}} > 0.14\,\mathrm{eV}$. This is indeed fulfilled according to the measurements for the top and base terrace sizes in the mound growth regime discussed in Sect. 4.3.3, which yield $\Delta E_{\mathrm{S}}^{\mathrm{eff}} = 0.18\pm0.01\,\mathrm{eV}$. For mound growth at 300 K, $\Delta E_{\mathrm{S}}^{\mathrm{eff}} > 0.08\,\mathrm{eV}$ would be needed, which according to the analysis in Sect. 4.4.3 is not satisfied. Mound formation at 300 K is no longer driven by preferential nucleation on top terraces. Instead, it provides a realization of the first scenario for mound formation, which will be discussed in more detail in the following section.

4.5 Mound Formation with Weak Barriers

The criterion (4.59) for mound growth due to enhanced second layer nucleation can be rearranged to include the length scales l_D and l_{ES} associated with the energies E_{D} and ΔE_{S}. Using the expressions (2.26) and (4.51) for $i^* = 1$, one finds

$$l_{\mathrm{ES}}/l_D > C_1 \approx 1.8 \ . \tag{4.61}$$

The *same* criterion (though presumably with different constants C_{i^*} on the right hand side) is obtained for larger values of the critical island size i^* [168]. It generalizes in a straightforward way the distinction between *weak* and *strong* step edge barriers introduced in Sect. 4.3.2 in the context of nucleation on a single island: Only strong barriers, in the sense of $l_{\mathrm{ES}} \gg l_D$, affect the growth of the first few layers.

In the weak barrier regime, $l_{ES} \ll l_D$, mounds form only gradually and require deposition of much larger film thicknesses. This regime is in the focus of the present section. The available theoretical description of the growth and shape of the mounds is less detailed than in the wedding cake regime of Sect. 4.2. For the most part we will employ a rather coarse-grained continuum picture based on the notion of the growth-induced surface current, which is introduced in the following subsection.

4.5.1 Diffusion Bias and the Growth-Induced Current

To understand the mechanism which destabilizes the growing surface in the weak barrier regime, we recall that, in addition to increasing the adatom density on top of an island, step edge barriers also introduce an asymmetry in the attachment of adatoms to the ascending and the descending step bordering a vicinal terrace (Sect. 4.3.1). The preferential attachment to the ascending step implies that the diffusion on the terrace is effectively *biased* in the uphill direction (Fig. 4.23). The uphill mass flow will tend to amplify small bumps that inevitably arise on the surface due to the fluctuations in the deposition flux, and the transport of mass away from the valleys will delay the closing of trenches. Both effects lead eventually to mound formation. It is worth emphasizing the generality of this mechanism: All that is needed is some process that couples the geometric asymmetry of the vicinal surface – the inequivalence between the uphill and downhill directions – to the motion of the adatoms [106,378]. The step edge barrier provides such a process, but other, less obvious possibilities exist as well (see Sect. 4.9.7).

For a quantitative description it is useful to introduce the notion of a growth-induced surface diffusion current [106,348,378,379]. Considering again a vicinal terrace, the current can be defined as the net number of diffusion jumps in the uphill direction (towards the ascending step) that cross a line of unit length parallel to the step per unit time. Within the theory of Sect. 4.3.1, the local uphill diffusion current is $D\partial n/\partial x$. Averaging this across the terrace and using the density profile (4.30,4.31) one obtains, for a terrace of width $l \gg a$,

$$j_{\mathrm{ES}} = \frac{D}{l}[n(l) - n(0)] = \frac{F l l_{\mathrm{ES}}}{2(4l + l_{\mathrm{ES}})} = \frac{Fl}{2}(2p_+ - 1) \; . \tag{4.62}$$

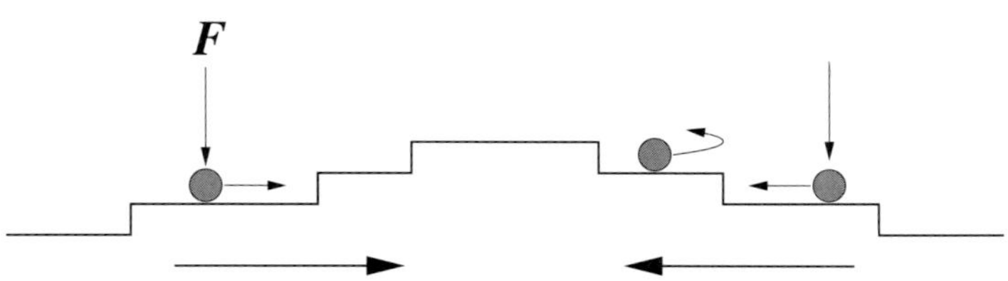

Fig. 4.23. Sketch of diffusion bias on a vicinal surface. The preferential attachment of adatoms to ascending step edges, and reflection at descending step edges, implies a net motion along the surface in the direction of increasing height. Note that this does *not* require that adatoms move to higher atomic layers

The last equality expresses the fact that each adatom attaches either to the ascending step (with probability p_+) or to the descending step (with probability $1 - p_+$), and travels a net distance $l/2$ from its point of arrival to the step.

In the following, it will prove useful to consider the current as a function of the *surface slope* m. More precisely, we take m to denote the slope of a one-dimensional, continuous height profile which represents the surface morphology on scales large compared to the typical terrace size [379]. For a train of steps of height d separated by terraces of width l, the magnitude of the slope is $|m| = d/l$, and $m > 0$ ($m < 0$) if the steps are ascending (descending). The uphill direction of the growth-induced current then implies that the sign of j_{ES} is equal to that of m, that is, j_{ES} must be an *odd* function of m. In particular, $j_{\mathrm{ES}}(0) = 0$, because there cannot be any diffusion bias on a horizontal surface with an equal number of ascending and descending steps.

The expression (4.62) does not satisfy this requirement: It approaches a nonzero limit for $l \to \infty$, which implies a discontinuity at $m = 0$. Equation (4.62) becomes invalid at small slopes (large terrace widths), because then islands begin to nucleate on the terrace, which capture adatoms and thus reduce the net uphill flux. In order of magnitude, this happens when the terrace width becomes comparable to the distance l_D between first layer islands on a flat surface[30] [380]. For $l > l_D$, only the atoms deposited within a distance l_D from the preexisting steps contribute to the net surface current. These form a fraction l_D/l of all atoms, and each of them travels a distance of the order of l_D to the step. The net current can then be estimated as $j_{\mathrm{ES}} \approx F(l_D/l)l_D(2p_+(l = l_D) - 1)$, which vanishes linearly in the slope for $m \to 0$. An interpolation formula which connects the regimes $l < l_D$ and $l > l_D$ is given by [348]

$$j_{\mathrm{ES}}(m) = \frac{F l_{\mathrm{ES}} l_D m}{2d(4 + l_{\mathrm{ES}}/l_D + |m| l_{\mathrm{ES}}/d)(1 + |m| l_D/d)} \,. \tag{4.63}$$

The qualitative shape of this function is illustrated by curve (I) in Fig. 4.24. We would like to emphasize that the interpolation across the region $|m| < a/l_D$ is the *only* way in which the continuum theory based in the growth-induced current takes account of the fact that nucleation occurs on flat regions of the surface. The development of a continuum theory which treats nucleation in a more explicit, and realistic manner remains an important challenge for the future.

To fully explore the consequences of the growth-induced current for the morphological evolution of the surface, expressions like (4.63) have been incorporated into two-dimensional phenomenological continuum equations which describe the competition between the growth instability and smoothening by surface energetics in a quantitative way. The reader interested in these theoretical developments is referred to Sect. 4.9.8. In the following subsections

[30] See also Sect. 4.7.4.

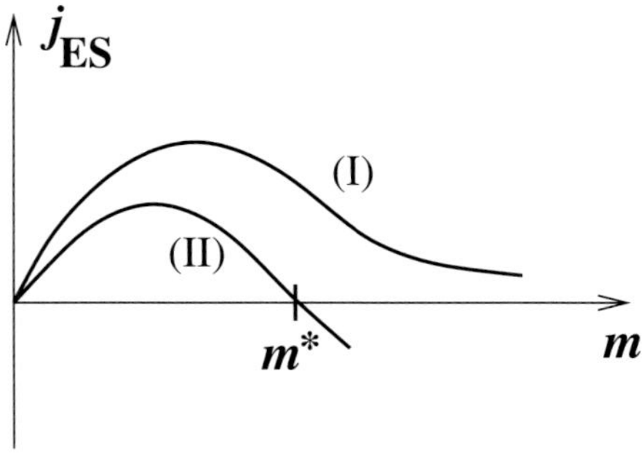

Fig. 4.24. Sketch of the growth-induced surface current with (II) and without (I) slope selection. The current for negative slopes follows from the requirement that $j_{ES}(m)$ is an odd function of m

we content ourselves with extracting some elementary (and fundamental) implications of the diffusion bias, specifically the selection of the initial mound wavelength (Sect. 4.5.2), the mound shape (Sect. 4.5.3), and the final mound slope (Sect. 4.6.1).

4.5.2 Weak Barriers: Onset of the Instability

Here we apply the concept of the growth-induced surface current to describe the growth instability in the weak barrier regime, $l_{ES} \ll l_D$. Anticipating that the lateral scale of surface features in this regime will be much larger than the typical size l_D of atomically flat terraces, we employ a continuum approximation, where the height of the surface above a substrate point x is given by a continuous function $h(x,t)$ [379]. For the present discussion it is sufficient to consider only one-dimensional surface profiles.

The morphology evolves in response to the mean deposition flux F, and the total mass current $j(x,t)$ along the surface. This is expressed through the continuity equation

$$\frac{\partial h}{\partial t} = -\Omega d \frac{\partial j}{\partial x} + \Omega d F \,, \qquad (4.64)$$

where Ωd is the atomic volume. A flat growing surface is given by $h(x,t) \equiv \Omega d F t$. To examine its stability, we superimpose a small height perturbation $\varepsilon(x,t)$ and expand the current $j = j_{ES}$, evaluated as a function of the *local* slope $m = \partial h/\partial x$, to leading order in $\partial \varepsilon/\partial x$. This yields a diffusion equation for ε with a *negative* diffusion coefficient [see (4.67,4.68)], which proves that the surface is unstable. However, the instability is unphysical in that the surface develops violent modulations on arbitrarily small length scales.

To cure this deficiency, we need to take into account additional contributions to the surface diffusion current that smoothen the surface on small scales. Near thermal equilibrium the theory of Mullins can be used [381], in which the smoothening current is proportional to the gradient of the adatom chemical potential. According to the Gibbs–Thomson relation, the chemical potential near an interface of curvature κ is enhanced by an amount [222]

$$\Delta\mu = \gamma\kappa \; , \tag{4.65}$$

where γ is the interfacial free energy[31]. In our one-dimensional setting the local curvature is $\kappa \approx -\partial^2 h/\partial x^2$, and consequently the smoothening current takes the form $j_{\mathrm{smooth}} \sim \partial^3 h/\partial x^3$, with a coefficient proportional to γ and to Dn^{eq}. Under the far–from–equilibrium conditions typical for mound growth this mechanism is not relevant, because adatom detachment from steps is virtually absent, and correspondingly the equilibrium adatom density is very low [382]. However, other smoothening mechanisms involving the adatoms from the deposition flux are conceivable, and can be argued to yield a smoothening current of the same general form [348]. Without specifying these mechanisms in detail, we therefore write the total current as

$$j = j_{\mathrm{ES}} + K_4 \frac{\partial^3 h}{\partial x^3} \; . \tag{4.66}$$

Inserting this into (4.64) and expanding around the flat surface, we now obtain the evolution equation

$$\frac{\partial \varepsilon}{\partial t} = K_2 \frac{\partial^2 \varepsilon}{\partial x^2} - K_4 \frac{\partial^4 \varepsilon}{\partial x^4} \tag{4.67}$$

for the perturbation, where

$$K_2 = -\frac{\mathrm{d} j_{\mathrm{ES}}}{\mathrm{d}m}(0) = -\frac{1}{8}\Omega F l_{\mathrm{ES}} l_D \; . \tag{4.68}$$

Equation (4.67) can be solved by decomposing the perturbation into Fourier components. The contribution with wavenumber q (wavelength $2\pi/q$) is given by

$$\varepsilon_q(x,t) = \varepsilon_0 \mathrm{e}^{\mathrm{i}qx + \omega(q)t} \; , \tag{4.69}$$

where ε_0 is the initial perturbation amplitude and ω is the growth rate of the perturbation. The surface is unstable with respect to perturbations of wavenumber q if $\omega(q) > 0$. Inserting (4.69) into (4.67) we find that the growth rate is given by the expression

$$\omega(q) = -K_2 q^2 - K_4 q^4 \; . \tag{4.70}$$

Since $K_2 < 0$, this is positive for $|q| < \sqrt{-K_2/K_4}$. Thus perturbations with wavelengths larger than $\lambda_{\mathrm{min}} = 2\pi\sqrt{-K_4/K_2}$ grow exponentially. The fastest growing mode – which dominates the morphology at early times – is characterized by the wavenumber q^* at which the growth rate (4.70) is maximal.

31 The two-dimensional version of the Gibbs–Thomson relation was used in Chap. 3, see (3.9).

The corresponding wavelength is $\lambda^* = \sqrt{2}\,\lambda_{\min}$, and the time constant for the growth of this mode is $\tau^* = \omega(q^*)^{-1} = 4K_4/K_2^2$.

The value of the coefficient K_4 in (4.66) can now be determined through a consistency argument [383]. We require (i) that K_4 is independent of l_{ES} (this is reasonable since the smoothening current involves diffusion on the terraces, not across steps), and (ii) that the dominant wavelength λ^* becomes of the order of the island spacing l_D when the strong barrier regime is approached, $l_{\mathrm{ES}} \to l_D$ (recall that in the strong barrier regime, the initial mound size is set by the island spacing in the first layer). Using (4.68) this leads to the relation

$$K_4 \approx \Omega F l_D^4 \,, \tag{4.71}$$

which was first proposed on dimensional grounds [353]. With (4.71) the expressions for the initial mound separation λ^* and the growth time τ^* take the form

$$\lambda^* \sim (l_D/l_{\mathrm{ES}})^{1/2} l_D, \quad \tau^* \sim (\Omega F)^{-1}(l_D/l_{\mathrm{ES}})^2 \,. \tag{4.72}$$

These equations are the central result of this subsection. They show two important characteristics of the weak barrier regime: The mound instability first appears at a length scale which is larger than the submonolayer island spacing l_D, and the onset of the instability is delayed – it occurs only after the deposition of $\sim (l_D/l_{\mathrm{ES}})^2$ layers, which can be a large number if $l_{\mathrm{ES}} \ll l_D$.

In metal homoepitaxy this scenario is typically not observed. Experiments reporting mound formation on metal surfaces have found that the instability sets in during the growth of the first few layers, at a length scale which coincides with the distance between first layer islands (see Sect. 4.6.2 for further discussion). This fact can be used to derive rough lower bounds on the step edge barrier in these systems. For example, for Fe grown on Fe(100) at 293 K [382] the initial island spacing is 3.5 nm. The criterion (4.61) then yields $l_{\mathrm{ES}} > 6.3$ nm, which implies, using (4.33) and assuming equal preexponential factors for ν and ν', that $\Delta E_{\mathrm{S}} > 0.08$ eV. Similarly one obtains $\Delta E_{\mathrm{S}} > 0.11$ eV for Cu(100) [384] and $\Delta E_{\mathrm{S}} > 0.27$ eV for Rh(111) [121].

A semiconductor system that appears to behave in accordance with the weak barrier scenario is Ge/Ge(001) [385]. As can be seen in Fig. 4.25, the mound pattern becomes discernible only after the growth of 200 nm, with a lateral length scale of about 70 nm which is much larger than the island spacing $l_D \approx 5$ nm at this temperature. From the growth of the roughness amplitude the characteristic time τ^* can be estimated, and it is concluded that the ratio l_D/l_{ES} varies between 8 at 373 K and 26 at 500 K. The weak temperature dependence of l_{ES} is consistent with an additional step edge barrier of $\Delta E_{\mathrm{S}} = 0.024$ eV.

The onset time τ^* can also be observed in the growth of a film on a rough substrate. Depositing thick InAs buffer layers onto substrates with considerable initial roughness, a minimum in the surface width was observed at a layer thickness of about 570 nm [386]. A similar roughness minimum was reported for Ag growth on preroughened Ag(111) [342]. A theoretical analysis

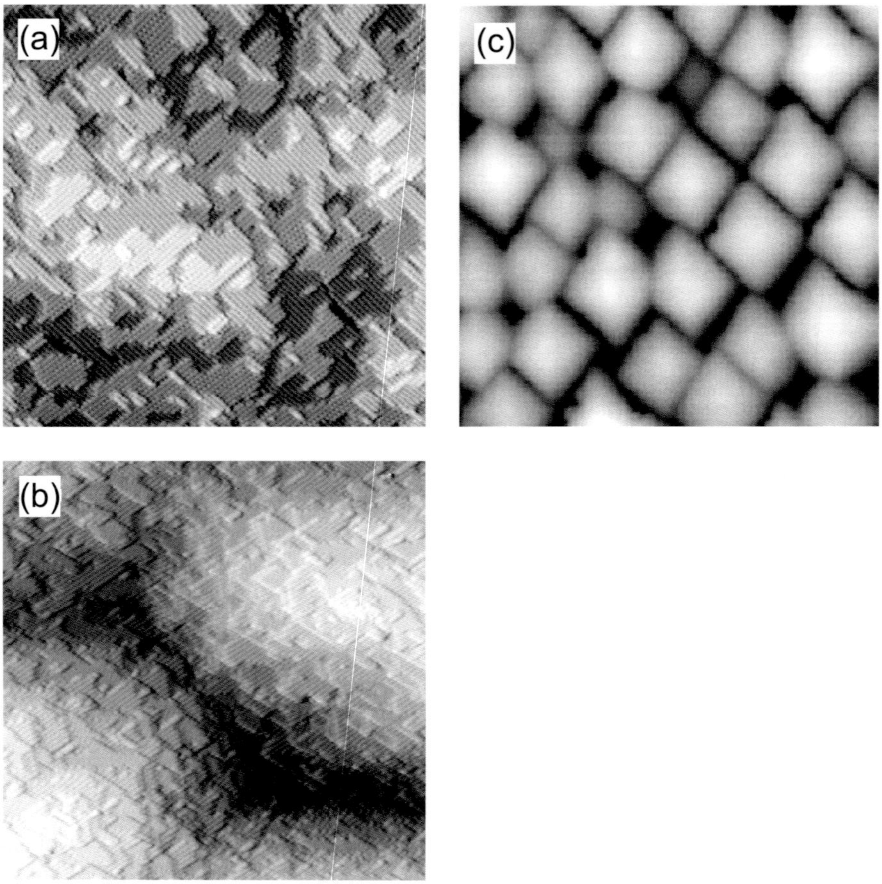

Fig. 4.25. STM images of mound formation on Ge(001). (**a**) 700 Å × 700 Å image of a film of mean thickness $\bar{h} = 100$ Å. (**b**) 700 Å × 700 Å image at $\bar{h} = 2000$ Å. (**c**) 7000 Å × 7000 Å image at $\bar{h} = 10000$ Å. Mounds are first visible in (**b**), and the pattern is fully developed in (**c**) (Fig. 1 of [385])

in the framework of continuum growth equations shows that this behavior can be explained by the dominance of the second, smoothening term on the right hand side of (4.67) in the inital stage of growth [387]. If the substrate roughness spectrum contains sufficient weight at small wavelengths, the rapid smoothening of these small scale features overcompensates the (slow) increase of roughness at the scale λ^* induced by the growth instability[32]. The two effects balance, and the surface width attains a minimum, at a time of the order of τ^*.

[32] In the experiment of [386], the dominating roughening mechanism may in fact have been deposition noise rather than the growth instability due to step edge barriers, see [388].

4.5.3 Weak Barriers: Mound Shapes

Apart from delaying the onset of the instability, a decrease of the ratio l_{ES}/l_D also changes the shape of the mounds. This is illustrated in Fig. 4.26, which shows mound shapes obtained from simulations of a one-dimensional growth model described in [353] (see also [389]). For $l_{\mathrm{ES}} > l_D$ the main effect of decreasing the step edge barrier is to increase the size of the top terrace, while the overall shape follows the cut-off error function profile given by (4.20) (compare to Fig. 4.9). The top terrace size cannot become larger than the lateral size of the mounds which, in this regime, is of the order of the island spacing l_D. Rewriting the expression (4.54) for the size of the top terrace in the form

$$\frac{R_{\mathrm{top}}}{l_D} \sim \left(\frac{l_D}{l_{\mathrm{ES}}}\right)^{1/5} \tag{4.73}$$

the condition $R_{\mathrm{top}} < l_D$ is seen to be equivalent to the strong barrier condition (4.61), and therefore a qualitative change in the mound shapes is expected at the transition to the weak barrier regime.

Simulations of the one-dimensional model of [353] produce dome-shaped mounds for $l_{\mathrm{ES}} \ll l_D$, see Fig. 4.26. The mounds appear at the initial wavelength λ^* given by (4.72). Subsequently the lateral length scale coarsens until an upper critical wavelength λ^{**} is reached, at which point coarsening ceases and deep crevices develop between the mounds. In an approximation neglecting the stochastic nature of nucleation, the upper critical wavelength is estimated to be [353]

$$\lambda^{**} \sim \frac{l_D^2}{l_{\mathrm{ES}}} \, . \tag{4.74}$$

Within the continuum theory, the distinct up–down asymmetry of the mounds is attributed to an additional, symmetry-breaking contribution

$$j_{\mathrm{SB}} = \frac{F\Omega}{6m^3}\frac{\partial m}{\partial x} \tag{4.75}$$

to the surface current (4.66), where $m = \partial h/\partial x$ [353, 390]. In contrast to the terms in (4.66), the expression (4.75) does not change sign under the transformation $h \to -h$. Its physical origin lies in the dependence of the density of adatoms on the local terrace size, or surface slope.

The low temperature growth on Pt(111) discussed in Sect. 4.4.3 appears to provide an example for mound formation in the weak barrier regime. Based on the estimates for the effective step edge barrier $\Delta E_{\mathrm{S}}^{\mathrm{eff}} \approx 0.01 - 0.02\,\mathrm{eV}$, the ratio l_D/l_{ES} is around 10. Correspondingly, one-dimensional cuts through the experimental morphology represented by Fig. 4.20 are compatible with the simulated one-dimensional mound shapes somewhere between Fig. 4.26a and b. As the morphological evolution at 300 K was only investigated up to 10 ML, no precise statements concerning the existence of lower and upper critical wavelengths λ^* and λ^{**} are possible. Nevertheless, coarsening of the lateral length scale is definitely seen at 300 K up to deposition of 10 ML.

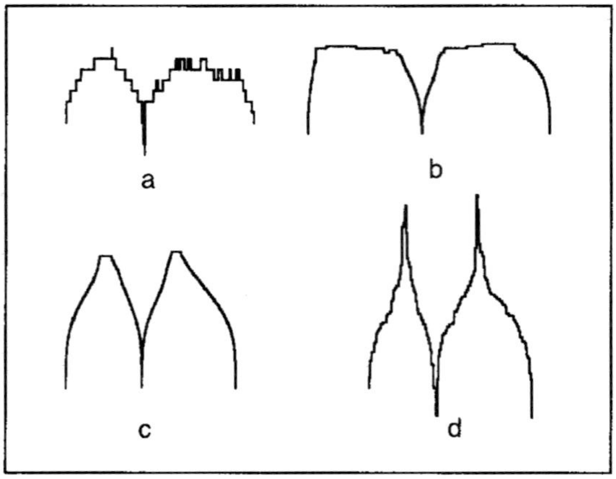

Fig. 4.26. Mound shapes generated using a one-dimensional model with $l_D/l_{\mathrm{ES}} =$ (**a**) 10, (**b**) 1, (**c**) 0.01 and (**d**) 0. In (**a**) the vertical scale is exaggerated by a factor of 5 (Fig. 2 of [372])

4.6 Long Time Evolution of the Mound Morphology

In this section we ask how the patterns induced by the suppression of interlayer transport evolve asymptotically for long times. There are two distinct aspects to this question: The evolution of the mound shapes, specifically the possible selection of the slope at the hillsides (*slope selection*), and the evolution of the lateral length scale of the pattern (*coarsening*).

4.6.1 Slope Selection

The existence of an uphill surface current implies that mass is transported from the valleys to the mound tops. At least for profiles of a fixed wavelength, it follows that the sides of the mounds *steepen* as a function of time. Correspondingly, a downhill current, if existent, flattens the morphology. Let us now suppose that $j_{\mathrm{ES}}(m)$ changes sign not only at $m = 0$, but also at a pair of nonzero slopes $m = \pm m^*$ (see Fig. 4.24). Then profiles with mound slopes of magnitude less than m^* steepen, while profiles with slopes larger than m^* flatten, and consequently the slope m^* is *selected* by the surface evolution. This can be seen also by repeating the stability analysis of Sect. 4.5.2 for a tilted surface segment with slope m. The coefficient K_2 in the perturbation equation (4.67) is then generally given by the negative derivative $-\mathrm{d}j_{\mathrm{ES}}/\mathrm{d}m$ evaluated at slope m, so $K_2 > 0$ for $m = \pm m^*$. We conclude that additional zeros of the function $j_{\mathrm{ES}}(m)$ with $\mathrm{d}j_{\mathrm{ES}}/\mathrm{d}m < 0$ correspond to stable slopes which will dominate the mound morphology at long times [378]. The detailed analysis of the full nonlinear surface evolution equations described in Sect. 4.9.8 confirms this conclusion.

It was noted early on that additional zeros in the current must exist for fundamental crystallographic reasons [378]. At the latest, the current has to vanish when new low index facets form the sides of the mounds. As an example, for growth on an fcc(111) surface, three new (111)–facets separated by polar angles of 120° appear at a slope of 54.7°. Since the surface current vanishes on the original surface for $m = 0$, it must vanish also for the facet slopes. Moreover, if the current was destabilizing on the original surface, the same must be true on the facets, which are crystallographically indistinguishable from the substrate. As the surface current is a continuous function of the slope, stabilizing and destabilizing zeros alternate. It follows that an additional, stabilizing zero has to intervene before the new (111)–facets are reached.

The experimental evidence indicates, however, that slope selection is mostly determined by kinetic, rather than crystallographic factors. The expressions (4.62) and (4.63) for the surface current decrease as $1/|m|$ for $m \to \pm\infty$ but do not vanish at any nonzero slope. Kinetic slope selection thus requires additional atomistic processes that generate a downhill contribution to the surface current, and are able to compensate the uphill current due to the step edge barriers. A plausible candidate is the funneling process described in Sect. 2.4.1. In the present context, funneling implies that an atom depositing directly onto the step edge is deflected towards the lower terrace, and hence does not feel the step edge barrier (Fig. 4.27). The fraction of such atoms increases with increasing step density. The corresponding downhill current is a linearly increasing function of the slope, which eventually cancels the (decreasing) contribution from the step edge barrier.

A quantitative analysis of this effect has been carried out for a one-dimensional model of a (100) surface [391]. The selected slope is found to be

$$m^* = \frac{2\nu/\nu' - 3}{8\nu/\nu' - 9}\,, \tag{4.76}$$

where it is assumed that $\nu/\nu' > 3/2$. This expression has several interesting consequences. First, the selected slope increases with decreasing temperature,

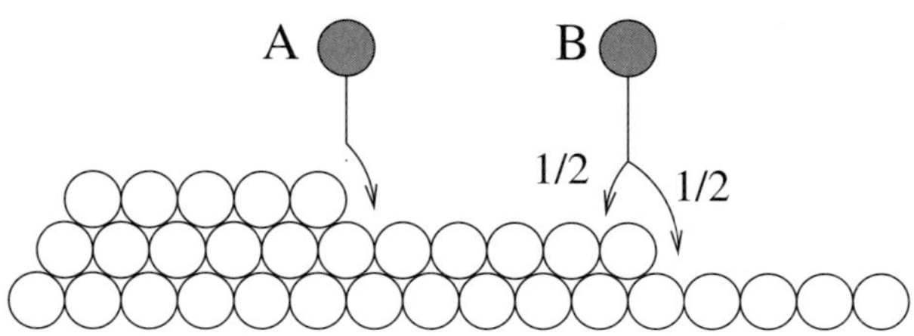

Fig. 4.27. Funneling rules used in the derivation of (4.76) [391]. Atom A impinges to the right of the midpoint of the edge atom, and is therefore deflected to the lower terrace. Atom B impinges exactly on top of the edge atom, and is deflected downward with probability 1/2

reaching a limiting value $m^* = 1/4$ for $\nu \gg \nu'$. Owing to the exponential temperature dependence of ν/ν', the limiting slope is approached rapidly. Second, the selected slope vanishes when $\nu/\nu' \to 3/2$. This value can be reached only if the preexponential coefficients satisfy $\nu_0/\nu_0' < 3/2$. The condition $\nu/\nu' = 3/2$ then defines a critical temperature

$$T_c = \frac{\Delta E_{\mathrm{S}}/k_B}{\log(3\nu_0'/2\nu_0)} \tag{4.77}$$

above which no mound formation is possible. Equation (4.76) is consistent with simulations and experiments on Fe(100) [382,391]. The general trend of the selected slope decreasing with increasing temperature is also confirmed by the experiments on Cu(100) reported in Table 4.2.

The opposite behavior has been observed in KMC simulations of a model for metal(100) homoepitaxy [392]. Its origin is illustrated in Fig. 4.28: While the downhill current due to funneling is essentially independent of temperature, the uphill current decreases with decreasing temperature and eventually falls below the downhill contribution; the simulations thus predict the disappearance of the mounds at *low* temperatures. This can be understood on the basis of the expression (4.63) for the (uphill) growth-induced current. For small slopes $j_{\mathrm{ES}}/m \approx F l_{\mathrm{ES}} l_D / [2d(4 + l_{\mathrm{ES}}/l_D)]$. Since l_{ES} increases, and l_D decreases with decreasing temperature, at low temperatures the strong barrier regime $l_{\mathrm{ES}}/l_D \gg 1$ is reached where $j_{\mathrm{ES}} \sim l_D^2$, which decreases with T. This effect is not captured in the analysis of [391], because it does not include nucleation of islands on the terraces of the sloped surface.

For mound formation on Pt(111) no slope selection is observed. Indeed, Figs. 4.6, 4.7 and 4.8 prove the absence of slope selection up to deposited amounts of 300 ML. This is not in contradiction to the above statement that slope selection must take place in the asymptotic stage of mound growth. It simply implies that this asymptotic stage has not yet been reached after deposition of 300 ML. On the fcc(100) surfaces with a comparatively much larger adatom diffusion barrier E_{D}, the initial island separation is much smaller.

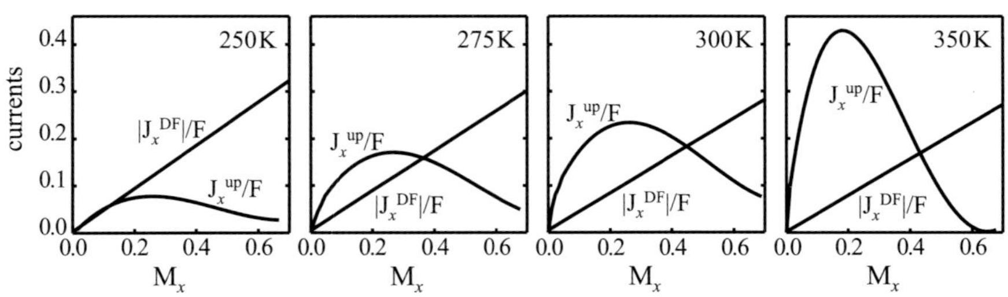

Fig. 4.28. Uphill (J_x^{up}) and downhill (J_x^{DF}) contributions to the surface current measured in KMC simulations of an (100) surface. The crossing point of the two contributions determines the selected slope (Fig. 11 of [392] reprinted with permission from Elsevier Science)

Thus large step densities arise earlier during mound growth and slope selection is reached at smaller coverages.

4.6.2 Coarsening

The mound morphologies appearing in metal homoepitaxy are often observed to *coarsen* with increasing film thickness (Fig. 4.29). This implies that the lateral feature size λ, extracted from the height–height correlation function (4.18), grows with the total coverage Θ in a way that is consistent with a power law relationship,

$$\lambda \sim \Theta^{1/z} , \tag{4.78}$$

which defines the *coarsening exponent*[33] $1/z$. Similarly the surface width W introduced in Sect. 4.2 grows as

$$W \sim \Theta^{\beta} , \tag{4.79}$$

where β is the dynamic roughening exponent[34]. The typical mound slope $\bar{m}$ is then

$$\bar{m} = m_0 \frac{W}{\lambda} , \tag{4.80}$$

where m_0 is a morphology dependent constant, which is for example $\sqrt{48} \approx 6.9$ for a one dimensional sawtooth profile, $\sqrt{72} \approx 8.5$ for a regular array of pointed, two-dimensional pyramids, and $\sqrt{16\,\pi} \approx 7.1$ for the minimal slope of the wedding cake profile (4.17). The slope increases (the morphology *steepens*) if $\beta > 1/z$, and remains constant if $\beta = 1/z$; the latter behavior corresponds to the slope selection scenario described in Sect. 4.6.1. The wedding cake model described in Sect. 4.2 corresponds to the case $1/z = 0$, $\beta = 1/2$, in which the mounds steepen without coarsening.

Table 4.2 summarizes some important characteristics of several experimental coarsening studies. The measured coarsening exponents are in the range $1/6 \leq 1/z \leq 1/3$. In most cases an initial regime of increasing mound slopes is seen followed by a regime with approximate slope selection (Fig. 4.30). For Ag/Ag(100) it has been observed that the steepening regime is more extended when the initial mound size is larger, which implies that slope selection is delayed at higher temperature [393].

Coarsening is a familiar feature of phase ordering phenomena such as Ostwald ripening, where it can be attributed to the tendency of inhomogeneous systems to reduce the excess free energy contained in the boundaries between coexisting phases [394]. The local thermodynamic driving force is

---

[33] In this notation z is a *dynamic exponent* relating a time scale – the coarsening time t_{coarse} – to a length scale – the mound size – as $t_{\mathrm{coarse}} \sim \lambda^z$. A different kind of dynamic exponent will be introduced in Sect. 4.8. In the literature the coarsening exponent is often denoted by n.

[34] This is not to be confused with the static roughness exponent defined in Sect. 4.8.

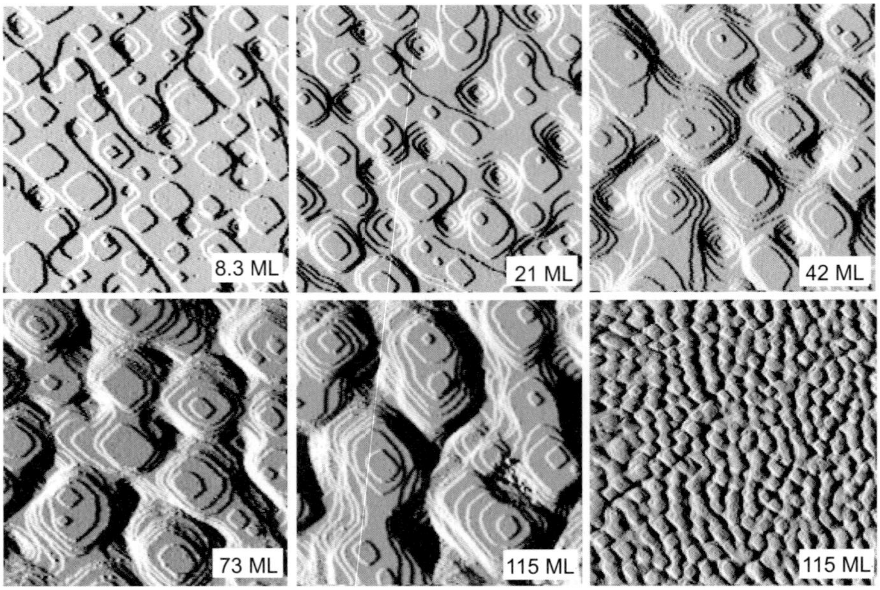

Fig. 4.29. STM images of mound formation on Cu(100) at 299 K. The first five images show an area of 1000 Å × 1000 Å, while the last image has an area of 5000 Å × 5000 Å (Fig. 3 of [384])

the *curvature* of the phase boundaries. The Gibbs–Thomson relation (4.65) implies a mass flow from small domains (with large curvature) to large domains (with small curvature), and correspondingly an increase of the mean domain size. It is, however, by no means obvious that this scenario can be applied to coarsening mounds on a growing crystal surface, which is (i) far from thermal equilibrium and (ii) subject to severe kinetic restrictions in the pathways for lateral mass transport.

The example of the Pt(111) surface illustrates the second point [399]. Comparison of the growth morphology at 440 K, which shows no coarsening, and the erosion morphology[35] shown in Fig. 4.31, which does coarsen at and above 600 K, suggests the onset of an atomistic process responsible for the mass transport between mounds in the range $440\,\mathrm{K} < T < 600\,\mathrm{K}$. The compact island shapes after growth at 440 K show that atoms arriving at the edge may diffuse as step adatoms along the edge to kink positions already at this lower temperature. However, the incorporation into kink positions is irreversible at 440 K as proven by the absence of step fluctuations at the growth temperature, after the deposition flux has stopped. The absence of ripening of submonolayer island arrays proves that the creation of adatoms by detachment from steps is negligible up to 650 K [398]. Consequently, coars-

[35] For the purposes of the present discussion, ion erosion will be regarded simply as the negative equivalent of growth. The conditions under which this equivalence holds are described in [398, 400].

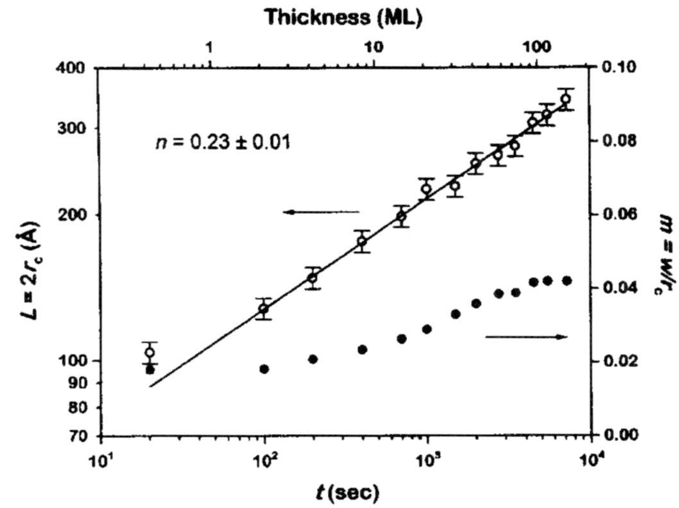

Fig. 4.30. Mound separation L and mound slope $2W/L$ as a function of time, recorded for the mound morphology shown in Fig. 4.29 (Fig. 4 of [384])

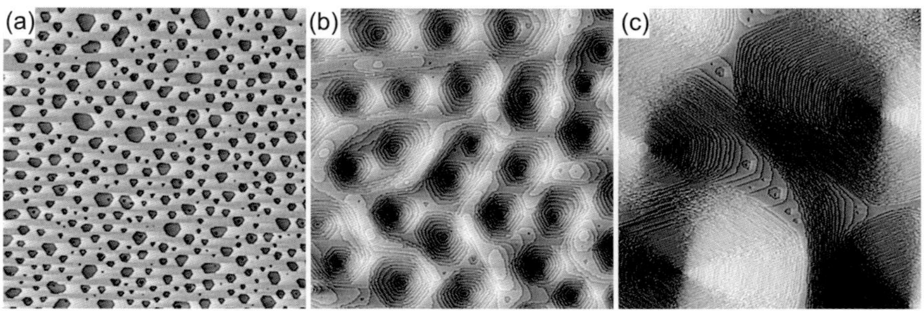

Fig. 4.31. STM images of pit formation on Pt(111) by erosion with 1 keV Xe^+ ions at 650 K. The removed amounts are (**a**) 0.26 ML, (**b**) 6.2 ML and (**c**) 454 ML. The topograph size is 1600 Å× 1600 Å

ening by exchange of atoms between steps via adatom formation and terrace diffusion can be ruled out at temperatures of 600 K – 650 K. We therefore conclude that coarsening proceeds by thermal step atom formation at kinks and subsequent diffusion of these step adatoms along the step edge. This conclusion is consistent with the hierarchy of activation energies as shown in Table 3.1[36]. Step adatom generation at kinks and step edge diffusion are sufficient to induce mound coarsening even in the absence of adatom creation by

[36] According to DFT calculations, step adatom diffusion including diffusion around island corners has an activation energy of 0.9 eV–0.99 eV, step adatom generation and full mobility along the island step has an activation energy of 1.41 eV–1.49 eV, while adatom creation either by direct kink detachment or by step adatom generation followed by step adatom detachment has an estimated ac-

Table 4.2. Experimental characteristics of mound formation in metal homoepitaxy and pit formation in ion erosion. The columns display values for the temperature (T), the total number of deposited or removed layers (Θ), the maximal mean feature size reached at the end of the growth or erosion process ($\lambda_{\max}$), the coarsening exponent ($1/z$), the typical mound or pit slope ($\bar{m}$) and the noise parameter $\mathcal{R}$ defined in (4.81). In cases where the latter two quantities change during surface evolution, the values representative of the latest stage of growth or erosion are given. Unless it is explicitly reported, the slope has been extracted from the surface width and the feature size through the relation (4.80) with $m_0 = 6.9$. In the experiment of [395] on Cu(100), no estimates of the coarsening exponent were reported. The numbers in the corresponding column refer to the roughening exponent β, which is expected to provide an upper bound on $1/z$.

Surface	T [K]	Θ	$\lambda_{\max}$ [nm]	$1/z$	$\bar{m}$	$\mathcal{R}$
Au(111) [396]	290–330	100	40	0.27 ± 0.02	0.17	0.003
Cu/Cu(100) [395]	160	60	6.1	≤ 0.26	0.36	0.038
	200	60	29	≤ 0.56	0.18	0.003
Cu/Cu(100) [384]	300	145	30	0.23 ± 0.01	0.15	0.006
Fe/Fe(100) [382]	293	600	7.3	0.16 ± 0.04	0.25	0.11
Fe/Mg(100) [397]	400–450	2000	18	0.23 ± 0.03	0.50	0.017
Pt/Pt(111) [333]	440	300	80	0.02 ± 0.02	0.28	0.0008
Pt(111) [398]	600–650	600	130	0.28 ± 0.02	0.28	0.0004
Rh/Rh(111) [121]	725	600	100	0.33 ± 0.03	0.07	0.0025
Ag/Ag(100) [393]	190	60	7.2	0.2–0.3	0.35	0.036
	260	60	18	0.18	0.16	0.012

detachment of step atoms or interlayer transport, because they allow atoms to move along step edges shared by two mounds in response to differences in *step* curvature [401]. Figure 4.32 illustrates the mechanism for the case of erosion.

The absence of coarsening in Pt(111) growth at 440 K can thus be attributed to the fact that step adatom generation at kinks is not yet activated. The most efficient pathway for mass transport between mounds is not yet turned on at this temperature. A similar picture may explain also the decrease of the coarsening exponent for Fe(100) between 400 K and 293 K. In [382], the small exponent observed at 293 K was associated with the absence of detachment of step atoms at this temperature; however, as we argued above, detachment is not a necessary condition for coarsening.

The quantitative evaluation of the time required for the merger of two mounds through edge diffusion within a near-equilibrium picture yields $t_{\text{coarse}} \sim \lambda^4$ [401], which reproduces Herring's classic result for the sintering of particles through surface diffusion [402]. The prediction for the coarsening ex-

tivation energy of 1.72 eV. In the model of Sect. 3.1 the corresponding energies are 0.9 eV, 1.5 eV and 1.8 eV.

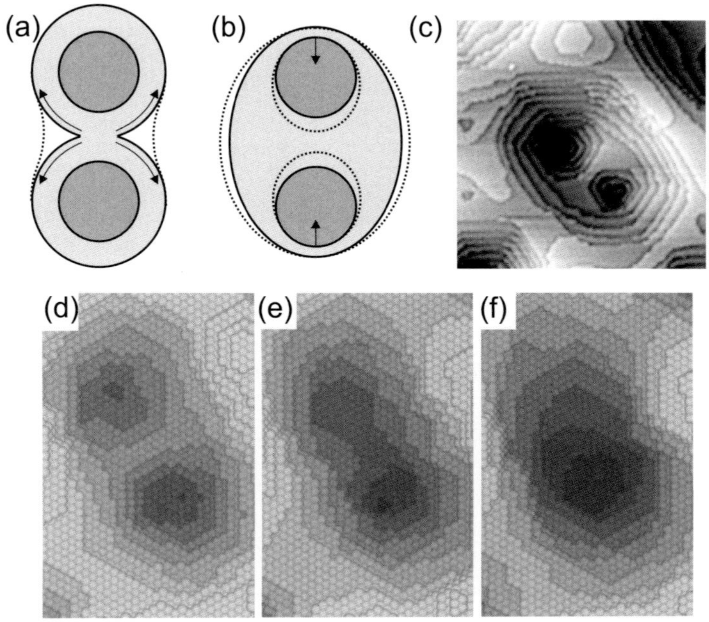

Fig. 4.32. (**a**), (**b**) Schematic sketch of coarsening by step edge diffusion. (**a**) Immediately after coalescence of two vacancy islands a strong edge diffusion current moves material away from the locations of high curvature. (**b**) During shape rearrangement in response to the step line tension, the coalesced vacancy island approaches the two bottom vacancy islands, starting to move them towards their common center of mass. If coalescence of the bottom islands occurs prior to nucleation on both of their bottoms, rapid rearrangement of all vacancy islands takes place and the pits coalesce. Otherwise the process repeats itself. (**c**) STM topograph of two sub-pits in the process of coalescence. Topograph size 500 Å× 500 Å, 6.2 ML removed at 600 K. (**d**)-(**f**) Snapshots of a KMC simulation showing a pit coalescence event. Removed amounts are (**d**) 3.2 ML, (**e**) 3.85 ML, and (**f**) 4.2 ML (Fig. 4 of [398])

ponent is thus $1/z = 1/4$. To understand this result, consider a configuration of two touching mounds which are about to merge (the analogous configuration for the case of erosion is shown in Fig. 4.32). The two-dimensional Gibbs–Thomson relation (3.9) implies a chemical potential difference $\Delta\mu$ between the outer, convex part of the step and the inner, concave coalescence region, which is of the order of the step curvature, $\Delta\mu \sim 1/\lambda$. The corresponding chemical potential gradient drives a mass current of order $\Delta\mu/\lambda \sim 1/\lambda^2$ into the coalescence region. The merger is completed when an area of order λ^2 has been filled, which therefore takes a time proportional to λ^4.

Kinetic Monte Carlo work has also emphasized the importance of step edge diffusion for mound coarsening [403–405]. However, for efficient step edge diffusion, which implies (i) that the kink spacing on the steps is large compared to the step distance and (ii) that diffusion around corners is facile, the numerically observed coarsening exponent is $1/z \approx 1/3$ rather than $1/4$; exponents close to $1/4$ are found if edge diffusion is hindered because of the violation of one of the two conditions (i) and (ii), though smaller exponents have also been reported [406]. Within the estimate of the merging time t_{coarse} sketched above, the value $1/z = 1/3$ is obtained if the difference in adatom density (or chemical potential) between the outer and inner parts of the step is independent of the mounds size, rather than proportional to $1/\lambda$ [403, 405]. This has been attributed to the nonequilibrium conditions at the step [403], i.e. the fact that the step adatom population is fed mainly by deposition rather than by step adatom generation at kinks; in the simulations step adatom generation (process 8 in Fig.3.2) is typically not allowed. Clearly this interpretation contradicts the experimental observation that coarsening sets in only when step adatom generation becomes efficient. Thus, while the central role of edge diffusion for mound coarsening seems to be established, a full understanding of the resulting coarsening law is still lacking.

Mass transport through step edge diffusion is not properly represented in the conventional continuum theory of mound growth described in Sect. 4.9.8, which assumes a mass current proportional to the gradient of a chemical potential that is linearly related to the surface curvature, without distinguishing between the mass transport along or across steps. Nevertheless the continuum theory recovers the coarsening exponent $1/z = 1/4$ for uniformly curved mound morphologies, because this result relies on little more than dimensional analysis.

As we explain in more detail in Sect. 4.9.8, simple dimensional analysis of the growth equations fails in the presence of slope selection due to the appearance of sharp ridges and crevices which concentrate most of the surface curvature. The coarsening exponent is then predicted to depend on the crystal symmetry of the surface, taking the value $1/z \approx 1/4$ for surfaces of fourfold symmetry, and $1/z = 1/3$ in the case of sixfold or higher symmetry (including the limiting case of an isotropic surface) [407]. This is consistent with the measured coarsening exponents for Cu(100) [384] and Fe(100) [397] at higher temperatures, which are close to $1/4$, and with the value for growth on Rh(111) [121], which is $1/3$. However, it is inconsistent with the erosion experiments on Pt(111) [398] and Au(111) [396], which yield exponents closer to $1/4$ than to $1/3$. In view of the fact that the theoretical analysis relies on an up–down symmetry between mounds and pits, which real surface morphologies do not possess (see Sect. 4.9.8), and that it assumes the smoothening Gibbs–Thompson current to be proportional to the gradient of the global surface curvature rather than the step curvature, these discrepancies may be not too surprising. Kinetic Monte Carlo simulations find no systematic dependence of the coarsening exponent on the symmetry of the underlying lattice [403, 405].

An atomistic feature which becomes prominent at low temperatures, and which is not covered by our previous considerations, is the shot noise in

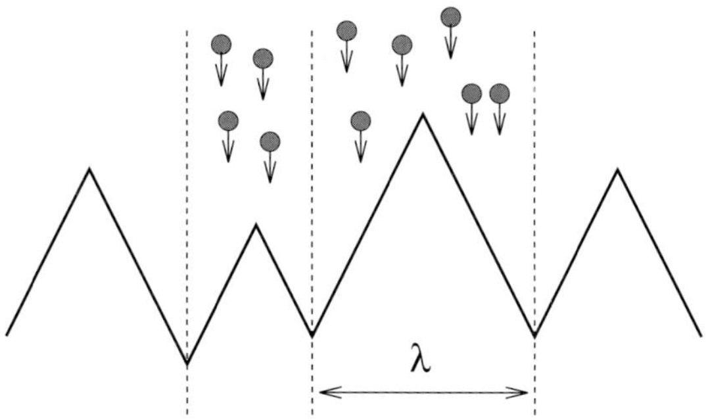

Fig. 4.33. Schematic of noise-induced mound coarsening. Each mound receives the flux incident in its capture region, as indicated by the dashed vertical lines. Due to the fluctuations in the flux, mounds move up or down relative to their neighbors. This shifts the boundaries between the capture regions and may lead to the disappearance of mounds

the deposition beam. In fact noise by itself can cause coarsening, even in the absence of mass transport between mounds [401, 408]. The basic idea is illustrated in Fig. 4.33. Consider an array of roughly equal sized mounds of area $A = \lambda^2$ and height $H = m\lambda/2$, where m is the mound slope. During a time t, a number FAt of atoms is deposited onto a mound, with a statistical fluctuation of $\pm\sqrt{FAt}$. If there is no mass transport between neighboring mounds, this translates into a relative height fluctuation $\delta H = d(\Omega/A)\sqrt{FAt}$. Coarsening then occurs if, by chance, a mound overgrows a less fortunate neighbor. The condition for this to happen is that $\delta H \approx H$, which implies the relationship

$$\mathcal{R} = \frac{\sqrt{\Theta}}{m(\lambda/a)^2} \approx 1 \tag{4.81}$$

between the mound size, the mound slope and the film thickness.

The dimensionless parameter $\mathcal{R}$ is a measure of the importance of noise in the growth process. When $\mathcal{R} \approx 1$, the coarsening is noise-driven and the mound size evolves according to

$$\lambda/a \approx m^{-1/2}\Theta^{1/4} . \tag{4.82}$$

Under conditions of slope selection, $m = const.$, the coarsening exponent is $1/z = 1/4$, while in general the combination of (4.78,4.79,4.80,4.81) yields the exponent relation

$$\beta + \frac{1}{z} = \frac{1}{2} . \tag{4.83}$$

This expresses a competition between coarsening and roughening (or steepening) [406]: The larger β, the smaller $1/z$, with the limiting case $\beta = 1/2$, $1/z = 0$ corresponding to purely statistical growth.

If, on the other hand, the noise parameter $\mathcal{R} \ll 1$, the mound size is much larger than the value (4.82) expected from noise-driven coarsening, and the beam fluctuations are effectively irrelevant. Inspection of Table 4.2 shows that all experiments operate in this low noise regime. The strongest noise effects are expected for Fe(100) at 293 K, for which $\mathcal{R} = 0.11$.

4.7 Growth Instabilities of Vicinal Surfaces

The orientation of a vicinal surface is close to (in the *vicinity* of) a high symmetry direction of the crystal lattice. Such a surface therefore consists of several lattice spacings wide, low index terraces separated by steps, usually of monolayer height.

During growth on a vicinal surface, the attachment of freshly deposited adatoms to the preexisting steps competes with nucleation of islands on the terraces. Nucleation is expected to be negligible if the distance l between the preexisting steps is small compared to the island spacing l_D,

$$l \ll l_D \tag{4.84}$$

(see also Sect. 4.5.1). The surface then maintains its vicinal shape, and growth occurs through step propagation or *step flow*. In the following the conditions for step flow will be assumed to hold. The question about the long term stability of this growth mode is discussed in Sect. 4.7.4.

On a perfectly ordered vicinal surface, as it would appear in thermal equilibrium at low temperatures, the steps are straight and equally spaced. Correspondingly, the morphological instabilities of stepped surfaces are of two kinds: Either the individual steps develop a *meander*, beyond their thermal or kinetic roughness, or several steps form *step bunches*, regions of high step density separated by large terraces (Fig. 4.34).

The main topic of this section is a generic step meandering instability in homoepitaxial growth, which was first predicted theoretically by Bales and Zangwill [409]. It is caused by the asymmetry between ascending and descending steps which the step edge barrier introduces; steps become wavy whenever the flux reaching the step from the lower terrace is larger than the flux from the upper terrace[37]. Meandering instabilities which have been associated with the Bales–Zangwill mechanism were so far identified on surfaces vicinal to Pt(111) [410] and Cu(100) [411–413]. Step bunching during homoepitaxy has been observed on several semiconductor surfaces [414, 415], but a simple generic mechanism has not been suggested. It is likely that growth-induced step bunching requires additional features such as anisotropic surface diffusion [416], a precursor species [417], adsorbates [418–421] or highly mobile dimers [422]. A combination of step bunching and step meandering during

[37] Even in the absence of a step edge barrier, such an asymmetry occurs for purely geometric reasons when an island grows from a large capture area [299]. This observation links the Bales-Zangwill instability to the diffusional island shape instabilities described in Sect. 3.6.

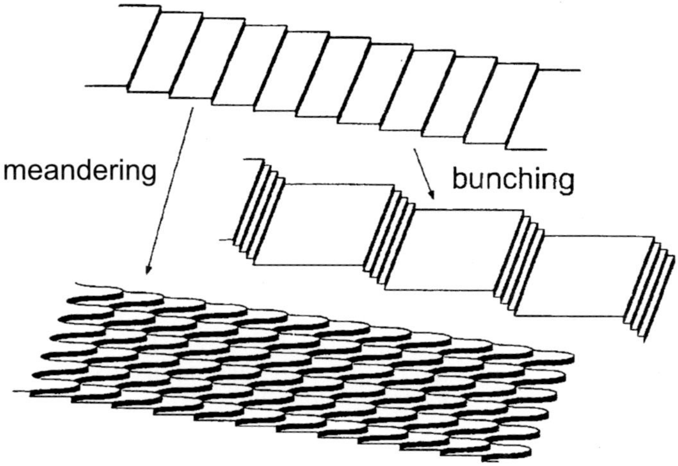

Fig. 4.34. Schematic of instability modes of a vicinal surface

growth has been observed, e.g., on Si(100) [414] and Cu(100) [423]. The two phenomena may be coupled through mass currents along the step edges [424], see Sect. 4.7.3 and Sect. 4.9.7.

In this section the stability of growing vicinal surfaces will be analyzed in terms of step motion. A complementary continuum viewpoint is sketched in Sect. 4.9.9.

4.7.1 Stability of a Step Train

From a theoretical perspective, the step flow growth mode is attractive because it can be described in terms of step motion without the need to treat island nucleation. The problem simplifies further if the steps are assumed to be straight. Then the advancement speed of the j-th step in a step train can be written as the sum of the contributions f_- and f_+ from the upper and lower terraces, each of which is a function of the corresponding terrace width (Fig. 4.35). Denoting the position of the j-th step by x_j, the evolution equations then read

$$\frac{\mathrm{d}x_j}{\mathrm{d}t} = f_+(x_{j+1} - x_j) + f_-(x_j - x_{j-1}) \; . \tag{4.85}$$

A train of equally spaced steps moving at speed $v_0 = f_+(l) + f_-(l)$ evidently satisfies these equations. To probe its stability, we consider a small perturbation of the form

$$x_j(t) = jl + v_0 t + \varepsilon_j(t) \tag{4.86}$$

and linearize (4.85) in the ε_j. The solutions of the linearized equations are of the form $\varepsilon_j(t) \sim \exp[i\phi j + \omega(\phi)t]$, where the growth rate ω is given in terms

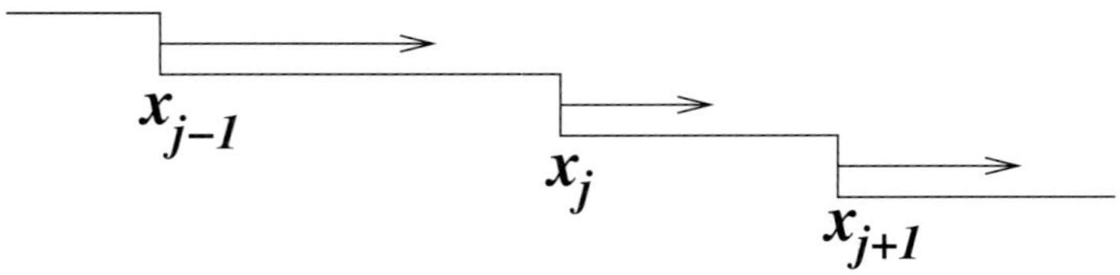

Fig. 4.35. Schematic of a growing step train with straight steps

of the phase shift ϕ by the expression

$$\omega(\phi) = -(1 - \cos\phi)(f'_+(l) - f'_-(l)) + i \sin\phi(f'_+(l) + f'_-(l)) \, . \tag{4.87}$$

Stability requires the real part of ω to be negative for all ϕ, which implies

$$\frac{\mathrm{d}}{\mathrm{d}l}(f_+(l) - f_-(l)) > 0 \, . \tag{4.88}$$

Since $f_\pm$ are increasing functions of l, this is roughly equivalent to the statement that a step train is stable if the steps are fed primarily from the lower terrace, in the sense that $f_+ > f_-$ [17]. This is easy to understand intuitively: Under this condition a step trailing a particularly wide terrace accelerates, and the uniform step spacing is restored. When (4.88) is violated the step train is unstable towards step bunching. The largest growth rate is then attained for $\phi = \pi$, hence step pairs form in the initial stage of the instability[38].

While the above analysis applies generally to growing or sublimating vicinal surfaces, we now specialize to a surface growing in the absence of evaporation. Then the functions $f_\pm$ are simply given by $f_+ = p_+ a^2 F l$, $f_- = (1 - p_+) a^2 F l$, with the attachment probability p_+ to the ascending step computed in Sect. 4.3.1. Using the expression (4.32) in (4.88) it is easy to check that a growing step train is stable whenever $\nu' < \nu$. A step bunching instability during growth would require a "negative" step edge barrier, in the sense of $\nu' > \nu$. Similar arguments show that normal step edge barriers generically do cause step bunching during sublimation [17,350].

The stabilization of the equidistant step train by the step edge barrier may be interpreted in terms of an effective, growth-induced step–step repulsion. This repulsion is very efficient, in the sense that the resulting terrace width fluctuations can be far smaller than in thermal equilibrium [377, 425].

4.7.2 Step Meandering: The Bales–Zangwill Instability

Bales and Zangwill made the remarkable observation that the very same mechanism that stabilizes a growing vicinal surface against step bunching

[38] This need no longer be true if long ranged step–step interactions are taken into account.

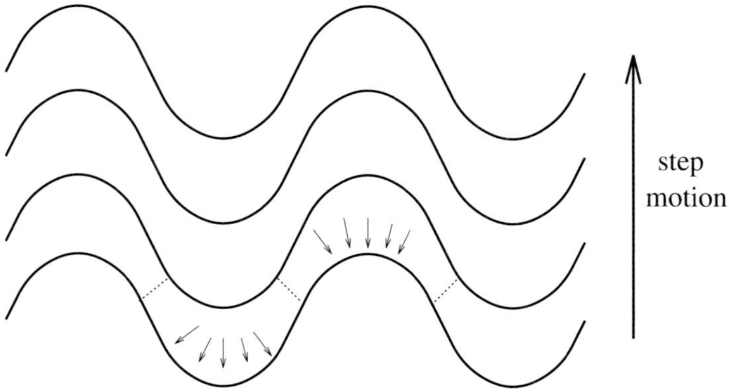

Fig. 4.36. Schematic of the Bales–Zangwill mechanism for step meandering. The dotted lines are drawn perpendicular to the lines of constant adatom density

also makes the steps susceptible to a meander instability [409]. Figure 4.36 illustrates the phenomenon on a qualitative level. To account for the mutual repulsion between the steps, it is assumed (and will be confirmed by the quantitative analysis) that they meander in phase. The terraces can then be subdivided into lots, as indicated by the dotted lines. Each lot receives the same number of atoms per unit time, which attach primarily to the corresponding segment of the ascending step. Because of the meander, the indented segments of the step are longer than the protruding ones. Since both capture the same flux, the protrusions propagate faster and the deformation is amplified.

For the quantitative stability analysis, we use the coordinate system shown in Fig. 4.37. The position of the j-th step is described by a function $\zeta_j(y,t)$. Between the steps the stationary diffusion equation (4.24) is to be solved subject to the boundary conditions (4.28,4.29) at $x = \zeta_j(y,t)$. The thermodynamic cost of step deformations enters the boundary conditions through the expression

$$n^{\mathrm{eq}} = n_0^{\mathrm{eq}} \left(1 + \frac{\Omega \tilde{\delta} \kappa_{\mathrm{st}}}{k_{\mathrm{B}} T} \right) , \tag{4.89}$$

where $\tilde{\delta}$ is the step stiffness introduced in Sect. 3.2.1, n_0^{eq} is the equilibrium adatom density at a straight step, and κ_{st} is the step curvature. Equation (4.89) follows from the two-dimensional Gibbs–Thomson relation (3.9) for small step curvatures.

Once the boundary value problem defined by (4.24,4.28,4.29,4.89) has been solved for a given configuration of steps, the local normal velocity $v_n^{(j)}$ of each step can be computed from the total mass flux reaching the step from the two adjoining terraces, as well as through diffusion *along* the step. This yields

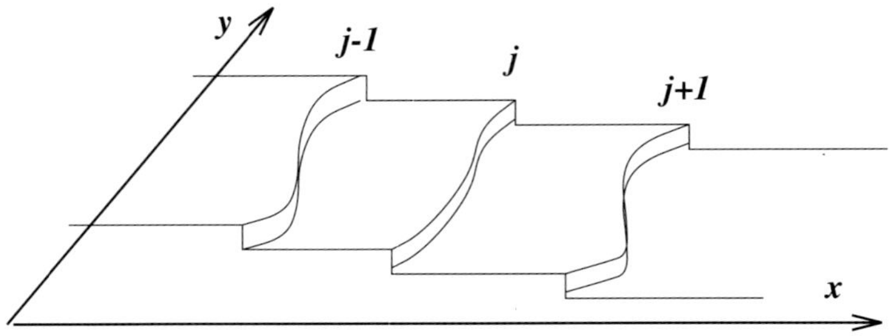

Fig. 4.37. Geometry of the vicinal surface used in the stability analysis

$$v_n^{(j)} = \Omega D \left[\mathbf{n} \cdot \nabla n|_+ - \mathbf{n} \cdot \nabla n|_- \right] + \frac{\partial}{\partial s} \sigma_{\mathrm{st}} \frac{\partial}{\partial s} \Omega \tilde{\delta} \kappa_{\mathrm{st}}^{(j)} \,, \tag{4.90}$$

where σ_{st} is the mobility for migration along a curved (that is, kinked) step introduced in Sect. 3.7.1, and s denotes the arc length of the step.

The calculation now proceeds, in principle, as in Sect. 4.7.1. The general form of the perturbed step train is

$$\zeta_j(y,t) = jl + v_0 t + \varepsilon_j(y,t) \,. \tag{4.91}$$

To linear order in the ε_j, the solution of the coupled equations can be decomposed into normal modes of the form $\varepsilon_j(y,t) \sim \exp[\mathrm{i}\phi j + \mathrm{i}qy + \omega(\phi,q)t]$. Here q denotes the wavenumber of the step deformation, corresponding to a meander wavelength $2\pi/q$. The real part of the growth rate $\omega(\phi,q)$ turns out to be maximal for the in-phase mode $\phi = 0$ [426]. This is a consequence of the kinetically induced step repulsion described in Sect. 4.7.1: The in-phase mode is a compromise which allows the deformed steps to keep the terrace width as uniform as possible.

Since the incipient morphology will be dominated by the fastest growing mode, we may assume $\phi = 0$ in the following. Moreover, for the present purposes it is sufficient to consider long wavelength deformations, with a meander wavelength large compared to the mean step spacing l. In this limit the expression for the growth rate reads [427]

$$\omega(0,q) = \frac{\Omega F l^2 f_{\mathrm{ES}}}{2} q^2 - (D\Omega n_0^{\mathrm{eq}} l / k_{\mathrm{B}} T + \sigma_{\mathrm{st}}) \Omega \tilde{\delta} q^4 \,. \tag{4.92}$$

It is worthwhile to pause to appreciate how (4.92) reflects the essential physics of the meandering instability. The positive term proportional to q^2 describes the destabilization of the straight step by the attachment asymmetry. The strength of the destabilization is proportional to the flux F, and to the factor $f_{\mathrm{ES}} = l_{\mathrm{ES}}/(4l + l_{\mathrm{ES}} + 2a)$, which is a dimensionless measure of the strength of the step edge barrier (compare to Sect. 4.3.1). The negative term proportional to q^4 describes the thermal relaxation of the step towards the (straight) equilibrium shape. The smoothening is driven by the step stiffness $\tilde{\delta}$, and it operates through two kinetic channels [302]: Detachment–reattachment processes over the terrace, with a rate proportional to the terrace diffusion coefficient and the terrace width, and step edge diffusion with a rate proportional to σ_{st}.

On the level of the linear stability analysis, the behavior of the step is similar to that of the one-dimensional surface profiles analyzed in Sect. 4.5.2. For sufficiently small q the quadratic term in (4.92) wins over the quartic term, and hence the step is subject to a long wavelength instability for arbitrarily small flux[39]. The range of unstable wavelengths is bounded below by $2\pi/q_c$, where q_c is the wavenumber at which the two terms on the right hand side of (4.92) balance. As in Sect. 4.5.2, the dominant meander wavelength λ_{BZ} corresponds to the maximum of (4.92), which yields $\lambda_{\mathrm{BZ}} = 2\pi\sqrt{2}/q_c$. Explicitly,

$$\lambda_{\mathrm{BZ}} = 4\pi \sqrt{\frac{(D\Omega n_{\mathrm{eq}}^{(0)} l/k_{\mathrm{B}}T + \sigma_{\mathrm{st}})\tilde{\delta}}{Fl^2 f_{\mathrm{ES}}}} \,. \tag{4.93}$$

For later reference we note that $\lambda_{\mathrm{BZ}} \sim F^{-1/2}$.

4.7.3 Step Meandering: Experiments and Alternative Mechanisms

The first quantitative experimental test of the prediction (4.93) was carried out by Ernst and collaborators for surfaces vicinal to Cu(100) [412,413]. They investigated the growth of Cu on the Cu(1,1,17) and Cu(0,2,24) surfaces, which consist of Cu(100) terraces separated by steps aligned along the close packed $\langle 110 \rangle$ and the open $\langle 100 \rangle$ directions, respectively[40]. The step–step distance was $l = 21.7 \pm 0.03$ nm in both cases. On both surfaces a clear, in-phase meander pattern was seen to develop after the deposition of a few monolayers (Fig. 4.38). The wavelength $\lambda_\perp$ of the pattern was studied as a function of temperature and flux. The temperature dependence was found to follow an Arrhenius law, $\lambda_\perp \sim \exp[-E_\lambda/k_{\mathrm{B}}T]$, with activation energies $E_\lambda = 0.092$ eV [Cu(1,1,17)] and $E_\lambda = 0.132$ eV [Cu(0,2,24)]. The ratio $\lambda_\perp/l$ of meander wavelength to step spacing varied in the range $2 \leq \lambda_\perp/l \leq 19$, hence the condition $\lambda_\perp \gg l$ assumed in the derivation of (4.93) is reasonably well satisfied[41].

To compare the experimentally determined activation energies with the theoretical prediction, we need to identify those contributions to (4.93) that show activated behavior. We focus here on the close packed $\langle 110 \rangle$ steps, for which fairly detailed information on the energetics and kinetics is available

[39] In the presence of desorption, which is the case originally considered by Bales and Zangwill [409], the instability sets in only above a critical flux.

[40] In [412], the Cu(0,2,24) surface was erroneously identified as Cu(1,1,17). The experimental results described below are taken from [413].

[41] This condition can be relaxed without affecting the conclusions that follow.

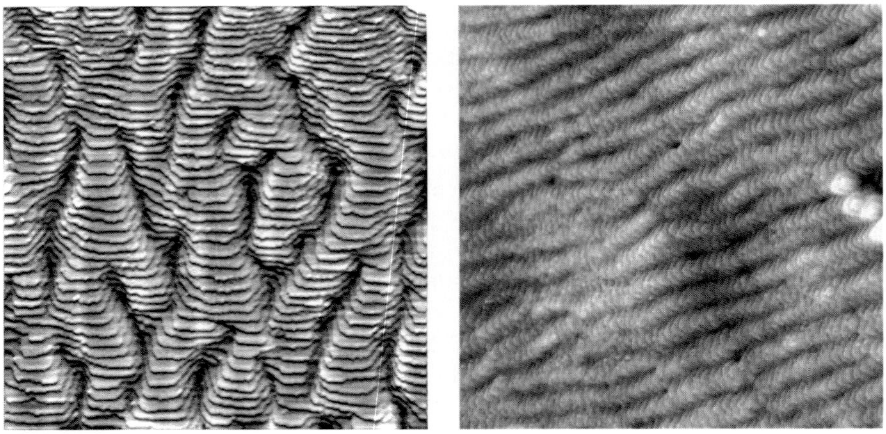

Fig. 4.38. Left panel: Step meandering on the Cu(1,1,17) surface after deposition of 18 ML of Cu at 280 K with flux $F = 5 \times 10^{-3}$ ML/s. The image size is 1000 Å × 1000 Å. Right panel: Step meandering on the Cu(0,2,24) surface after deposition of 20 ML at 250 K with flux $F = 3 \times 10^{-3}$ ML/s; image size is 1300 Å × 1300 Å (left panel is courtesy of H.J. Ernst, right panel is Fig. 6 of [413])

[172]. According to (3.22), the step stiffness $\tilde{\delta}$ is inversely proportional to the step diffusivity (3.21), hence $\tilde{\delta} \sim \exp[\epsilon/k_B T]$ with a kink energy $\epsilon \approx 0.13$ eV for the $\langle 110 \rangle$ step. In the temperature range of interest, 276 K $\leq T \leq$ 385 K, step smoothening occurs primarily by step edge diffusion. The activation energy E_{st} governing diffusion along a kinked step can be obtained from the observation of temporal step fluctuations in thermal equilibrium [314], which yields[42] $E_{st} \approx 0.91$ eV. Provided the step edge barrier is sufficiently strong, such that $f_{ES} \approx 1$ independent of temperature, we thus arrive at an activation energy of $E_{BZ} = (E_{st} - \epsilon)/2 \approx 0.39$ eV, which is much larger than the experimental value. Assuming a weak step edge barrier makes matters worse, because then the additional barrier ΔE_S is added to the activation energy. The discrepancy between theory and experiment is further emphasized by considering the flux dependence of the meander wavelength [413]. Over the range 2.7×10^{-3} ML/s $\leq F \leq 1.1 \times 10^{-2}$ ML/s, a power law dependence $\lambda_\perp \sim F^{-0.21 \pm 0.08}$ was found, which is inconsistent with the predicted behavior $\lambda_\perp \sim F^{-1/2}$.

One has to conclude, then, that the Bales–Zangwill scenario is *not* operative in the Cu(1,1,17) system. An important clue to a possible alternative meandering mechanism is provided by the observation that the experimentally determined wavelength, in its dependence on flux as well as on temperature,

[42] Recall from Sect. 3.7.1 that the activation energy of the prefactor of the step fluctuation variance (3.40) is $(3\epsilon + E_{st})/4$. For this energy the value 0.324 ± 0.008 eV was obtained in [314]. Note that the resulting estimate of E_{st} differs from that used in [412].

approximately matches the expected spacing $l_{1\mathrm{d}}$ between *one-dimensional* nuclei forming on the close packed step during the very early stages of growth [412,413]. This length scale can be estimated from one-dimensional nucleation theory (see Sect. 2.4.3). In order of magnitude, $l_{1\mathrm{d}} \sim (aD_\mathrm{e}/F_{1\mathrm{d}})^{1/4}$, where D_e is the coefficient of one-dimensional diffusion along a *straight* (rather than kinked) step edge, and $F_{1\mathrm{d}} = Fl$ is the effective flux impinging onto a unit length of the step edge [see (3.27)]. A more precise calculation yields [372]

$$l_{1\mathrm{d}} \approx \left(\frac{12aD_\mathrm{e}}{Fl} \right)^{1/4} . \qquad (4.94)$$

For the activation energy of D_e the estimate $E_\mathrm{e} = 0.45 \pm 0.08$ eV has been obtained from step fluctuation measurements [428]. Equation (4.94) then implies an effective activation energy of $E_\lambda = E_\mathrm{e}/4 \approx 0.11$ eV for the meander wavelength, which is close to the experimental value. Also the flux dependence $\lambda_\perp \sim F^{-1/4}$ of (4.94) is consistent with the experiment on Cu(1,1,17).

There are two conceivable scenarios for the way in which the one-dimensional nucleation length (4.94) might come into play here. First, it could simply set the dominant lengthscale for the initial step fluctuations, which are then acted upon and amplified by the Bales–Zangwill instability. Under the conditions of the experiment, $l_{1\mathrm{d}}$ is considerably larger than λ_BZ. According to (4.92), fluctuations at wavelength $l_{1\mathrm{d}}$ are then indeed amplified, though at a much lower rate than those at wavelength λ_BZ. Studies of a nonlinear extension of the Bales–Zangwill theory indicate, however, that initial wavelengths larger than about $3\lambda_\mathrm{c} \approx 2\lambda_\mathrm{BZ}$ are *not* preserved during the evolution of the meander; rather, as the meander amplitude grows, the system reverts to a wavelength near λ_BZ [429].

The second possibility is that not only the initial wavelength, but also the instability mechanism differs from that envisioned by Bales and Zangwill. A plausible candidate for an alternative mechanism is the one-dimensional analogue of the mounding instability described in Sects. 4.2 and 4.5: If the step is viewed as a one-dimensional surface, then the existence of additional energy barriers suppressing the diffusion around kinks – a one-dimensional *kink Ehrlich–Schwoebel effect* (KESE) – should be sufficient to induce a morphological step instability [424,430,431]. The theory of the mounding instability developed above carries over directly to the one-dimensional situation. In particular, if the one-dimensional analog of the Ehrlich–Schwoebel length (4.33) is large compared to the nucleation length $l_{1\mathrm{d}}$, then the meander wavelength will be set by $l_{1\mathrm{d}}$ (compare to Sect. 4.2).

The KESE scenario has received support from KMC simulations of a model of the Cu(100) surface based on EMT energy barriers[43] [433]. Within EMT, a kink rounding barrier of the same magnitude as E_e is found; a similar

[43] In contrast, the experimentally observed instability of the ⟨100⟩ steps is not reproduced in these simulations. The origin of the meander instability at the ⟨100⟩ steps is unclear at present, see [432] for further discussion.

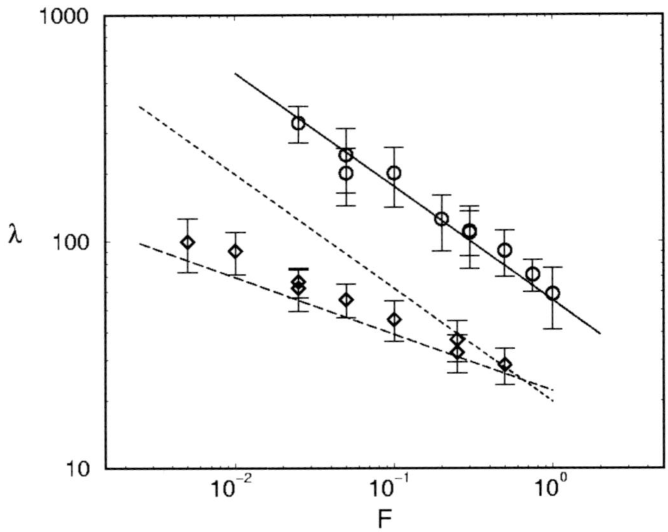

Fig. 4.39. Flux dependence of the meander wavelength from KMC simulations of a simple cubic solid–on–solid (SOS) model. Circles show data for model I (no suppression of detachment relative to edge diffusion) and diamonds show data for model II (with suppressed detachment). The full line is the prediction (4.93) of the Bales–Zangwill theory, and the long–dashed line represents the one-dimensional nucleation length (4.94). All thermodynamic and kinetic quantities entering (4.93) and (4.94) can be calculated analytically for the SOS model; there are no free parameters. The short–dashed line shows the Bales–Zangwill wavelength (4.93) evaluated for model II. For low fluxes the ratio between the Bales–Zangwill prediction and the wavelength observed in the simulations amounts to a factor which is larger than $\sqrt{2}$. Since wavelengths smaller than $\lambda_{\mathrm{BZ}}/\sqrt{2}$ are not amplified according to (4.92), this definitely rules out the relevance of the Bales–Zangwill mechanism for model II (Fig.4 of [432])

estimate can be derived from a comparison with step fluctuation measurements [316, 432]. This places the system in the strong KESE regime, and indeed the generated step configurations resemble closely the morphology of the one-dimensional wedding cake model described in Sect. 4.2. An important difference between the KESE and BZ scenarios is that in the former case the instability does not involve any mass transport across the terraces. Consequently the meander develops independently at different steps, and, in contrast to the BZ stability analysis, there are no phase correlations between the steps in the initial stage of the instability. This feature is clearly seen in the simulations of [433].

The foregoing discussion leaves open whether the dominance of the KESE over the BZ mechanism is generic, or whether the BZ instability in its original form may be observable in other systems. A partial answer is provided by KMC simulations of a simple cubic lattice model, in which both types of instabilities are found for different choices of parameters [432]. In one case (*model I*) the edge diffusion barrier E_e is set equal to the barrier for the detachment of adatoms from a straight edge, while in the second case (*model II*) detachment is suppressed through an additional energy barrier. Since kink rounding can occur only by detachment from the step, model II also displays a KESE. In model I, edge diffusion, along with its possible destabilizing effects, is essentially absent. The simulations show *quantitative* agreement with the BZ prediction (4.93) for model I, and with the one-dimensional nucleation length (4.94) for model II (Fig. 4.39). This suggests that the BZ instability could be realized under conditions where edge diffusion and detachment occur at comparable rates.

4.7.4 The Ultimate Fate of Step Flow

Under the simple step flow condition (4.84), the probability for the nucleation of an island on a terrace is small. Because of the stochastic nature of nucleation, this does not imply that nucleation events do not occur, only that they are rare. Theoretical analysis and KMC simulations show that islands continue to appear on the terraces even when (4.84) is satisfied [380]. The reduction of the adatom density due to the capture at the steps leads to a new scaling regime for the island density, in which the exponent in the scaling law (2.19) takes the value $\chi = i^*$.

The question then arises whether rare nucleation events on the terraces may trigger a global destabilization of a growing vicinal surface. The appearance of single layer islands on the terraces does not necessarily endanger the step flow growth mode, because these islands are quickly swept up by the advancing step edge. This healing process is no longer effective if second layer islands nucleate before the first layer islands have been engulfed by the step. A more stringent criterion for the integrity of step flow growth is therefore that the step spacing should be small compared to the critical island radius R_c for second layer nucleation. In the absence of step edge barriers $R_c \sim l_D$ and the criterion reduces to (4.84). For the case of strong step edge barriers, R_c can be estimated along the lines of Sect. 4.3.3. The island density N in (4.51) has to be replaced by l^{-2}, since l sets the distance to the step edges with which the growing island competes (other islands can be assumed to be far away). Thus for $i^* = 1$ we have $R_c \sim (a\Omega\nu' l^2/F)^{1/7}$, and the condition that $l \ll R_c$ becomes

$$l \ll (a\Omega\nu'/F)^{1/5} . \qquad (4.95)$$

The length scale on the right hand side is of the same order of magnitude as the size of the top terrace of a mound, see (4.54).

A detailed study of the nucleation-induced breakdown of step flow was carried out for a one-dimensional model with $\nu' = 0$ (the model of Sect. 4.2 applied to a vicinal surface) [377]. In this case the condition (4.95) is always violated. The problem can be analyzed using the approach of Sect. 4.3.2. The interarrival time between deposition events on a terrace of length l is $\Delta t =$

$1/Fl$. Since atoms are easily adsorbed at the ascending step, the residence time τ is of the order of the diffusion time $\tau_D \sim l^2/D$, and the adatom density is of the order of $\bar{n} = F\tau \sim Fl^2/D$. The condition (4.84) ensures that $\tau \ll \Delta t$ always, and the mean number of adatoms on a terrace is small compared to unity. The rate of nucleation on a terrace is then

$$\omega = \tau/(\Delta t)^2 \sim F^2 l^4/D \, . \tag{4.96}$$

For $\nu' = 0$, the nucleus of a multilayer mound forms whenever an atom is deposited on top of a previously deposited adatom, which has not yet joined a step. The probability for this to happen is $\bar{n}$, and consequently the rate of mound formation per terrace is

$$\omega_{\mathrm{m}} \sim \bar{n}/\Delta t \sim F^2 l^3/D \, , \tag{4.97}$$

which is smaller than the nucleation rate (4.96) by a factor of l. Mounds appear randomly on the surface and their concentration grows initially as $\omega_{\mathrm{m}} t$. The transition to the mound morphology is completed when the spacing between mounds has reached the (one-dimensional) island distance l_D, after a time of the order of $\tau_{\mathrm{m}} \sim (l/l_D)\omega_{\mathrm{m}}^{-1} \sim (D/F)^{3/4} l^{-2}$. Apart from the asymmetry induced by the overall vicinality, the morphology for times $t > \tau_{\mathrm{m}}$ is indistinguishable from the wedding cakes in Fig. 4.5. During the transition period the surface is well described as a random mixture of mounds and regions growing in the step flow mode, somewhat similar to the transitory mound morphology described in Sect. 4.4.4.

For two-dimensional surfaces the meandering instabilities described above provide an alternative pathway to the destabilization of step flow. As the meander amplitude increases, it becomes possible for different parts of a deformed step to touch and to form closed loops. This is most likely to happen in the fjords of the meander, where vacancy islands form which act as nucleation centers for pits and prevent the further propagation of the succeeding steps. This scenario has been observed in KMC simulations of growth on vicinal surfaces in the presence of step edge barriers [410, 434]. Figure 4.40 shows the surface morphology before and after the global destabilization. At 80 ML, the surface displays the ripple structure of a well-developed meander instability, while at 320 ML an essentially isotropic mound morphology is seen. On closer inspection, isolated pits can be detected already at 80 ML, which presumably trigger the destabilization. A similar transition has been found in the numerical solution of two-dimensional continuum equations for vicinal surfaces [410] (see also Sect. 4.9.9). A possible experimental realization has been reported in GaAs homoepitaxy [415].

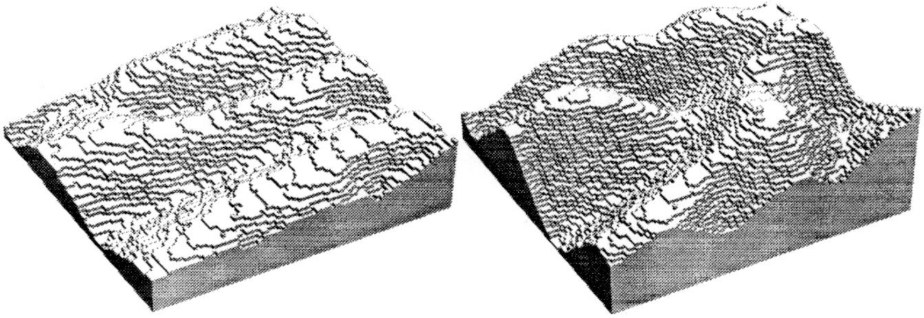

Fig. 4.40. Breakdown of step flow in a KMC simulation of growth on a vicinal surface. The transition from the meander morphology in (**a**) to the mound morphology in (**b**) is triggered by the formation of vacancy islands in the fjords of the meander. An example of such a vacancy island is seen in the upper part of the foremost valley in (**a**) (Fig. 1 of [415])

4.8 Self–Affine Growth

In recent years, the concepts of fractal geometry [435] have been used extensively to characterize the disorderly structures that form in growth far from equilibrium [32, 36, 106]. For our purposes, fractals can be defined as objects that show *statistical scale invariance*, which means that they look the same (in a statistical sense) under arbitrary levels of magnification[44]. The notion of self-affinity arises naturally in the description of randomly rough surfaces. A self-affine surface is statistically invariant under a scale transformation which involves different scale factors along the surface and perpendicular to it. If the surface fluctuations (relative to the mean height $\bar{h}$) are described by a one-dimensional function $h(x)$, as in the example of Fig. 4.41, the affine scale transformation takes the form

$$h(x) \quad \Rightarrow \quad b^{-\alpha} h(bx) \tag{4.98}$$

where b is the scale factor, and α is the *roughness exponent* characterizing the surface[45]. In words, statistical invariance under (4.98) has the following meaning: If the lateral coordinate x is rescaled by an arbitrary factor b, then rescaling the height coordinate h by a factor b^α produces a profile with the same statistical properties as the original surface. For a self-similar surface

[44] In practice, the range of possible magnifications is of course limited by the atomic scale on the one side, and large scale inhomogeneities (the edges of the sample, grain boundaries etc.) on the other. Actual experimental evidence for scale invariance is usually restricted to 1 – 2 decades of spatial magnification [436, 437].

[45] The roughness exponent is sometimes called the Hurst exponent, and denoted by H, h, ζ or χ.

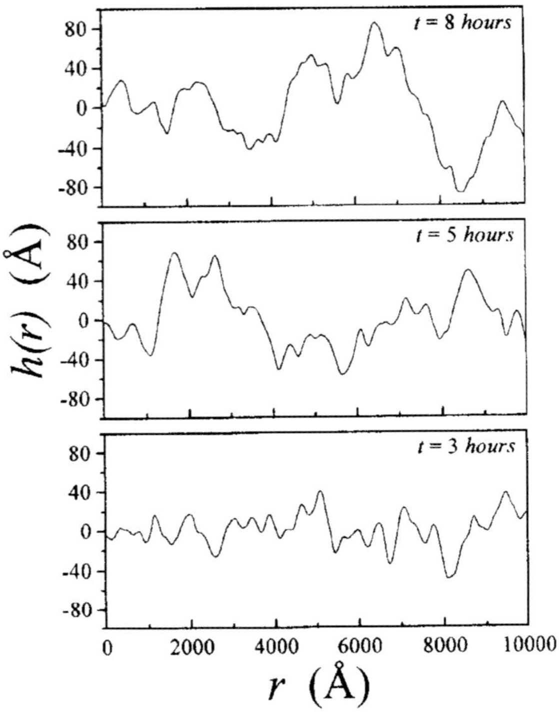

Fig. 4.41. AFM profiles of amorphous silicon films grown by thermal evaporation around room temperature, at a deposition rate of $0.8 \pm 0.2\,\text{Å}\,\text{s}^{-1}$. The deposition times are indicated in the panels. The figure illustrates the buildup of random roughness with increasing film thickness. The analysis of the height difference correlation function (4.99) for these profiles yields the estimate $\alpha = 0.96 \pm 0.02$ for the roughness exponent (Fig. 2 of [438])

the two scale factors are the same, i.e. $\alpha = 1$, while self-affine surfaces have[46] $\alpha < 1$.

Invariance under (4.98) implies that the height–difference correlation function

$$\tilde{G}(r) = \langle (h(x+r) - h(x))^2 \rangle = 2(G(0) - G(r)) \tag{4.99}$$

satisfies the homogeneity relation $\tilde{G}(r) = b^{-2\alpha}\tilde{G}(br)$. Setting $b = 1/|r|$ it follows that

$$\tilde{G}(r) \sim |r|^{2\alpha}\,. \tag{4.100}$$

The power law relationship (4.100) is the hallmark of self-affine scaling, and it is most often used to determine the roughness exponent α experimentally.

[46] For a surface with $\alpha > 1$ the fluctuations in the slope $\partial h/\partial x$ would grow without bound with increasing magnification, so that a description in terms of a single-valued height function is no longer meaningful [106].

For growth processes it is natural to extend the notion of scale invariance also to the temporal dimension. The transformation (4.98) then generalizes into

$$h(x,t) \;\Rightarrow\; b^{-\alpha}h(bx, b^z t)\,, \tag{4.101}$$

which defines the *dynamic exponent* z. Here t denotes the deposition time, which is usually taken to be proportional to the film thickness, and it is assumed that the surface is flat ($h = 0$) at $t = 0$. If invariance under (4.101) holds, then the height difference correlation function takes on the scaling form

$$\tilde{G}(r,t) = |r|^{2\alpha}\tilde{\mathcal{G}}(|r|/t^{1/z})\,. \tag{4.102}$$

The scaling function $\tilde{\mathcal{G}}$ approaches a constant for small arguments, so that (4.102) reduces to (4.100) for long times, and vanishes as $\tilde{\mathcal{G}}(x) \sim x^{-2\alpha}$ for $x \to \infty$. The latter limit has the following interpretation: At a finite deposition time t, the height fluctuations at sufficiently distant points must be uncorrelated; hence $G(r,t) \approx 0$ for $r \gg t^{1/z}$, and $\tilde{G} \approx 2d^2W^2$ becomes independent of r. It follows that the surface width increases with time according to

$$W(t) \sim t^{\alpha/z}\,, \tag{4.103}$$

which can be used, together with (4.100), to determine z. At time t the correlations extend up to a distance of order $t^{1/z}$, which defines the correlation length

$$\xi(t) \sim t^{1/z} \tag{4.104}$$

of the roughening process. An example for the experimental observation of the power laws (4.103) and (4.104) is shown in Fig. 4.42.

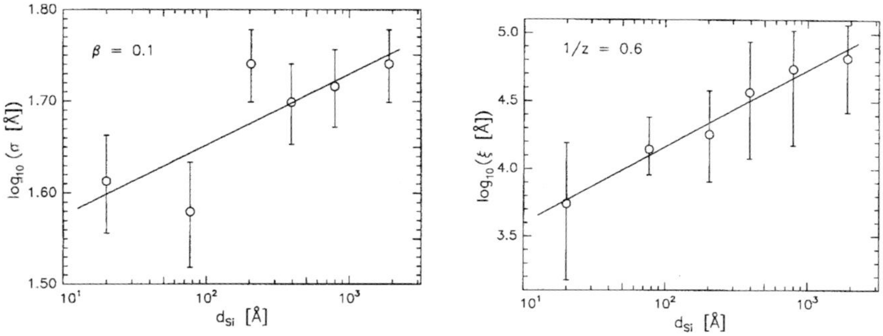

Fig. 4.42. Surface width (left panel) and correlation length (right panel) during the deposition of amorphous silicon films, plotted as a function of film thickness. The measurements were performed using specular and diffuse x–ray scattering in the region of total external reflection. Based on these data, the self-affine scaling exponents were estimated to be $\alpha/z = 0.1 \pm 0.05$ and $1/z = 0.6 \pm 0.2$ (Figs. 5 and 6 of [439])

4.8.1 Universality Classes of Kinetic Roughening

The theoretical modeling of self-affine growth is based on Langevin equations for the height fluctuations $h(\boldsymbol{r}, t)$ of the growing film. The approach is close in spirit to the continuum theory of mound formation that is employed in Sects. 4.5 and 4.9.8, with two important differences: First, the Langevin equations contain random noise terms describing the fluctuations in the deposition beam; second, the deterministic (non-random) terms in the Langevin equations are relaxational, i.e. they are chosen such that the surface would become flat in the absence of the deposition noise. In contrast to mound formation, which has the character of a morphological instability, self-affine roughness arises from the competition between random flucutuations and various smoothening processes. To distinguish it from mounding, the term *kinetic roughening* will be reserved for this type of growth-induced roughness.

Langevin equations for kinetic roughening are of the generic form

$$\frac{\partial h}{\partial t} = \mathcal{K}(\nabla h, \nabla^2 h, ...) + \delta F(\boldsymbol{r}, t) . \tag{4.105}$$

Here $\mathcal{K}$ is a (polynomial) function of the derivatives of h, which encodes the relaxational processes mentioned above, and $\delta F(\boldsymbol{r}, t)$ is a random function of space and time describing the fluctuations in the deposition flux[47]. The terms appearing in $\mathcal{K}$ can be obtained from a small-gradient expansion of a macroscopic surface evolution equation [106], or they can be written down on the basis of symmetry arguments [32, 440]. The notion of *universality* allows one to include only a few leading terms in such an expansion. This is possible because the Langevin equation (4.105) is constructed to describe only the asymptotic, large scale and long time fluctuations of the surface. Through a technique known as power counting (which is essentially a form of dimensional analysis), different terms in the equation can be ordered according to their relevance in the limit of large scales and long times[48]. When the most relevant terms are contained in the equation, the addition of further, less relevant terms does not change asymptotic properties, such as the values of the scaling exponents α and z.

A Langevin equation containing the most relevant terms appropriate to describe a particular growth situation defines a *universality class* of kinetic roughening. The most important universality classes are summarized in Table 4.3. They are arranged in order of decreasing relevance, in the following sense: If the terms appearing on the right hand sides of two of the equations are combined, then the asymptotic properties of the resulting equation are

[47] Technically speaking, δF is taken to be a Gaussian stochastic process which can be fully characterized by its mean and covariance function. The latter is assumed to be short ranged in space and time.

[48] A more rigorous analysis of the relevance of different terms requires renormalization group techniques, see [32, 165].

Table 4.3. Universality classes of kinetic roughening. The abbreviations in the first column refer to the Kardar–Parisi–Zhang (KPZ) [440], the Edwards–Wilkinson [441], the conserved KPZ [379,442,443] and the (noisy) Mullins [442,444] equations. In the second column the coefficients K_2 and K_4 are positive, while λ_2 and λ_4 may take both signs. In the third column d_s denotes the dimensionality of the surface. The scaling exponents for the linear EW and Mullins equations are derived e.g. in [32, 106]. The expressions given for the CKPZ equation are the leading terms in an expansion in $4 - d_s$ [445]. The scaling exponents for the KPZ equation are for the physical surface dimensionality $d_s = 2$, and are taken from recent numerical simulations [446] and an analytic conjecture [447].

Class	Langevin equation	α	z
KPZ	$\partial h/\partial t = K_2\nabla^2 h + \lambda_2(\nabla h)^2 + \delta F$	≈ 0.40	≈ 1.60
EW	$\partial h/\partial t = K_2\nabla^2 h + \delta F$	$(2 - d_s)/2$	2
CKPZ	$\partial h/\partial t = -\nabla^2[K_4\nabla^2 h + \lambda_4(\nabla h)^2] + \delta F$	$\approx (4 - d_s)/3$	$\approx (8 + d_s)/3$
Mullins	$\partial h/\partial t = -K_4(\nabla^2)^2 h + \delta F$	$(4 - d_s)/2$	4

those of the equation with a higher position in the table. This ordering coincides with increasing values of the dynamic exponent z [448]. According to (4.104) a smaller value of z implies a faster spreading of correlations, and it is plausible that at long times the fastest growing correlation length dominates the behavior.

4.8.2 Conserved and Nonconserved Growth

A central distinction among the universality classes in Table 4.3 is between *conserved* and *non-conserved* growth [379]. In conserved growth, the non-random processes included on the right hand side of (4.105) conserve the volume of the growing film, and hence $\mathcal{K}$ has to be of the form

$$\mathcal{K} = -\nabla \cdot \boldsymbol{J}(\nabla h, \nabla^2 h, ...) \,. \tag{4.106}$$

The EW, CKPZ and Mullins equations are conserved growth equations. A general consequence of (4.106) is the relation

$$z = d_s + 2\alpha \tag{4.107}$$

between the scaling exponents α and z, and the surface dimensionality d_s [442, 449, 450] (see Sect. 5.6 for a derivation). Conserved growth equations are appropriate to describe epitaxial crystal growth without desorption and vacancy formation (*ideal MBE* [449]), because in that case the volume of the growing film is determined only by the number of deposited atoms, which is independent of anything that happens on the surface.

Since the Laplacian $\nabla^2 h$ is the lowest order derivative term compatible with the form (4.106), the Edwards–Wilkinson (EW) equation defines the most robust conserved universality class [378]. EW growth results when the coefficient K_2 of the Laplacian is positive, whereas a negative coefficient implies mound formation (see Sect. 4.5). The other two conserved universality classes (CKPZ and Mullins) apply if, for reasons of symmetry or because certain kinetic processes are suppressed, K_2 is zero or very small. For the physical dimension $d_s = 2$ the EW roughness exponent is $\alpha = 0$, which corresponds to logarithmic roughness in space and time, i.e.

$$\tilde{G}(r) \sim \log(|r|), \quad W(t) \sim \sqrt{\log(t)}\,. \tag{4.108}$$

Fairly clearcut experimental evidence for EW behavior has been found in deposition by energetic cluster impact [451]. In this case the coefficient $K_2 > 0$ originates from a downhill mass current, which is associated with the asymmetric spreading of clusters impacting onto an inclined part of the surface.

The Kardar–Parisi–Zhang (KPZ) equation provides the generic large scale description of non-conserved self-affine growth [270,440,452]. The distinguishing feature of this equation is the nonlinear term $\lambda_2(\nabla h)^2$ on the right hand side, which cannot be brought into the form (4.106). The presence of this term enforces the identity [453]

$$\alpha + z = 2 \tag{4.109}$$

relating the two scaling exponents. The physical meaning of the nonlinearity is that the local growth rate of the film depends on the local surface slope. There are two kinds of deviations from the ideal MBE scenario which can induce such a slope dependence in homoepitaxial crystal growth [106]. The first is desorption, which becomes important at very high temperatures. As discussed further in Sect. 4.9.8, the desorption rate depends on the local step density, hence on the local slope, which translates into a corresponding slope dependence of the net growth rate[49].

The second mechanism sets in at very low temperatures, where vacancy formation begins to significantly reduce the film density below that of a perfect crystal. The growth rate V of the film (the rate at which the film thickness increases) is generally inversely proportional to the film density [455],

$$V = F/\rho\,, \tag{4.110}$$

where ρ denotes the number of film atoms per volume; for a perfect crystal $\rho = 1/(d\Omega)$. As bulk vacancies originate from overhangs, which in turn are generated e.g. at the sides of nanoprotrusions through restricted downward funneling [151, 153] (see Sect. 2.4.1), the concentration of vacancies depends

[49] A related scenario, in which material is continually exchanged between the surface and a fluidlike metastable precursor phase, has been suggested in the context of GaAs homoepitaxy. In that system experimental evidence for the presence of the KPZ nonlinearity has been derived from the evolution of the height profile of initially pitted surfaces [454].

on the local surface structure, in particular on the slope of the surface. Direct experimental evidence for this effect is available in the context of amorphous film growth, where the film density is known to depend on the slope of the surface relative to the direction of the incoming beam [456]; oblique incidence generally increases the defect concentration, hence ρ is a decreasing function of the magnitude of the surface slope. Through (4.110) this implies an increase of the local growth rate with increasing slope, and therefore a nonzero KPZ nonlinearity with a positive coefficient $\lambda_2 > 0$.

Asymptotic KPZ scaling has been established numerically for the *ballistic deposition* model, a caricature of low temperature deposition in which atoms approach the surface along linear (usually vertical) trajectories and stick irreversibly at the point of first contact with the film [36]. The gradual onset of vacancy formation in low temperature epitaxy, and the concomitant appearance of KPZ behavior, has been addressed in several simulation studies [151, 457, 458], but unambiguous identification of the asymptotic self-affine scaling exponents has been possible only for two-dimensional deposits with one-dimensional surfaces (see Fig. 4.43 for an example of such a simulation). It is difficult to reach the true long-time, large-scale regime in simulations of three-dimensional films because the coefficient λ_2 of the KPZ nonlinearity (as determined e.g. for the restricted downward funneling model at zero temperature [151]) is very small; due to the appearance of a nonequilibrium phase transition to smooth growth above three dimensions, the length and time scales beyond which asymptotic behavior set in are exponentially large in $1/\lambda_2$ [459]. This may well render the observation of KPZ scaling in low temperature epitaxial growth an elusive goal for the foreseeable future.

Experimental evidence for vacancy formation in metal homoepitaxy at low temperatures has been reported for Cu(100), Ag(100) and Ag(111) [460]. A typical value for the vacancy concentration at or below 100 K is 2 %, which is similar to the simulated two-dimensional film in the lower panel of Fig. 4.43. Vacancy formation appears to affect the roughness of the surface, but the nature of the coupling is not clear. For Cu(100) the surface roughness was observed to decrease with the onset of vacancy formation [461], while for Ag(111) the surface becomes extremely rough at low temperatures, and the distribution of exposed coverages φ_n becomes strongly non-Gaussian [462].

Several groups have turned to amorphous films in the quest for an experimental realization of KPZ growth. In an x–ray scattering study of sputter deposited W/Si multilayers, good agreement with the predictions of the EW equation was found [463]. This would indeed be expected, if the coefficient λ_2 is small, so that the asymptotic regime is not experimentally accessible. Amorphous silicon films have been investigated with a variety of methods, but the situation remains inconclusive. In the first study [25] the evolution of the surface was monitored using Ge marker layers. A very rapid roughness buildup followed by a regime of constant interface width was observed, but no power law of the type (4.103) could be detected. Two subsequent investiga-

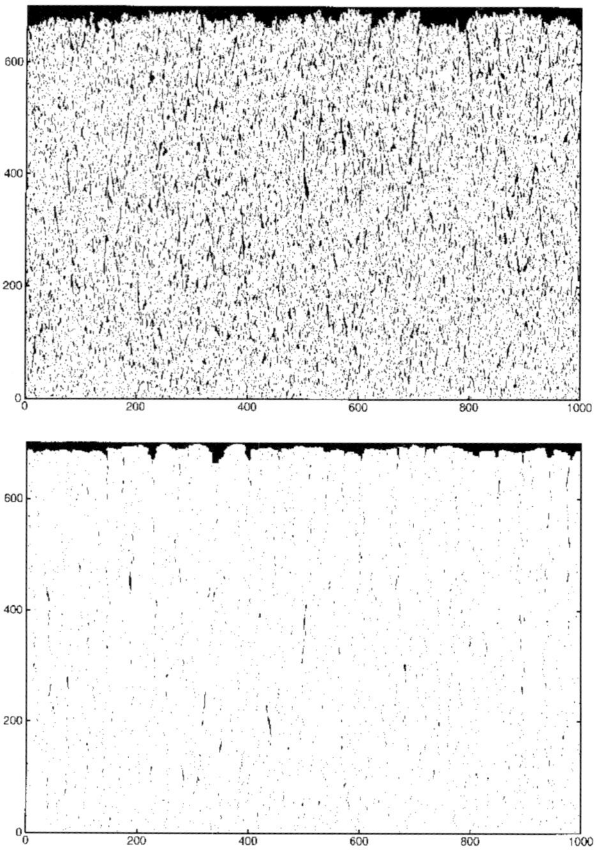

Fig. 4.43. Two-dimensional films generated with a growth model that allows for surface diffusion and the formation of bulk vacancies, but contains no additional barriers for interlayer transport. The dimensionless ratio $D/(F\Omega^2)$ is 5 in the upper panel and 200 in the lower panel. Note the coexistence of isolated point vacancy with elongated vacancy clusters. The latter represent grown-in surface crevices (Fig. 4a and b of [457])

tions using atomic force microscopy [438] (see Fig. 4.41) and x–ray scattering [439] (see Fig. 4.42) did arrive at estimates of the scaling exponents α and z, but assigned the observed behavior to different universality classes – Mullins [438] and KPZ [439], respectively.

4.8.3 Kinetic Roughening and Mound Formation

The paradigm of self-affine scaling has been applied in the analysis of growth-induced roughness for a wide range thin film systems. A systematic survey of experiments conducted up to 1995 can be found in [464], and many other examples are described in [36]. In a typical experiment of this kind, the roughness exponent α is estimated from a measurement of the height difference correlation function (4.100) using real space probes (STM or AFM) or diffuse scattering, and the ratio α/z is determined through (4.103) from the dependence of the surface width on the film thickness. The direct measurement of the dynamic exponent z from the variation of the correlation length (4.104) with film thickness is more difficult, and has been achieved only in a few cases (e.g. [439, 465, 466]).

By and large, the values of α and z obtained from vapor deposition experiments are consistent with the predictions of the conserved growth equations (CKPZ and Mullins) [464]; in particular, the roughness exponent is often found to lie in the range $\alpha \approx 0.7 - 1$. In general, however, the validity of this interpretation must be questioned because of the difficulty in distinguishing between kinetic roughening and mound formation on the basis of experimentally measured height correlation functions. The attentive reader may already have noticed that the self-affine scaling form (4.102) is very similar to the scaling form (4.131) that is used in the theory of mound formation. The lateral length scales that appear in the two expressions are conceptuatlly quite different – the wavelength $\lambda(t)$ is the *typical* length scale of the mound morphology, whereas the correlation length $\xi(t)$ in kinetic roughening is the *maximal* length scale up to which the morphology displays statistical scale invariance – but both length scales increase as a power law with film thickness[50]. Also the power law growth of the surface width, (4.79) and (4.103), is qualitatively similar in both cases.

Asymptotically, the height difference correlation function $\tilde{G}$ of a mounded surface with selected slopes should follow (4.100) with $\alpha = 1$, because the hillsides of the mounds have a fixed, nonzero misorientation relative to the average film height, so that the height difference between two points at distance r increases linearly with r. In practice, however, effective roughness exponents smaller than unity must be expected. Partly this reflects the contribution from the mound tops and valleys, where the surface slope is small, and partly the random roughness that is always superimposed onto the mound morphology. In a numerical study of mound formation with added deposition noise, a roughness exponent of $\alpha \approx 0.8$ was extracted [467]. Similarly, in an experimental investigation of GaAs and InP epitaxy, roughness exponents in the range $\alpha \approx 0.4 - 0.6$ were found for morphologies that by visual inspection show a clear mounded appearance [468].

These considerations show that the distinction between kinetic roughening and mound formation cannot be made only on the basis of measured scaling exponents. In addition, the visual appearance of the morphology and, as an objective measure, the *shape* of the height correlation function must be taken into account. A mounded surface is characterized by a (damped) oscillatory spatial correlation function, while for a randomly rough surface the correla-

[50] We follow here the widespread convention of using the same notation, $1/z$, for the exponents of the power law in both cases.

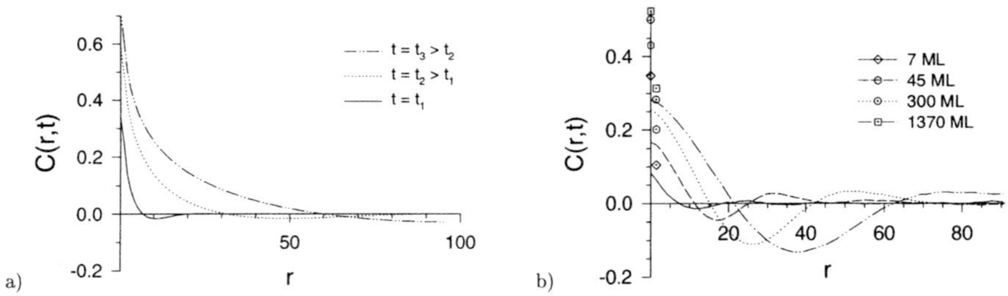

Fig. 4.44. Slope-slope correlation function $C(\boldsymbol{r},t) = \langle \boldsymbol{m}(\boldsymbol{r},t) \cdot \boldsymbol{m}(\boldsymbol{0},t)\rangle$ for (**a**) a kinetically roughened surface, generated by the noisy Mullins equation, and (**b**) a mounded surface generated by KMC simulations. In (**b**) the oscillations of the correlation function are seen to become more pronounced with increasing layer thickness. The symbols in (**b**) indicate the value of the correlation function at distance $r = 0$, while the lines are extrapolations of the correlation functions at finite distance. The difference between the two values is a measure of the random fluctuations in the slope (Fig. 11 of [467])

tions should decay monotonically. This criterion is not completely unambiguous, since also self-affine surfaces may show weakly oscillatory correlations; this has been established analytically for the Mullins equation [469]. However, in the case of mound growth the oscillations are much more pronounced, and they become more clearly developed with increasing film thickness, as the relative importance of fluctuations decreases [467]. This is illustrated in Fig. 4.44, where slope–slope correlation functions obtained from KMC simulations of mound formation are compared to the correlations generated by the noisy Mullins equation. Similar considerations apply to the distinction between rough and mounded surfaces on the basis of the intensity patterns in diffraction experiments [470–473] (see also Sect. 6.3).

4.9 Supplementary Topics

4.9.1 Growth Regimes and Surface Atomic Structure

It is an interesting question to ask whether general trends can be identified which relate the growth regimes in metal homoepitaxy to the atomic structure of the surface. Reviewing the available experimental observations of the specularly reflected intensity of thermal energy atoms, electrons or X–rays together with some additional microscopic information, a simple answer appears indeed possible[51].

Generally, growth on fcc(001) surfaces gives rise to intensity oscillations in a broad temperature interval, which begins with the onset of adatom diffusion

[51] A review on the growth modes in metal homoepitaxy until 1996 may be found in [474].

at low temperatures and ends with the onset of step flow at high temperatures. Indeed, for Ag(001) [342,475,476], Cu(001) [41,477–480], Ni(001) [481], and Pd(001) [482] oscillations are observed, which are strongly damped at the lowest temperatures, well developed at intermediate temperature, and fade away when step flow is approached. The same behavior is encountered also for growth on the Fe(001) surface [483], which exhibits the same surface structure and symmetry as fcc(001), even though Fe is a bcc material. Supported by investigations that combine scattering probes with real space observations [483], one may conclude that homopepitaxy on (001) surfaces generally displays an initial regime of layer–by–layer growth, which may be limited to only one or a few layers.

In contrast, for fcc(111) surfaces a strong barrier regime ($l_{\mathrm{ES}} \gg l_D$) with immediate mound formation appears to be a general phenomenon. For Ag(111) a continuous decay of the scattering intensity is observed at all temperatures up to the onset of step flow [475, 484, 485], in agreement with real space observations of immediate mound formation [203, 340]. For Cu(111) the same behavior is observed[52] [487–489]. For Pt(111) such a monotonic decay exists around 400 K (see Fig. 4.2c) [324], and it is associated with immediate mound formation (see Figs. 4.3 and 4.6) [333]. For Rh(111) at temperatures below 500 K and above 700 K immediate mound formation appears to occur [121]. Finally, for Pd(111) STM topographs indicate immediate mound formation at 400 K [490]. As the fcc(111) surfaces are the densest packed surfaces of face centered cubic crystals, one may expect the growth behavior on the densest packed surfaces of bcc crystals to also exhibit immediate mound formation. In fact, for the (110) surfaces of the bcc materials Fe and W, immediate mound formation does take place [343, 491]. The densest surfaces of fcc and bcc metals thus appear to be the only surfaces that show a regime of immediate mound formation, i.e. an interval of growth temperatures without initial LBL growth.

The origin of this apparently different growth behavior for different crystallographic faces remains unclear to date. The densest packed fcc(111) and bcc(110) surfaces are – compared to other more open surfaces – characterized by (i) the largest adatom diffusion coefficients and (ii) the largest step free energies due to the poor coordination of step adatoms. The impact of the first factor may be evaluated by using the criterion (4.60) for immediate mound formation. It is clear that for a fixed additional effective step edge barrier $\Delta E_{\mathrm{S}}^{\mathrm{eff}}$, a large adatom diffusion coefficient (small E_{D}) tends to make mound formation less likely. Thus large adatom diffusion coefficients on the densest packed surfaces do not favor the observed immediate mound formation. As step free energies are largest on the dense packed surfaces, one may

[52] Only by spot profile analysis low energy electron diffraction (SPALEED) strong oscillations were found [486] during the growth of Cu on Cu(111). These results are critically discussed in [487].

speculate that an adatom descending from such a step edge will experience the relatively largest differences in bonding by moving from the terrace to the transition state at the step prior to being incorporated. Consequently the step edge barriers on the densest packed surfaces should be larger than on the (001) surfaces and, as expressed by (4.60), a large additional step edge barrier favors immediate mound formation. Indeed, the experimentally determined step edge barriers tend to be significantly larger on the fcc(111) and bcc(110) surfaces, compared to (001) surfaces. As an example, for Ag(111) the measured step edge barrier values range from 0.12-0.15 eV, [239, 366, 369] (see Sects. 4.3.3 and 4.9.6) while for Ag(001) the value is 0.03 eV [322]. For Ag(001) the low additional edge barrier was associated with the existence of a low energy exchange path for the descent of an adatom into the step edge. Following this exchange path DFT calculations find no additional step edge barrier within the accuracy of the method [374]. Possibly such low energy exchange processes are a quite general feature for steps on fcc(001) surfaces.

In the preceding analysis it was tacitly assumed that the presence of oscillations in the reflected scattering signal excludes immediate mound formation, in agreement with the TEAS and STM results for growth on Pt(111) discussed in Sect. 4.1. Comparing scattering signals for growth on Ag(111) [475, 484, 485] with transmission electron microscope and STM results [203, 340] one arrives at the complementary conclusion that immediate mound formation excludes scattering signal oscillations. On the other hand, for growth on Pt(111) it was also pointed out that damped oscillations do not exclude the onset of mound formation after a few deposited layers. In fact, comparing real space mound formation observations on Fe(001), Cu(001) and Ag(001) with the corresponding oscillatory scattering signals, it appears to be the rule rather than an exception that damped oscillations are indicative of initial LBL growth followed by a mound formation regime. As an example, in Fig. 4.45 the reflection high energy electron diffraction (RHEED) signals at 200 K, 300 K and 400 K during homoepitaxial growth on Ag(001) are compared to the temperature dependent film roughness after deposition of 25 ML and the growth morphology at 190 K. Obviously the presence of 10 RHEED oscillations at 200 K (Fig. 4.45a) does not exclude the existence of mounds after deposition of 25 ML at the same temperature (Fig. 4.45e). The surface width after deposition of 25 ML actually exhibits a maximum around 200 K (Fig. 4.45d). The relation between the damping of RHEED oscillations and surface roughening due to weak step edge barriers or deposition noise will be further discussed in Sect. 5.6.

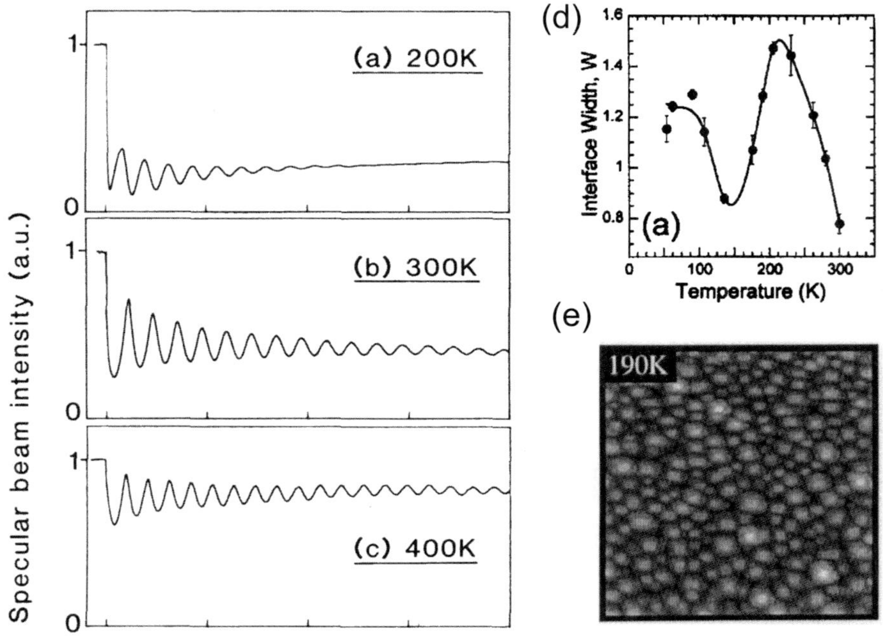

Fig. 4.45. Comparison of (**a**)–(**c**) RHEED intensity oscillations during growth on Ag(001) (Fig.1 of [475]) with (**d**) the temperature dependent surface width after deposition of 25 ML on Ag(001) (Fig. 2a of [322]) and (**e**) the corresponding growth morphology at 190 K ((**e**) is part of Fig. 1 of [151])

4.9.2 Ripple Formation in Homoepitaxial Growth

By reducing the symmetry of the deposition arrangement to a twofold symmetric situation (rotation by 180° allowed) or to a situation with a single mirror symmetry (reflection at one mirror plane allowed), instead of mounds ripples are observed. A generic case of the second kind is the growth on vicinal surfaces discussed in Sect. 4.7, where ripples form parallel to the steps (step bunching) or perpendicular to the steps (step meandering). Here two examples for ripple formation on singular surfaces will be described.

The first situation is realized by use of a twofold symmetric substrate and deposition at normal incidence [492]. The substrate is the Ag(110) surface, which exhibits channels along the $[1\bar{1}0]$ direction, as illustrated in the ball model of Fig. 4.46. After deposition of about 30 ML Ag, spot profile analysis low energy electron diffraction (SPALEED) contour plots exhibit, apart from the specular reflected peak, two sidepeaks in $[1\bar{1}0]$ at 130 K and in [001] at 210 K (see Fig. 4.47a and b). These sidepeaks indicate ripples oriented along [001] at 130 K and along $[1\bar{1}0]$ at 210 K. As seen from the STM topographs in Fig. 4.47c and d, the ripple orientation appears to be determined already during submonolayer growth, where elongated islands form.

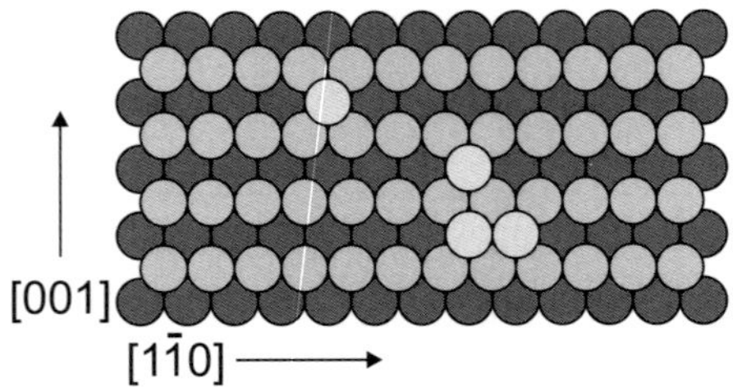

Fig. 4.46. Ball model of an fcc(110) surface, which forms channels along the [1$\bar{1}$0]–direction.

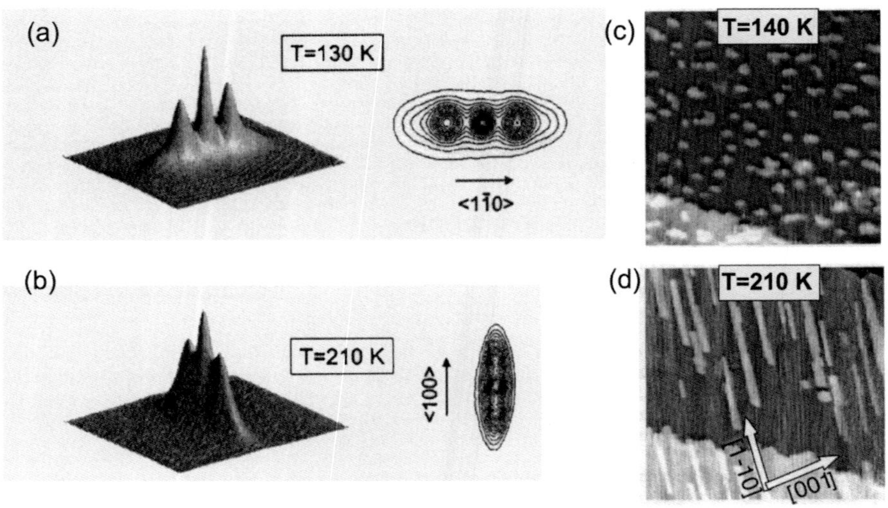

Fig. 4.47. (**a**), (**b**) Two-dimensional SPALEED profiles measured near the in-phase diffraction condition and centered at the (00) LEED beam after evaporation of 30 ML Ag on Ag(110) at the temperatures indicated. The arrows refer to real space surface orientations and have a length of 10% of the surface Brillouin zone (Figs. 1a and 1c of [492]). (**c**), (**d**) STM topographs after deposition of 0.16 ML Ag on Ag(110) at the temperatures indicated. Topograph sizes 1000 Å× 1000 Å (part from Fig. 2 of [493] reprinted with permission from Elsevier Science)

The switch in the island orientation with increasing temperature can be understood as follows. At low temperatures only diffusion along the channels is possible and the weak bonds between atoms in different channels allow the alignment of atoms in different channels along [001]. Increasing the temperature makes the weak cross-channel bonds unstable but still allows for the formation of the strong in-channel bonds. This leads to the formation of adatom islands aligned along [1$\bar{1}$0]. The change in the orientation of islands and ripples can be reproduced in KMC simulations which incorporate anisotropic bonding and diffusion [493, 494]. At intermediate temperatures (around 170 K) the experiments indicate the existence of a pattern of rectangular mounds. A continuum theory of the type described in Sect. 4.9.8 shows

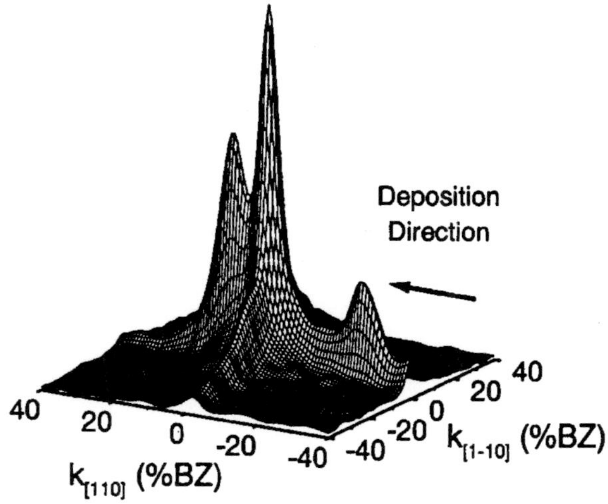

Fig. 4.48. Two-dimensional SPALEED profile after deposition of 40 ML Cu on Cu(001) at 250 K at an angle of 80° from the surface normal (Fig. 1 of [158])

that the transition between the two rippled states should generically proceed through such an intermediary, rectangular phase [495].

The second situation is realized by grazing incidence Cu deposition (at about 80° from the surface normal) on the fourfold symmetric Cu(001) surface [158, 159]. The existence of a single mirror plane in the deposition geometry is reflected by a growth morphology of asymmetric ripples. Figure 4.48 shows a SPALEED contour plot after deposition of about 40 ML at 250 K. The two sidepeaks are facet spots and indicate the formation of $(\bar{1}\bar{1}1)$ and (113) facets on the illuminated and shaded sides of the ripples, respectively. The slope angle on the illuminated side of the ripple is $\approx 56°$, much larger than on the shaded side ($\approx 25°$). Again the tendency for ripple formation manifests itself already in submonolayer deposition experiments, where rectangular islands elongated normal to the deposition flux are observed. As discussed in Sect. 2.4.1, the formation of rectangular islands in grazing incidence deposition is due to steering effects. Adatoms arriving on top of an island are distributed isotropically to the descending island steps. The extra steering flux on the island top is thus distributed isotropically, while the flux deficit behind an island leads to a reduced island growth speed in the direction of the depositing beam.

4.9.3 Derivation of the Step Edge Boundary Condition

In this subsection we provide a microscopic derivation of the boundary condition (4.29) for the adatom diffusion field at the descending step edge. To avoid the complications associated with the two-dimensional lattice geome-

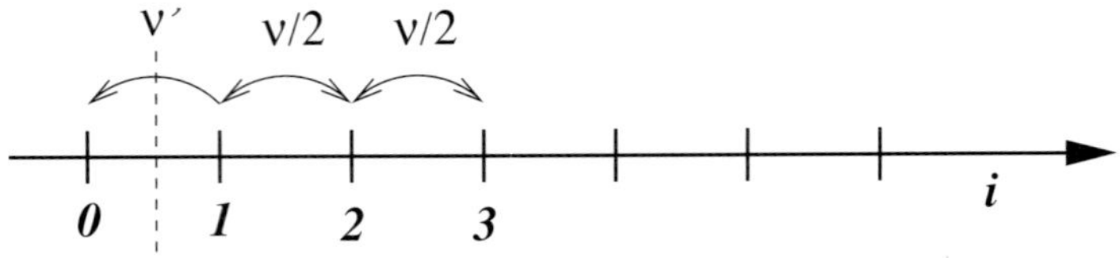

Fig. 4.49. Sketch of the one-dimensional geometry used in the derivation of the step edge boundary condition. Site 1 is the last site on the island, and site 0 is a fictitious additional site introduced to formulate the boundary condition

try, we use the one-dimensional setup shown in Fig. 4.49. The lattice sites are labeled by i, and $i = 1$ is the boundary site (the last site on the island). Adatoms are deposited at rate $\hat{F}$ per site, perform a simple random walk with jump rate ν on the island, and descend from site 1 at rate ν'. In the steady state, the adatom density $\hat{n}_i$ at sites $i \geq 2$ satisfies the balance equation

$$\frac{\mathrm{d}\hat{n}_i}{\mathrm{d}t} = \frac{\nu}{2}(\hat{n}_{i+1} - 2\hat{n}_i + \hat{n}_{i-1}) + \hat{F} = 0 \ , \tag{4.111}$$

which is the discrete version of the stationary diffusion equation (4.24). At the boundary site $i = 1$ we have instead

$$\frac{\mathrm{d}\hat{n}_1}{\mathrm{d}t} = \frac{\nu}{2}\hat{n}_2 - \left(\frac{\nu}{2} + \nu'\right)\hat{n}_1 + \hat{F} = 0 \ . \tag{4.112}$$

Introducing an additional, fictitious site $i = 0$ outside of the island, (4.112) can be rearranged into

$$\frac{\nu}{2}(\hat{n}_2 - 2\hat{n}_1 + \hat{n}_0) + \hat{F} = \left(\nu' - \frac{\nu}{2}\right)\hat{n}_1 + \frac{\nu}{2}\hat{n}_0 \ . \tag{4.113}$$

Setting the left hand side of (4.113) to zero extends the validity of the discrete diffusion equation (4.111) to the site $i = 1$. But then the right hand side of (4.113) has to vanish also, which fixes the auxiliary variable $\hat{n}_0$ to be

$$\hat{n}_0 = (1 - 2\nu'/\nu)\hat{n}_1 \ . \tag{4.114}$$

The quantity $(1 - 2\nu'/\nu)$ is the probability for the adatom to be reflected at the island edge. It interpolates between the limiting cases of a perfectly reflecting boundary ($\nu' = 0$, $\hat{n}_0 = \hat{n}_1$) and a perfectly absorbing boundary ($\nu' = \nu/2$, $\hat{n}_0 = 0$). Thus in this one-dimensional geometry, the complete absence of a step edge barrier implies $\nu' = \nu/2$. For a general, two-dimensional lattice, the corresponding condition reads $\nu' = (z_{\mathrm{down}}/z_{\mathrm{all}})\nu$, where z_{down} and z_{all} denote the number of diffusion channels for adatom descent, and the total number of diffusion channels leaving the edge site, respectively; in the case of several inequivalent edge sites an appropriate average has to be taken.

To make contact with the continuum description, we now rewrite (4.114) as

$$\nu' \hat{n}_1 = \frac{\nu}{2}(\hat{n}_1 - \hat{n}_0) \approx \frac{\nu}{2} a \frac{\mathrm{d}n}{\mathrm{d}x} . \quad (4.115)$$

Since $D = a^2 \nu/2$ in one dimension, this reduces to (4.29) (note that in (4.29) the descending step edge is at the right end of the terrace).

4.9.4 Second Layer Nucleation and Rate Equations

The first calculation of the rate of second layer nucleation [354] was based on the rate equation (2.13), in which the average adatom density n_1 is replaced by its local value $n(\boldsymbol{r})$ at a point $\boldsymbol{r}$ on the island. The *local* nucleation rate $J(\boldsymbol{r})$, which counts the number of nucleation events per time and lattice site, is then given by

$$J(\boldsymbol{r}) = \Omega^{-1} \sigma_{i^*} D \mathrm{e}^{E_{i^*}/k_\mathrm{B}T} (\Omega n(\boldsymbol{r}))^{i^*+1} . \quad (4.116)$$

Inserting the density profile (4.34) and integrating over the island area, the total nucleation rate on a circular island is obtained. Comparison with the exact expression (4.42) shows that the rate equation approach overestimates (4.42) by a factor of the order of $l_\mathrm{ES}/R \gg 1$ [334]. In addition, (4.116) incorrectly predicts a uniform spatial distribution of nucleation events for $l_\mathrm{ES} \gg R$, whereas in fact the distribution is distinctly peaked near the center of the island [360].

To elucidate the origin of this discrepancy, it is useful to view the encounter of the two atoms as a first passage problem in the four-dimensional space spanned by their coordinates. The joint diffusion of the atoms stops either when the two reach adjacent lattice sites (nucleation) or when one of them escapes, whatever happens first. The detailed analysis of this process [360–362] reveals the nature of the approximation based on (4.116): Setting the nucleation rate proportional to the square of the adatom density amounts to treating the atoms as *noninteracting*, in the sense that they are allowed to continue their diffusion even after they have met; in this way a single pair can accumulate several (spurious) nucleation events, and the nucleation rate is overestimated.

For a quantitative comparison, recall from Sect. 2.4.3 that the nucleation probability p_nuc per atom can be written as the product of the (mean) adatom density and the number $\mathcal{N}(\tau)$ of *distinct* sites the atom encounters during its residence time. Using (4.38) this amounts to writing the total nucleation rate on the island as

$$\omega = (\Delta t)^{-1} p_\mathrm{nuc} = F \Omega \mathcal{N} \frac{\tau}{\Delta t} . \quad (4.117)$$

In the strong barrier limit, $\mathcal{N}$ equals the total number of sites on the island, and (4.117) reproduces (4.42). To compare (4.117) to the rate equation approximation, we multiply the local nucleation rate (4.116) with the island area and obtain, in order of magnitude,

$$\omega_{\text{rate}} \sim R^2 D(\bar{n})^2 \sim F\Omega\mathcal{N}_{\text{all}}\frac{\tau}{\Delta t} , \tag{4.118}$$

where (4.38) has been used, and the number $\mathcal{N}_{\text{all}} = \nu\tau$ of *all* sites visited by an adatom during its residence has been introduced. Thus the expressions (4.117) and (4.118) differ, in general, by a factor $\mathcal{N}_{\text{all}}/\mathcal{N} > 1$. For weak barriers $\mathcal{N}_{\text{all}} \sim \Omega R^2$ and $\mathcal{N} \sim \Omega R^2/\log(\Omega R^2)$, hence the discrepancy amounts to a logarithmic correction similar to those discussed in Sect. 2.4.3, see (4.44).

4.9.5 Second Layer Nucleation with Reversible Aggregation

The extension of the theory developed in Sect. 4.3.2 to reversible aggregation with $i^* > 1$ is difficult, because the lifetimes of the unstable clusters interfere with the basic time scales (4.36,4.37,4.39). The number of relevant time scales becomes so large that distinct regimes in which they are well separated can no longer be identified [357]. An interesting qualitative feature is that the confinement of the adatoms on top of the island can effectively decrease the size of the critical nucleus, because the constituents of the unstable clusters cannot escape. For example, if the dimer dissociation time τ_{dis} satisfies

$$\tau_{\text{dis}} \gg (\tau_{\text{D}}/\tau)\Delta t \tag{4.119}$$

the critical nucleus size is shifted from $i^* = 2$ to $i^* = 1$ [334].

A fairly complete analysis is possible in a simplified "noninteracting particle model", in which the binding energies of all unstable clusters are set to zero [168, 357]. Then the only additional time scale entering the problem is the *encounter time*

$$\tau_{\text{enc}} \sim R^{2i^*}/\nu \tag{4.120}$$

required for the $i^* + 1$ atoms forming a stable nucleus to meet, once they are present on the island. The expression (4.120) can be understood as follows [168]. The coordinates of the $i^* + 1$ adatoms form a $2(i^* + 1)$-dimensional region of linear size R. Since the point of encounter can lie anywhere on the island, nucleation amounts to finding a two-dimensional submanifold within this region. Thus the total number of points to be searched is of the order of R^{2i^*}, which takes a time given by (4.120).

The total nucleation rate can then be estimated as

$$\omega \sim (\Delta t)^{-1}\, p_{i^*+1}\, p_{\text{enc}} , \tag{4.121}$$

where $p_{\text{i}^*+1} \sim (\tau/\Delta t)^{i^*}$ is the probability that i^* atoms are present on the island when the $i^* + 1$'th atom arrives. The encounter probability p_{enc} is of order unity when $\tau_{\text{enc}} \ll \tau$, and of order τ/τ_{enc} when $\tau_{\text{enc}} \gg \tau$. In the latter, *encounter limited* nucleation regime, which exists only for $i^* > 1$, the nucleation rate coincides in order of magnitude with that predicted by the rate equation approach (see Sect. 4.9.4).

The estimate (4.121) is still based on the fluctuation-dominated "lonely adatom" picture of second layer nucleation sketched in Sect. 4.3.2: It is assumed that each nucleation event can be associated with the deposition of the last $(i^* + 1)$'th atom, and that the time-averaged number of adatoms on the island is much smaller than $i^* + 1$. It is important to keep in mind that this assumption is subject to a consistency requirement [168,357]. When describing, for example, the process of second layer nucleation for an array of growing first layer islands (as in Sect. 4.3.3), it has to be checked that the relations between the relevant time scales assumed in the derivation of the nucleation rate are still valid when the island has reached the critical size at which nucleation actually occurs. Within the noninteracting particle model one finds that the critical island radius always violates the relation (4.40) when $i^* \geq 3$; the general condition reads $\chi > 2/(i^* + 1)$, where χ is the island density scaling exponent introduced in (2.19) [168]. The encounter time (4.120) becomes so large in this case that many more than $i^* + 1$ atoms have to be present on the island before nucleation takes place. The fluctuation-dominated regime gives way to a *mean field* nucleation regime in which the rate equation approach based on (4.116) becomes *quantitatively* accurate [357]. For nucleation on the top terrace of a mound, the fluctuation-dominated scenario can be realized for arbitrary values of i^* [168].

4.9.6 Further Experimental Signatures of the Step Edge Barrier

This subsection collects additional approaches to the experimental determination of the step edge barrier, complementing the discussion in Sect. 4.3.3.

Interlayer Diffusion at Kinks. An atomistic experiment emphasizing the importance of kinks for interlayer transport was performed for Co adatoms on the Pt(111) surface [496]. Due to the chemical contrast achieved in atomic resolution STM images of the surface after incorporation of deposited Co atoms into steps, it was unambiguously shown that Co adatoms descend most easily at kinks, rather than at straight A- or B-steps. This finding is in good agreement with the general trend found in calculations, that kinks are positions of minimal step edge barrier (compare Table 4.1), and the model developed for the effective step edge barrier in Sect. 4.4.

Surface Morphology. A number of studies have attempted to estimate ΔE_S through a direct comparison of statistical measures of the surface morphology, such as the width W or the layer coverages θ_n, with model predictions usually obtained from KMC simulations. For example, in [341] the surface width of a 2 ML Ag film on Ag(111) was measured by LEED in the temperature range 130 – 200 K. Above 130 K, a reduction of the width relative to the statistical growth limit (4.12) was observed, and by comparison with a KMC simulation the additional step edge barrier was estimated to be $\Delta E_S \approx 0.05$ eV. This approach has also been applied to Fe(100) [497, 498].

A general difficulty, which is well illustrated by a recent detailed study for Ag(100) [151], is that the surface width is affected by a multitude of kinetic processes, such that the specific influence of interlayer transport cannot be easily extracted.

An approximate analytic model for the layer coverages can be derived by generalizing the multilayer rate equations (4.9) to include the (configuration-dependent) exchange of matter between the layers [335]. Meyer et al.[366] integrated such equations subject to a boundary condition corresponding to the nucleation of a new layer at a critical coverage θ_c (see Sect. 4.2). A fit to layer distributions obtained from STM images at different total coverages yields the estimate $\Delta E_S = 0.15 \pm 0.02$ eV for Ag/Ag(111). Similar investigations have been reported for Ag(100) [499] and Cu(100) [500].

Nucleation and Growth Near Preexisting Steps. On a vicinal terrace, the step edge barrier was shown to imply an asymmetry in the attachment probabilities between the ascending and the descending step [(4.32)]. This asymmetry becomes readily observable in a growth regime where island nucleation on the terrace competes with capture at steps. Because it is a less effective sink for adatoms, the *denuded zone* in which nucleation is suppressed is smaller near the descending step than near the ascending step[53] [502].

Similarly, if only a single island is growing on the vicinal terrace, it will absorb a larger fraction of the deposited atoms, the more the attachment to the descending step is hindered. Under suitable conditions the size of such islands is reflected in the total step density on the surface, which in turn can be related to certain diffraction signals. Both approaches have been applied to the Ag/Ag(111) system to estimate the preexponential factor ν_0' for interlayer diffusion [370, 503]. Their common drawback is that they are insensitive to large step edge barriers, because the descending step has no influence on the adatom distribution once it is completely blocked (i.e., once $l_{ES} \gg l$ in (4.32)). In this sense they are complementary to the second layer nucleation technique described in Sect. 4.3.3, which is most precise for strong barriers.

Decay of Monolayer Islands. In thermal equilibrium, arrays of adatom and vacancy islands on a surface undergo the process of Ostwald ripening, in which small islands shrink at the expense of larger ones in order to reduce the total step free energy [504]. Through a judicious choice of island geometry it is possible to direct the current of detaching adatoms in such a way that it has to cross a step, thus probing the step edge barrier. This is illustrated in Fig. 4.50, which contrasts the decay of an adatom island and a vacancy island, both of which are placed inside a larger vacancy island. The decay of the vacancy island is slower because the adatoms filling it have to cross a descending step. The decay process can be modeled quantitatively by solving the diffusion

[53] A careful LEEM study of nucleation on surfaces vicinal to Si(111) revealed a slight shift of the probability distribution of nucleation sites towards the *ascending* step, indicating that attachment is in fact more facile at the *descending* step for this system [501].

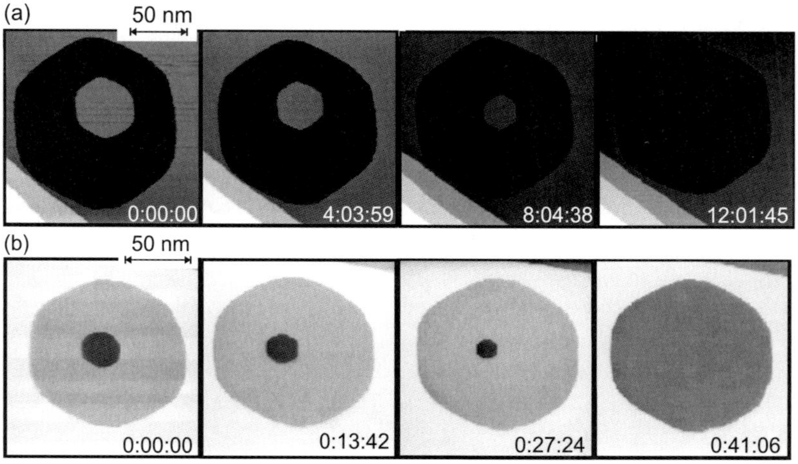

Fig. 4.50. Decay of adatom and vacancy islands on Ag(111). (**a**) shows an adatom island at 300 K, (**b**) a vacancy island at 360 K. The decay of the vacancy island is slowed down by the step edge barrier (Figs. 2 and 3 of [239])

equations (4.24,4.28,4.29) for two concentric, circular steps [239]. Comparison with the measured decay trajectories for a range of temperatures yields the estimates $\Delta E_S = 0.13 \pm 0.04$eV and $\nu_0'/\nu_0 = 10^{-0.6\pm0.5}$ for Ag/Ag(111).

A similar analysis has been carried out for Cu adatom islands on a stepped Cu(111) surface [241]. The asymmetry between the attachment to descending and ascending steps implies that islands near ascending steps decay faster. A comparison of the decay rates of differently positioned islands thus allows to estimate ΔE_S.

Decay of Multilayer Mounds. The thermal decay of a mound consisting of a stack of roughly concentric islands is driven by the line tension of the island edges. The boundaries of the upper, smaller islands are more strongly curved than those of the lower, larger islands, and the corresponding differences in the edge chemical potential induce a downhill mass current which flattens the mound. The kinetics of this process is evidently strongly affected by the step edge barrier.

In an experimental investigation of mound decay on Cu(111), it was found that the decay rate speeds up by several orders of magnitude when the boundary of the uppermost, shrinking island touches the descending step [505]. Such contacts occur occasionally because of the random walk motion of the island. Initially this phenomenon of *rapid mound decay* was interpreted as a deactivation of the step edge barrier when the distance between the step edges falls below a critical value of about six atom rows. Recent studies on Cu(111) and Ag(111) have instead converged on the view that not only the step edge barrier but, more importantly, the barrier for detachment from the step can be bypassed for steps in close proximity [506, 507]. The microscopic mechanism involves concerted two-atom exchange processes which allow atoms to descend without losing coordination by detachment and subsequent step crossing [508].

Rapid decay events can be avoided when a small island decays on top of a large one. The decay dynamics can then be analysed using the solution of the diffusion equation (4.24) for concentric, circular islands, and the step edge barrier entering the boundary condition (4.29) can be extracted. For Cu(111) this procedure lead to the estimates $\Delta E_S = 0.224 \pm 0.009$ eV and $\nu_0' = 3.5 \times 10^{12\pm1}\mathrm{s}^{-1}$ [509].

4.9.7 Diffusion Bias Without a Step Edge Barrier

The step edge barrier is probably the most common source of a diffusion bias, but it is not the only way in which the symmetry between ascending and descending steps can be broken.

An example of an alternative atomistic mechanism is the attraction of adatoms to an ascending step which was described in Sect. 2.4.1. It is quite plausible that this should have similar consequences as the usual repulsion from the descending step, and the corresponding uphill surface current can be calculated along the lines of Sect. 4.5.1 [510]. Also the steering effect discussed in Sect. 2.4.1 can be viewed as a source of an uphill current, since it causes the deposited atom to land uphill from the place where the extrapolated linear trajectory would reach the surface [154–156].

However, a diffusion bias can be generated even without any asymmetry in the motion of terrace adatoms near the step. This rather surprising insight was reached through computer simulations [431]. Figure 4.51 illustrates the effect. The key observation is that the motion of atoms along the step edges also contributes to the surface current [424]. Since the step edges undergo thermal and kinetic fluctuations, the local orientation of a step segment generally differs from the mean step direction. The motion along the step edge therefore has a component perpendicular to the mean step direction, that is,

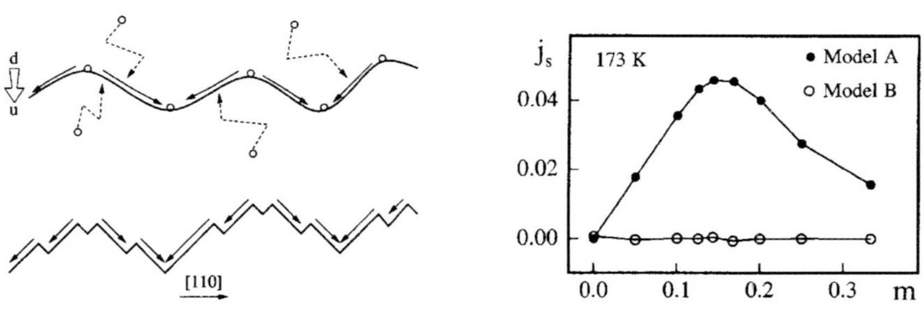

Fig. 4.51. Left panel: Illustration of the uphill current generated by step edge diffusion (Fig. 3 of [430]). Right panel: Numerically measured surface current in KMC simulations with (Model A) and without (Model B) facile edge diffusion (Fig. 3a of [431])

along the direction of the mean miscut. On average this current is directed from the protrusions of the step towards its indentations, and hence, from the point of view of the large scale vicinality, it flows in the uphill direction. In a continuum picture, this asymmetry can be attributed to the difference in step chemical potential between protrusions and indentations due to the two-dimensional Gibbs-Thomson effect (upper left panel in Fig. 4.51), but it can also be viewed as an effect of the microscopic step geometry (lower left panel). These arguments are essentially independent of the existence of kink barriers which prevent step atoms from moving around corners (see Sects. 3.7.1 and 4.7.2). In KMC simulations such a step fluctuation induced current has been measured (see right panel of Fig. 4.51), and the resulting formation and coarsening [511] of mounds has been observed[54].

An interesting feature of this fluctuation-induced current is that it becomes independent of tilt in the step flow regime [424]. For a vicinal surface with step spacing l, the flux onto each step, and hence the contribution of the step to the current, is proportional to l. As the step density equals $1/l$, the l-dependence cancels. Since the current must decrease when l become comparable with l_D and nucleation sets in, the current is a decreasing function l, and an increasing function of the surface slope d/l. According to the continuum theory of vicinal surfaces developed in Sect. 4.9.9 this implies a step bunching instability, which has been seen in KMC simulations [513].

The relevance of these rather subtle effects for real metal surfaces has not yet been explicitly demonstrated. We should like to point out, however, that there is at least one case – the Cu(100) surface – for which mound formation [395], step meandering [413] and possibly step bunching [423] instabilities have been observed, but attempts to directly verify the existence of a nonzero step edge barrier, along the lines of Sect. 4.9.6, have not been successful [172, 500]. Possibly this is an example for a system in which the exotic mechanisms described in this subsection play a role.

4.9.8 Continuum Theory of Mound Coarsening

In this subsection we elaborate on the two-dimensional continuum theory of mound growth, with the goal of arriving at a qualitative and semi-quantitative understanding of the coarsening laws which were mentioned in Sect. 4.6.2. Analogous to the one-dimensional theory sketched in Sect. 4.5.2, the two-dimensional surface configuration is described by a continuous height function $h(\boldsymbol{r}, t)$, where $\boldsymbol{r} = (x, y)$ is a point in the substrate plane. Here the height will be measured relative to the mean height, which grows at a constant rate given by the deposition flux F. The two-dimensional evolution equation then takes the general form[55]

[54] A distinct effect of edge diffusion which also promotes mound formation is that it makes islands more compact. This increases the adatom density on top of the islands, and leads to earlier second layer nucleation [512].

$$\frac{\partial h}{\partial t} = -\nabla \cdot \boldsymbol{j}_{\mathrm{ES}} - (\nabla^2)^2 h \,. \tag{4.122}$$

The growth-induced current $\boldsymbol{j}_{\mathrm{ES}}$ is a two-component vector field depending on the two-component slope vector $\boldsymbol{m} = \nabla h$. The second term on the right hand side derives from the two-dimensional version of the curvature-driven smoothening current in (4.66). We follow common practice and use the most simple, linear and isotropic form of this term. This is sufficient as long as the smoothening term is included only to make the equation well-behaved on short scales. More complicated expressions would be necessary to represent specific relaxation pathways such as diffusion along steps, see Sect. 4.6.2. A remark pertaining to other linear smoothening terms is made towards the end of this subsection.

Crystal Anisotropy. As discussed in Sect. 4.6.1, the zeros of $\boldsymbol{j}_{\mathrm{ES}}$ determine the stable slopes of the fully-developed mounds. Thus the expression for $\boldsymbol{j}_{\mathrm{ES}}$ must incorporate, in particular, the crystal symmetry of the surface [514]. To give an example, a possible choice for a surface of square symmetry reads [515]

$$\begin{aligned} j_{\mathrm{ES},x} &= m_x(1 - m_x^2 - b m_y^2) \\ j_{\mathrm{ES},y} &= m_y(1 - m_y^2 - b m_x^2) \,, \end{aligned} \tag{4.123}$$

which generates pyramidal mounds with selected slopes

$$\boldsymbol{m}^* = \frac{(\pm 1, \pm 1)}{\sqrt{1+b}} \tag{4.124}$$

for $-1 < b < 1$. The microscopic origin of the anisotropy of $\boldsymbol{j}_{\mathrm{ES}}$ is twofold [424]. First, because of its dependence on the step structure through, for example, the density of kinks, the effective step edge barrier depends on the orientation of the step in the plane. Second, the growth-induced current contains contributions from the diffusion of atoms *along* step edges [430, 431], which depend even more strongly on the step orientation. The latter issue is discussed further in Sect. 4.9.7. An approximate evaluation of these contributions shows that it is difficult to make contact with phenomenological expressions like (4.123), because the microscopic analysis predicts – in contrast to (4.123) – that the current remains anisotropic even in the limit of small slopes, $\boldsymbol{m} \to 0$. In mathematical terms, this implies a singularity in the function $\boldsymbol{j}_{\mathrm{ES}}(\boldsymbol{m})$ at $\boldsymbol{m} = 0$, which may point to a fundamental problem of the continuum theory at small slopes.

Recently the continuum theory has been extended to surfaces of rectangular symmetry, such as the Ag(110) surface discussed in Sect. 4.9.2 [495].

[55] Throughout this subsection dimensionful material parameters will be omitted from the equations. Surface evolution equations can always be brought into a dimensionless form like (4.122), for example by rescaling the spatial coordinates by the initial mound spacing and the slopes by the selected slope, see e.g. [410].

The reduced symmetry implies a large parameter space for the most general current function $\boldsymbol{j}_{\mathrm{ES}}$, and, correspondingly, a wide variety of patterns including straight and buckled ripples, and rectangular as well as rhomboidal pyramids. Remarkably, the theory predicts stationary patterns which are *not* associated with zeros of $\boldsymbol{j}_{\mathrm{ES}}$, but rather display persistent surface currents.

Driving Force of Coarsening. Since much of the coarsening phenomenology can be described without reference to crystal anisotropy, we now specialize to a class of *isotropic* models in which the current $\boldsymbol{j}_{\mathrm{ES}}$ always points in the direction of the slope. We can then write

$$\boldsymbol{j}_{\mathrm{ES}}(\boldsymbol{m}) = f(|\boldsymbol{m}|^2)\boldsymbol{m} \,, \tag{4.125}$$

where the function f depends only on the magnitude, but not on the direction of the slope. The form of f can be inferred from the expressions derived for one-dimensional situations, as described in Sect. 4.5.1 and 4.6.1. A dimensionless, somewhat simplified version of the current (4.63) leads to [516]

$$f_{\mathrm{I}}(m^2) = \frac{1}{1+m^2} \qquad \text{(model I)} \tag{4.126}$$

while the simplest expression accounting for slope selection (with the normalized selected slope $m^* = 1$) reads [514]

$$f_{\mathrm{II}}(m^2) = 1 - m^2 \qquad \text{(model II)} \,. \tag{4.127}$$

The anisotropic current (4.123) reduces to this form in the limit $b \to 1$. The evolution equations defined by (4.126) and (4.127) will be referred to as models I and II in the following.

The isotropic current (4.125) can be written[56] as the negative gradient (with respect to $\boldsymbol{m}$) of a *slope potential* defined by

$$\mathcal{V}_{\mathrm{ES}}(\boldsymbol{m}) = -\frac{1}{2}\int_0^{|\boldsymbol{m}|^2} du\, f(u) \,. \tag{4.128}$$

The slope potential allows to introduce a *Lyapunov functional* [517] for the evolution equation (4.122), that is, a functional of the surface morphology which is a strictly decreasing function of time. Defining

$$\mathcal{F}[h(\boldsymbol{r},t)] = \int d\boldsymbol{r} \left(\frac{1}{2}(\nabla^2 h)^2 + \mathcal{V}_{\mathrm{ES}}(\nabla h)\right) \,, \tag{4.129}$$

it is a simple matter to verify that [407, 518]

$$\frac{\mathrm{d}\mathcal{F}}{\mathrm{d}t} = -\int d\boldsymbol{r} \left(\frac{\partial h}{\partial t}\right)^2 < 0 \,. \tag{4.130}$$

[56] This is true also for some, but not all, anisotropic expressions for $\boldsymbol{j}_{\mathrm{ES}}$.

This is an extremely useful relation, because it suggests to view the far-from-equilibrium growth process in analogy to a process of relaxation to equilibrium, in which a quantity akin to a free energy is minimized.

The two terms in the integrand of (4.129) describe different aspects of this minimization: On the one hand, the value of the slope potential $\mathcal{V}_{\mathrm{ES}}$ should be minimized locally; on the other hand, the square of the surface curvature $\nabla^2 h$ should become small. The minimization of the slope potential drives the process of slope selection discussed in Sect. 4.6.1. It is evident from (4.128) that the potential has a minimum only if the function f goes through zero at some nonzero slope; otherwise, $\mathcal{V}_{\mathrm{ES}}$ decreases indefinitely with increasing $|\boldsymbol{m}|$, and the attempt to minimize it leads to unbounded steepening of the morphology. This proves the assertion of Sect. 4.6.1 that slope selection requires zeros in the growth-induced current.

The minimization of the surface curvature term in (4.129) provides, within the continuum equation (4.122), the driving force for coarsening. As will be shown in the following paragraphs, this implies that the coarsening behavior depends on the spatial distribution of the curvature on the suface.

Coarsening Laws. To extract the coarsening behavior from a nonlinear continuum equation such as (4.122), one commonly imposes a *scaling hypothesis* stating that the patterns at different times are similar, in a statistical sense, up to a rescaling by the average feature size [394]. In the present context this implies that the height–height correlation function (4.18) (as well as any other statistical measure of the morphology) depends on time only through the lateral mound size λ and the surface width W, and hence can be written in the form

$$G(\boldsymbol{r}, t) = W^2(t)\mathcal{G}(\boldsymbol{r}/\lambda(t)) , \tag{4.131}$$

where $\mathcal{G}$ is a time-independent scaling function. Innocuous as it may seem, this widespread hypothesis need not be correct. For the model (4.123) with square symmetry, it has been shown that the scaling hypothesis breaks down due to the existence of metastable checkerboard mound patterns, which coarsen only through the motion of dislocations ("roof tops") [407, 515]. The distance between the dislocations defines a second length scale, which is much larger than the mound size.

In the isotropic case of interest here such complications do not arise, hence the hypothesis (4.131) can be adopted. To put it to work, we consider the time evolution of the surface width. Since the mean height has already been subtracted, it is given by $W^2 = \langle h^2 \rangle$, where the angular brackets represent a spatial average. Multiplying (4.122) by h and integrating spatially we obtain [519]

$$\frac{1}{2} \frac{\mathrm{d}W^2}{\mathrm{d}t} = \langle \boldsymbol{j}_{\mathrm{ES}} \cdot \nabla h \rangle - \langle (\nabla^2 h)^2 \rangle . \tag{4.132}$$

This clearly demonstrates how the surface is roughened (the width increased) by the uphill component of the growth-induced current, and smoothened by the curvature term in (4.122). In the isotropic case $\boldsymbol{j}_{\rm ES}\cdot\nabla h = f(|\nabla h|^2)|\nabla h|^2 \geq 0$, and both terms on the right hand side of (4.132) have definite signs. Since the coarsening process can be viewed as a competition between the two terms, we expect them both, as well as their difference, to be of a similar order of magnitude[57]. This assumption is sufficient to fix the exponents z and β entering the scaling laws (4.78) and (4.79).

The scaling hypothesis (4.131) implies that the right hand side of (4.132) should be expressable in terms of λ and W. On dimensional grounds, the typical curvature of the surface is of the order of W/λ^2, and hence the right hand side of (4.132) can be estimated as W^2/λ^4. This immediately leads to $z = 4$ independent of β. However, the estimate implicitly assumes that the curvature is *uniformly* distributed over the surface, which is true for model I [(4.126)] without slope selection, but not for model II [(4.127)]. In the presence of slope selection, numerical integration of the evolution equation (4.122) shows that the surface breaks up into flat facets at the selected slope, $|\boldsymbol{m}| = 1$, which are separated by straight domain boundaries [407]. The width of these boundaries introduces another length scale into the problem, which invalidates naive dimensional analysis [394]. This length scale is independent of time, and coincides in order of magnitude with the initial mound size computed in Sect. 4.5.2. The surface curvature vanishes on the facets and is concentrated in the network of domain boundaries. The correct estimate of the right hand side of (4.132) is therefore $\langle(\nabla^2 h)^2\rangle \sim 1/\lambda$, which yields the scaling relation $2\beta - 1 = -1/z$. Since slope selection also implies $\beta = 1/z$, the result $z = 3$ follows for model II.

To complete the derivation for model I, we note that because of the unbounded steepening $f_{\rm I}(m^2) \approx 1/m^2$ for long times, and hence the first term on the right hand side of (4.132) is of order unity. Thus W^2 increases linearly in time, and $\beta = 1/2$. In summary, we have

$$z = 4, \quad \beta = 1/2 \quad \text{(model I)} \tag{4.133}$$

$$1/z = \beta = 1/3 \quad \text{(model II)}\ . \tag{4.134}$$

Equation (4.134) seems to be valid also in the presence of hexagonal crystal anisotropy, because the network of domain boundaries remains sufficiently disordered for the scaling arguments to apply. For square symmetry, as in (4.123), an analysis of the dislocation dynamics mentioned above suggests that $z \approx 4$, but for quite different reasons than in the case of model I [407]. For model II a weak form of the inequality $1/z \leq 1/3$ has been established with mathematical rigor [522].

The scaling arguments for model I can be extended to equations with a smoothening term of the general form $-(-1)^k(\nabla^2)^k h$ with $k > 2$ [523]. Such

[57] Exact calculations for one-dimensional growth equations show that this assumption may fail, because the two terms on the right hand side of (4.132) cancel almost completely [520, 521]. The simple scaling arguments presented here then only provide upper bounds on $1/z$ and β. This kind of behavior seems however to be specific to the one-dimensional geometry.

terms have been proposed to model situations in which thermal detachment from steps is impossible, although no clear identification of the associated microscopic processes has been provided [382]. The second term on the right hand side of (4.132) then becomes $\langle(\nabla^k h)^2\rangle$, which is estimated as W^2/λ^{2k}, leading to $z = 2k$. The estimate of the first term remains the same as above, so that $\beta = 1/2$ independent of k.

Up–down Symmetry. As written, (4.122) is symmetric under the transformation $h \to -h$, and correspondingly the morphologies it generates are up–down symmetric, that is, mounds and valleys have the same shapes. This is in strong disagreement with the mound patterns observed in experiments and KMC simulations, which display isolated hilltops separated by a connected network of valleys. Such morphologies can be modeled by supplementing (4.122) by a symmetry-breaking term similar to (4.75) [382]. In a gradient expansion, the lowest order term which accomplishes this is of the form $\nabla^2(\nabla h)^2$. For a one-dimensional evolution equation, it has been shown that such a term does not qualitatively change the coarsening behavior [520]. Since much of the analysis in this subsection relied on the introduction of the slope potential $\mathcal{F}$ defined in (4.128), which becomes impossible in the presence of a symmetry-breaking term, it is unclear if the same is true in two dimensions. In fact, it seems that the effect of the symmetry-breaking may be quite substantial: If the ridges are removed from the network of domain boundaries and only the valleys remain, it is no longer possible to localize most of the surface curvature in the domain boundaries, and the scaling arguments developed above have to be modified. In any case, it is worth keeping in mind that, because of the spurious up–down symmetry alone, (4.122) should not be expected to describe real mound morphologies in a quantitative manner.

Growth with Desorption. The up–down symmetry is also broken when desorption from the surface is allowed for. The probability for a deposited atom to desorb from the surface before being captured at a step evidently depends on the step density, and hence on the surface slope. The desorption rate is therefore an even function of ∇h. Adding such a function to the right hand side of (4.122) fundamentally changes the character of the evolution equation, because the slope-dependent desorption rate cannot be written as the divergence of a current. In [524] the effect of desorption on mound coarsening was studied in the framework of the evolution equation

$$\frac{\partial h}{\partial t} = -\nabla \cdot [(1 - (\nabla h)^2)(\nabla h)] - \frac{\varrho}{1 + (\nabla h)^2} - (\nabla^2)^2 h \,, \tag{4.135}$$

where $\varrho > 0$ is a dimensionless measure of the desorption rate. The numerical integration of (4.135) shows the emergence of conical mounds separated by a network of crevices (Fig. 4.52). The form of the desorption term on the right hand side of (4.135) implies that most desorption occurs from hilltops and valleys, where $\nabla h \approx 0$. Since the valleys form a network of lines, while the hilltops (the tips of the cones) are point-like objects, the growth rate in the

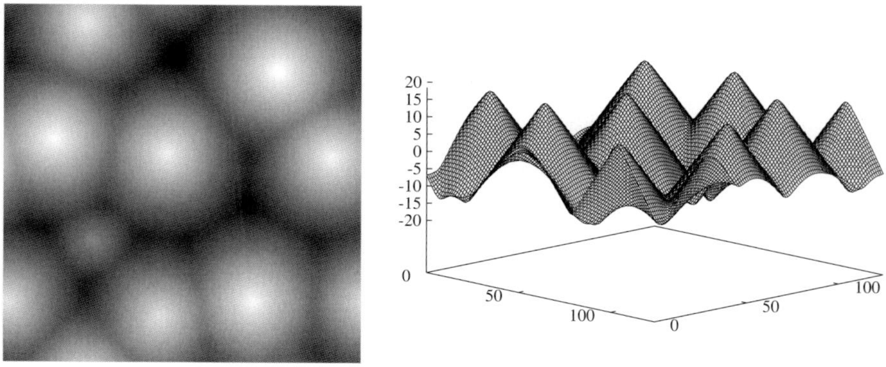

Fig. 4.52. Mounded surface generated by the evolution equation (4.135) with $\varrho \approx 0.32$. Conical mounds (right) form a cellular structure shown in gray-scale representation on the left (Fig. 4a of [524])

valleys is smaller than that at the hilltops by an amount of the order of $1/\lambda$, where λ is the lateral mound size. The surface width then increases according to $\mathrm{d}W/\mathrm{d}t \sim 1/\lambda$. Together with the slope selection property of (4.135) this implies $\beta = 1/z = 1/2$. We conclude that desorption leads to a significant speedup of coarsening.

4.9.9 Stability of Vicinal Surfaces: Continuum Approach

The continuum theory developed in the preceding section can be straightforwardly applied to the growth of a vicinal surface [410]. It is instructive to use it to discuss the stability of such a surface [106, 377]. To this end the evolution equation (4.122) is expanded to linear order around a surface of constant slope $\boldsymbol{m}_0$, that is, the ansatz

$$h(\boldsymbol{r}, t) = \boldsymbol{m}_0 \cdot \boldsymbol{r} + \varepsilon(\boldsymbol{r}, t) \tag{4.136}$$

is inserted into (4.122) and only terms linear in ε are kept (compare to the one-dimensional stability analysis in Sect. 4.5.2). Using the isotropic current function (4.125), the tilt vector $\boldsymbol{m}_0$ can be chosen along the x-axis without loss of generality. This results in an anisotropic, two-dimensional version of (4.67),

$$\frac{\partial \varepsilon}{\partial t} = K_2^{\parallel} \frac{\partial^2 \varepsilon}{\partial x^2} + K_2^{\perp} \frac{\partial^2 \varepsilon}{\partial y^2} - (\nabla^2)^2 \varepsilon \, . \tag{4.137}$$

In analogy to the one-dimensional case discussed in Sect. 4.5.2, the coefficients $K_2^{\parallel}$ and $K_2^{\perp}$ determine the stability of the surface with respect to undulations parallel and perpendicular to the tilt vector $\boldsymbol{m}_0$, respectively. These two instability modes correspond precisely to step bunching and step meandering: The surface undergoes a step bunching instability if $K_2^{\parallel} < 0$, and a step meandering instability if $K_2^{\perp} < 0$.

The two coefficients in (4.137) can be compactly expressed in terms of the *one-dimensional* surface current corresponding to (4.125), which is defined by $j_{\mathrm{ES}}(m) = f(m^2)m$. One finds that

$$\begin{aligned} K_2^{\|} &= -[f(m_0^2) + 2f'(m_0^2)m_0^2] = -\mathrm{d}j_{\mathrm{ES}}/\mathrm{d}m(m_0) \\ K_2^{\perp} &= -f(m_0^2) = -j_{\mathrm{ES}}(m_0)/m_0 \,. \end{aligned} \tag{4.138}$$

Thus step meandering requires a current that is directed uphill, in the sense that $j_{\mathrm{ES}}(m_0)/m_0 > 0$, while step bunching requires that the current increases in magnitude with increasing tilt if it is directed uphill, and decreases with increasing tilt if it is downhill. An application of these ideas is described in Sect. 4.9.7.

A classification of different step bunching mechanisms based on continuum theory has been proposed in [513]. For a proper of the step bunches, the stabilizing term in (4.122) has to be modified to take into account the repulsive step-step interactions.

5. Layer–By–Layer Growth and Growth Manipulation

When Shannon opened the shutter controlling the Ga oven to begin the growth sequence, we observed with amazement that the diffraction spots were pulsing in brightness.
John R. Arthur [525]

Technical applications of thin films and thin film systems often necessitate atomically sharp interfaces between the film and the vacuum or between different film layers. To mention a few examples: Magnetic coupling through non-magnetic layers depends sensitively on the spacer layer thickness [526]; tunneling devices require insulator spacer layers of precisely controlled thickness (implying negligible roughness) [527]; and spin dependent electron transport through interfaces is strongly correlated with the interface roughness [528–530]. Thus, generally it is desirable to grow thin films in a layer–by–layer fashion resulting in smooth interfaces. Occasionally, however, the opposite may be required, and a rough interface resulting from mound growth may be advantageous, e. g. for shadow deposition of nanomagnets [531]. Although in practical thin film applications mostly heteroepitaxial and polycrystalline films are used, investigations on how to influence the thin film growth morphology in homoepitaxy uncover the means supplied by growth *kinetics* towards the goal of creating thin film interfaces with purposely defined morphological properties.

5.1 Basic Concepts

In this section, concepts to deliberately change the morphology of a thin film by intervening in the atomistic processes during its formation are presented. In principle, two ways of modifying the growth processes towards layer–by–layer growth can be distinguished: Either segregating adsorption layers of *surfactants* are used, or methods are applied to decouple the nucleation of islands on coalesced layers from the nucleation on top of islands (second layer nucleation).

Traditionally, the term surfactant refers to substances that lower the surface energy, thereby improving the spreading and wetting properties of the film [532]. In the context of epitaxial growth, the first studies concerned with the effect of adsorbed layers on the growth of metals [533] and semiconductors [534] indeed did emphasize this point: An adsorption layer floating on the growing film and lowering its surface free energy will tend to improve the wetting of the film material to the substrate, thereby reducing film roughness. However, meanwhile the term surfactant is generally accepted to designate any substance adsorbed in monolayer or submonolayer concentration to the film surface, which effectively segregates during growth such that it floats on the surface (ensuring its continued action at the growth front and a clean, surfactant-free film interior), and which reduces the roughness of the growth front in comparison to the situation without the surfactant, be it for kinetic or energetic reasons. In the context of homoepitaxy discussed here, surfactant action is primarily investigated in view of its action on the growth kinetics, as even in the absence of surfactants thermodynamics favors the growth of smooth films (see Sect. 1.3).

Methods to decouple island nucleation on large terraces (coalesced layers) from the nucleation on top of islands (compare Sect. 4.3.2) were first applied to the homoepitaxial growth of Ag on Ag(111) [535]. The basic idea is to use pulses of flux, temperature, or particles creating nucleation centers with monolayer periodicity, such that an artificially high concentration of islands is nucleated at the beginning of each layer. This effectively reduces the island spacing l_D without significantly influencing the critical radius for second layer nucleation R_c. Applying these methods of growth manipulation, layer–by–layer growth can be achieved for growth situations in homoepitaxy [487, 535–537] or even in heteroepitaxy [538], where otherwise the growth kinetics would induce mound formation. The use of pulses of flux and temperature to manipulate growth was initially developed with the goal to synchronize nucleation on slightly miscut semiconductor surfaces with a broad initial terrace distribution [539].

The analysis of the methods for synchronized nucleation and growth manipulation will also provide us with an understanding of the morphological effects of *naturally* occurring surface modifications that appear with monolayer period at the growth front. Specifically, it will allow us to understand the high quality layer–by–layer growth observed for homoepitaxy on Pt(111) around 600 K on the basis of a reconstruction sweeping over the surface with monolayer period [332]. Likewise for surfaces with large unit cell reconstructions, such as the Si(111) 7×7 surface, a network of domain boundaries forms periodially which may have profound consequences on growth behavior [332, 540, 541].

A good starting point for the discussion of surfacant action and growth manipulation is the mound growth criterion (4.59), which states that mound formation sets in immediately if the average island separation l_D is larger than twice the critical radius R_c for second layer nucleation. The following discussion will be limited to the case $i^* = 1$, although all conclusions hold qualitatively also for situations with $i^* > 1$. As was described in Sect. 4.4.6, the criterion may be reformulated in terms of the fundamental activation energies E_S for step crossing and E_D for surface diffusion, as [compare to (4.60)]

$$E_\mathrm{S} - 5/6 E_\mathrm{D} > k_\mathrm{B} T \ln c_0 . \tag{5.1}$$

Assuming for the moment a fixed deposition rate F the quantity c_0 is a constant. Based on (5.1) we distinguish two principal scenarios of surfactant action.

In the first scenario a surfactant layer changes the activation energy E_D for surface diffusion without an effect on the step edge barrier E_S. According to (5.1), an increase of E_D tends to drive the system away from mound growth towards layer–by–layer growth. Indeed, in many cases an increase of the effective activation energy for surface diffusion in the presence of an adsorption layer could be inferred from a drastic enhancement of the island density N in the submonolayer regime [203, 279–281].

In the second scenario a surfactant layer changes the step edge barrier E_S without affecting E_D. This scenario may be realized e.g. by adsorption of a small fraction of a monolayer of mobile surfactant molecules, which preferentially bind to step edges. This scenario requires efficient *two-dimensional* segregation of the adsorbates, to ensure that they remain at the step edge while the islands grow [198]. The adsorption of a trace amount of CO in the homoepitaxy on Pt(111) is an example for such a situation (see Sects. 3.5 and 4.4.5). For CO on Pt(111), however, the step edge barrier is not lowered, but increased, and thus CO is an anti-surfactant or a mound growth agent [131].

There is a third scenario of surfactant action, which cannot be inferred from (5.1). The criterion (5.1) is based on (4.59), which assumes compact first layer islands. This assumption is not necessarily correct. Specifically, an adsorption layer may leave E_D and E_S unchanged but influence the mobility of atoms along the island steps, such that it changes the island shape. If the adsorption layer induces a transition from compact to ramified islands, it acts as a surfactant: The surfactant increases the fraction of step edge sites compared to terrace sites, thereby reducing the adatom lifetime τ on top of an island and thus the probability for second layer nucleation prior to coalescence. An example for a surfactant induced transition from compact to ramified island shapes is the action of Sb in the homoepitaxy on Ag(111) [203], which was already briefly mentioned in Sect. 3.5 (see Fig. 3.21b).

Growth manipulation relies on the creation of an artificially high island number density at the beginning of each deposited monolayer. Using (4.51), the mound growth criterion (4.59) can be brought into the form

$$N^{5/2} < \frac{\pi^3}{896} \frac{F}{\Omega a \nu'} . \tag{5.2}$$

The decoupling of nucleation and second layer nucleation is expressed in (5.2) by the fact that the island number density N on the left hand side is established during the nucleation phase, while F and ν' on the right hand side are relevant during second layer nucleation. Anything that creates a larger island number density will shift the system towards layer–by–layer growth. According to (2.26), the dependence of N on temperature T is much stronger than that on the deposition rate F_{nuc} during nucleation (here F_{nuc} refers to the deposition rate during nucleation, while F is the deposition rate during second layer nucleation). Practically, however, it is much easier to change the deposition rate F_{nuc} (e.g. with the help of a second pulsed evaporator) than to change the temperature during deposition. Rapid temperature changes require large heating and cooling powers and will give rise to a significant increase of the background pressure. The most efficient way of increasing N is to use ion pulses, which easily allow a rise of N by an order of magnitiude, without affecting the growth conditions and parameters (F, T, ν and ν') [281, 487, 535, 537].

In the following three sections, a number of experimental examples will be described which illustrate the concepts introduced here. Section 5.5 presents a simple theoretical description of growth manipulation by pulsed deposition, while Section 5.6 addresses the fundamental limits on the duration of layer–by–layer growth induced by the fluctuations in the deposition flux.

5.2 Layer–By–Layer Growth by Surfactants

Changing the growth morphology by adsorption prior to deposition is experimentally extremely simple. However, not every adsorbate is a surfactant, in the sense that it segregates efficiently and smoothens the growth morphology. As there is no simple way of predicting the action of an adsorbate for a specific growth system, finding a good surfactant is more or less a matter of trial and error. Moreover, the three surfactant effects separated in Sect. 5.1, (i) increase of E_{D}, (ii) decrease of E_{S} and (iii) transition from compact to ramified islands, are generally observed in combination. For a specific surfactant it is thus often difficult to establish the dominant mechanism that changes the growth morphology. In this section three examples of surfactants in metal homoepitaxy are discussed, for which this separation could be established at least partially: adsorbate layers of CO and O on Pt(111) [280, 281, 327], Sb on Ag(111) [203, 204, 484], and In and Pb on Cu surfaces [489, 542–545].

While these examples illustrate the basic mechanisms of surfactant action, the application of surfactants to alter growth behavior is obviously not limited to homoepitaxial metal systems. There are numerous studies of surfactant action in semiconductor growth systems, e.g. [540, 541, 546] (a review is given in [547]) and heteroepitaxial metal systems, e.g. [548–550].

CO on Pt(111). As was already pointed out in Sects. 3.5 and 4.4.5, a *small amount* of CO adsorbed during growth has a tremendous effect on the film morphology. Without significantly affecting E_{D} and the compactness of the

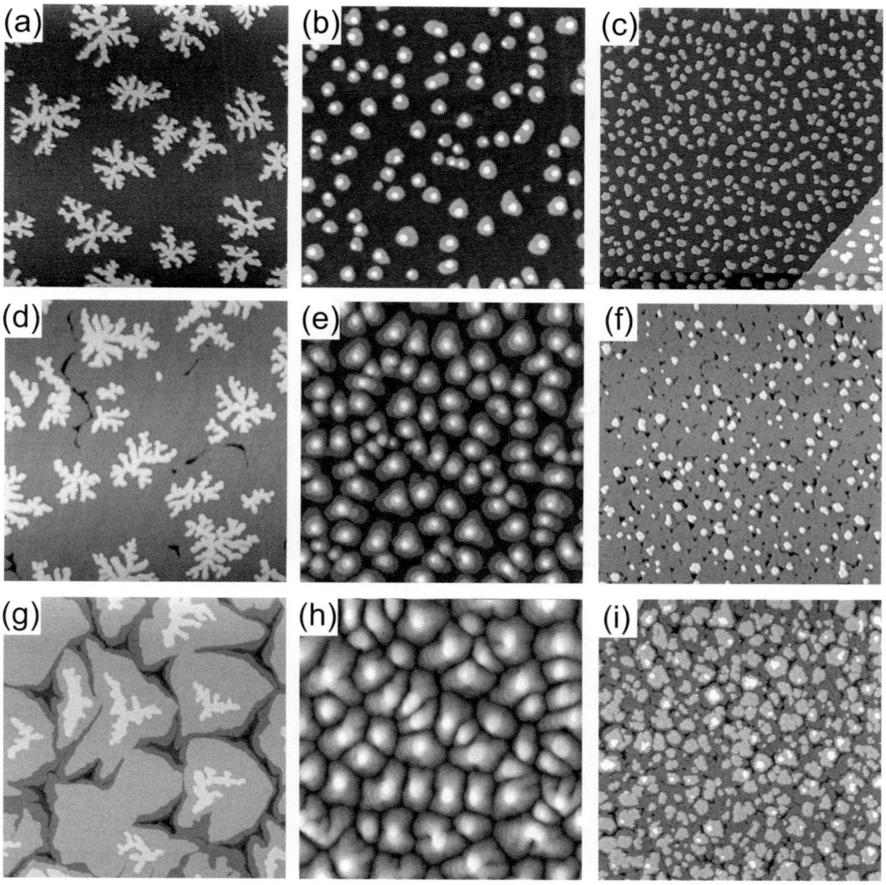

Fig. 5.1. Comparison of homoepitaxy at 300 K on clean Pt(111) (first column), on c(4×2)CO–Pt(111) with $\Theta_{\mathrm{CO}} = 0.5$ ML (second column) and on p(2×2)O–Pt(111) with $\Theta_{\mathrm{O}} = 0.25$ ML (third column). The deposited amounts are $\Theta \approx 0.2$ ML in (**a**)–(**c**) (first row), $\Theta \approx 1.0$ ML in (**d**)–(**f**) (second row) and $\Theta \approx 5.0$ ML in (**g**)–(**i**) (third row). The deposition rate is $F = 5 \times 10^{-3}$ MLs^{-1}, and topograph sizes are 1100 Å × 1100 Å ((**b**), (**c**), (**e**), (**f**), (**h**) and (**g**) from Fig. 35 of [281])

islands, it raises E_{S} and thus $\Delta E_{\mathrm{S}} = E_{\mathrm{S}} - E_{\mathrm{D}}$. On the other hand, CO adsorbed to *saturation* on Pt(111) forms an ordered c(4 × 2) overlayer with a coverage of $\Theta_{\mathrm{CO}} = 0.5$ [284]. As can be seen from the comparison of Figs. 5.1a and b, the CO overlayer increases the island number density by a factor of 5.5 compared to clean growth. Assuming $i^* = 1$ also in the presence of the CO overlayer and identical capture numbers σ and attempt frequencies ν_0, this corresponds to an increase in the effective adatom diffusion barrier E_{D} by 0.13 eV. Still, this increase in E_{D} is not sufficient to drive the balance in (5.1) towards layer–by–layer growth. On the contrary, as seen from the comparison between clean and CO overlayer growth after deposition of 1.0 ML (Figs. 5.1d and e) and 5.0 ML (Figs. 5.1g and h), despite the increased Pt island number density the CO overlayer growth is considerably rougher than growth on clean Pt(111). Quantitatively the roughness increase can be expressed by an increase of the growth number[1] g from 0.33 (clean substrate) to 0.96 (substrate with CO overlayer). Equation (5.1) then immediately shows that E_S must have increased by more than $5/6 \times 0.13\,\mathrm{eV} = 0.11\,\mathrm{eV}$. This is consistent with the findings in Sect. 4.3.3, where we saw that already a small amount of CO increases E_S by more than 0.18 eV. In conclusion, CO adsorption causes a drastic increase in E_S, which is able to overcompensate the reduced adatom mobility in a saturated CO overlayer.

O on Pt(111). Supply of oxygen from the background gas during homoepitaxial growth on Pt(111) was observed in TEAS experiments to drive the system towards layer–by–layer growth [327]. Therefore also a preadsorbed oxygen layer is likely to exhibit surfactant action. Indeed, preadsorption of the p(2 × 2) oxygen overlayer on Pt(111) gives rise to an even more drastic increase of the island number density than preadsorption of CO: N increases by a factor of 25 in comparison to the clean case (compare Figs. 5.1a and c). Assuming again that i^*, ν_0 and σ remain unchanged, this implies an increase in E_D from 0.26 eV in the clean case to an effective value of 0.51 eV. This shifts the balance in (5.1) strongly away from mound formation. Growth with a similar roughness as for the clean case would be maintained only by a simultaneous increase of the step edge barrier by 0.21 eV. Figures 5.1f and i show that such a strong compensating increase of E_S does not take place: The roughness of the 5.0 ML film grown on the O covered Pt(111) surface (Fig. 5.1i) is considerably smaller than that of the film grown on the clean substrate (Fig. 5.1g), as quantified by a growth number g of only 0.17 compared to 0.33 in the clean case.

To date, nothing more can be said about the action of O on the step edge barrier than that it increases E_S by significantly less than 0.21 eV. It also may leave it unchanged or even decrease its value. In previous experiments [280, 281] it was postulated that E_S is in fact decreased due to the O overlayer in comparison to clean growth. However, although the analysis of the experiments was in principle correct, the interpretation was marred by imperfect growth conditions. During the growth of the "clean" reference film, CO was still created and coadsorbed in significant amounts, decorating the island edges [131]. The apparent decrease of E_S due to the preadsorbed oxygen overlayer was therefore largely due to the fact that the overlayer keeps the surface and specifically the step edges free of CO.

A nice demonstration of the surfactant action of O is given in Fig. 5.2, where the morphologies after deposition of 5.0 ML on clean (Fig. 5.2a) and oxygen covered Pt(111) (Fig. 5.2b) at 400 K are compared. Oxygen preadsorption changes the clean growth of maximal roughness with a growth num-

[1] See (4.6) for the definition of the growth number g.

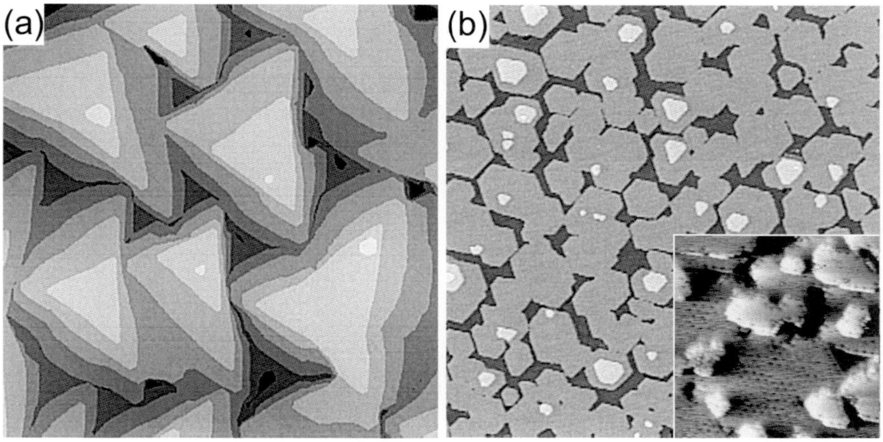

Fig. 5.2. Comparison of homoepitaxy at 400 K on (**a**) clean Pt(111) and on (**b**) p(2 × 2)O–Pt(111) with $\Theta_O = 0.25$ ML and 5 ML Pt deposited. The inset in (**b**) recorded after growth of 5 ML at 300 K demonstrates the presence of oxygen atoms (black dots) on the terrace and at step edges. Topograph sizes are 2130 Å × 2130 Å for the main figures and 140 × 140 Å for the inset ((**b**) and inset from Fig. 2d and Fig. 3 of [280])

ber $g = 0.65$ (compare Fig. 4.4) to nearly perfect layer–by–layer growth with $g = 0.07$. The inset in Fig. 5.2b was imaged 30 minutes after a film of 5.0 ML at 300 K was grown and shows the lateral distribution of oxygen, which appears as black dots. Besides patches of the p(2 × 2) overlayer also decoration of step edges by O atoms is observed, consistent with DFT calculations which find that oxygen binds preferentially to step edges [551]. Such images provide evidence that O effectively segregates on the film surface. In contrast to most other surfactants, oxygen can be completely removed from the film surface after growth through H_2 exposure at room temperature [280]. After the adsorbed O had been reacted away, no O could be detected by Auger spectroscopy, which implies that O is not significantly incorporated into the film during growth.

Sb on Ag(111). The first striking example for surfactant action on a metal surface was antimony supplied prior to homoepitaxial growth on Ag(111) [484]. Growth on the clean Ag surface at 300 K gives rise to an exponentially damped scattering signal without oscillations, characteristic for immediate mound formation [compare to (4.13)]. As can be seen in Fig. 5.3, preadsorption of 0.2 ML Sb induces strong oscillations in the scattering signal, characteristic of layer–by–layer growth. The oscillations are damped, but an additional supply of antimony after 20 ML of Ag are deposited leads to a recovery of the oscillation amplitude. Although Sb segregates effectively (otherwise a small preadsorbed amount could not influence growth for 20 ML),

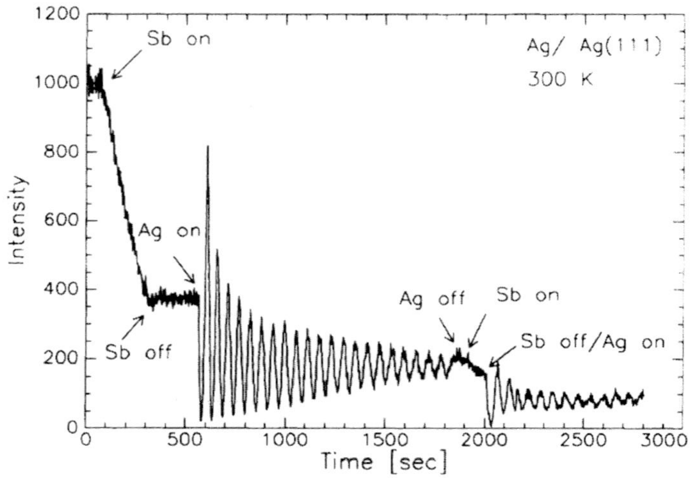

Fig. 5.3. X-ray reflection signal during homoepitaxial growth on Ag(111) at 300 K. Initially, 0.2 ML Sb were adsorbed (Fig. 2 of [484])

the damping indicates a slight loss of Sb during growth, presumably to the interior of the thin film.

Originally it was speculated that Sb lowers the step edge barrier on Ag(111). However, a subsequent STM study revealed a different picture [203]. Although nothing could be concluded about the magnitude of E_S, the STM topographs of Fig. 5.4 show two effects: First, Sb, which embeds into the topmost surface layer, gives rise to an increased island number density, and second, Sb causes a transition from compact to ramified island shapes in the first and all further layers. It is therefore likely that the effective increase of E_D and the shape transition trigger the onset of layer–by–layer growth. Ab initio calculations support this picture [204]. They show that Sb has its energetically favorable position in the surface layer. Thus for energetic reasons the topmost layer forms a surface alloy floating on the growing film. The substitutional Sb slightly repels a nearby Ag adatom. Even more important, the Ag adatom diffusion barrier is raised significantly by several tenths of an eV in the vicinity of a substitutional Sb atom. Thus, the Sb–Ag surface alloy gives rise to an increased effective diffusion barrrier E_D in agreement with the measurements. One may speculate that also for Ag step adatoms a nearby substitutional Sb atom increases the barrier for step edge diffusion, which would explain the transition from compact to ramified islands.

In and Pb on Cu. It was predicted on the basis of EMT calculations that indium should act as an efficient surfactant for the homoepitaxial growth on copper surfaces [542,543]. Experiments on Cu(001) and Cu(111) indeed show that In enhances the surface smoothness during growth and forms surface alloys which segregate efficiently [489,544]. On Cu(100), In was postulated to improve growth primarily by lowering the step edge barrier. This is supported

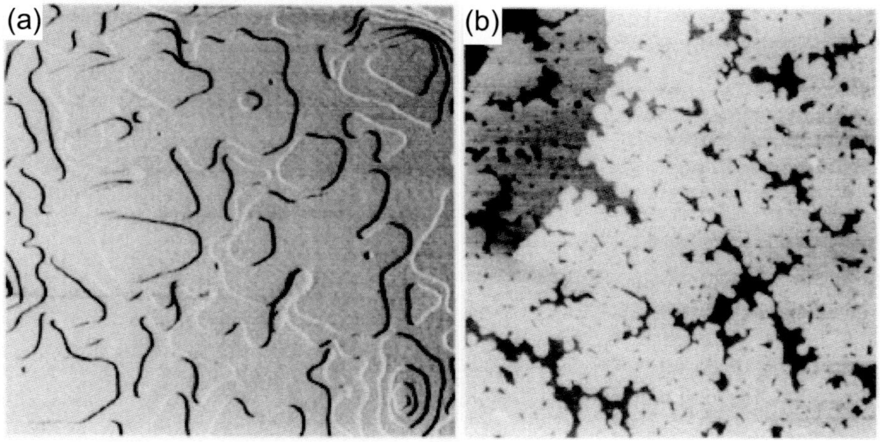

Fig. 5.4. (**a**) 2.7 ML Ag deposited on the clean Ag(111) substrate at room temperature. Topograph size 11000 Å × 11000 Å. (**b**) 1.7 ML Ag deposited at room temperature on Ag(111) precovered with 0.08 ML Sb. Topograph size 5700 Å × 5700 Å (Figs. 1b and 2c of [203])

by the experimental finding that In does not significantly change the island number density during growth [489]. On Cu(111), it was postulated that In changes the growth behavior largely by acting as a trap for Cu adatoms, thereby enhancing the island number density. Experimentally it was found that In in fact does induce a transition from immediate mound formation to layer–by–layer growth on Cu(111), but the mechanism of the growth mode change could not be clearly identified [544].

Predeposition of an atomic layer of Pb onto the Cu(111) surface changes growth at 300 K from immediate mound formation to layer–by–layer growth [545]. The preadsorbed Pb layer leads to an enhancement of the island number density by two orders of magnitude and thus to a strong rise in the diffusion barrier, which by itself may explain the onset of layer–by–layer growth. Annealing measurements and simulations indicate that the deposited Cu adatoms immediately penetrate through the lead cover layer and migrate on the Cu surface by exchange diffusion rather than by hopping motion as on the clean surface. Since an efficient surfactant must float in or on the outermost layer of the growth front, exchange mechanisms between surfactant and deposit atoms are a necessity for surfactant growth. The change of the diffusion mechanism of the deposit to exchange diffusion, however, is not a necessary condition for surfactant mediated growth, but is a specific feature that appears only in certain systems.

5.3 Layer–By–Layer Growth Through Growth Manipulation

A straightforward realization of the idea of decoupling the processes of first and second layer nucleation has been applied to homoepitaxy on Ag(111) [535]. As already explicated in Sect. 5.1, the kinetic pathway to growth manipulation is to create a high island number density on flat terraces at the beginning of the deposition of each monolayer, without affecting the parameters determining second layer nucleation [compare to (5.2)]. In Fig. 5.5a the reflected intensity of the thermal energy helium beam during homoepitaxy on Ag(111) at 260 K is seen to decay monotonically. Predeposition of 0.05 ML at 100 K leads to a high island density, and thus subsequent deposition at 260 K results in a growth behavior of the first monolayer with an intensity evolution characteristic for layer–by–layer growth (Fig. 5.5b). The intensity drops, reaches a minimum and then almost recovers its initial value. It may be concluded that the low temperature predeposition gives rise to a nearly perfect coalescence of the first layer prior to second layer nucleation.

The same effect is achieved by bombarding the substrate prior to growth with a pulse of energetic Ar^+ ions as shown in Fig. 5.5c. It is much simpler to expose the growing sample to a new ion pulse after the completion of each monolayer, compared to the tedious procedure of periodically cooling the sample in order to prepare a high density of nuclei for the next layer.

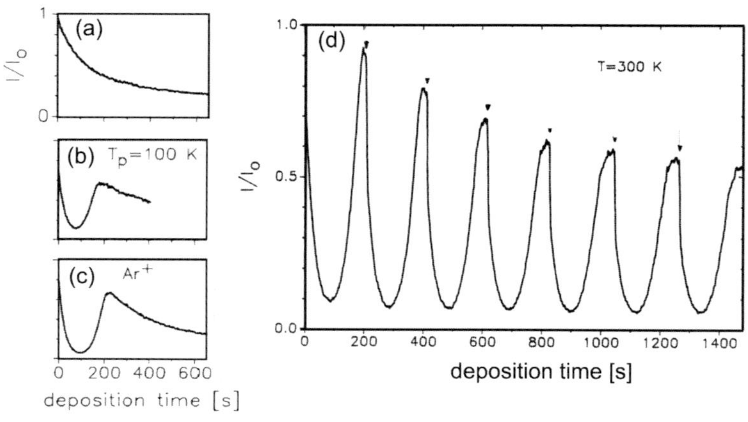

Fig. 5.5. Evolution of the He specular peak intensity during deposition of Ag on Ag(111). (**a**) Deposition on bare substrate with a deposition rate $F = 4.2 \times 10^{-3}\,\mathrm{ML/s}^{-1}$ at 260 K. (**b**) Substrate prepared with a high density of nuclei by predeposition of 0.05 ML of Ag at 100 K and with a rate $F = 4.3 \times 10^{-2}\,\mathrm{ML/s}^{-1}$ followed by deposition under identical conditions as in (**a**). (**c**) Substrate bombarded by 600 eV Ar^+ ions with a fluence corresponding to sputter removal of 0.014 ML, followed by deposition under identical conditions as in (**a**). (**d**) Deposition at 300 K with $F = 4.7 \times 10^{-3}\,\mathrm{ML/s}^{-1}$. At the start and in the maxima a short ion pulse with bombardment parameters as in (**c**) was given (Figs. 2a, 2e, 2f and 3 of [535])

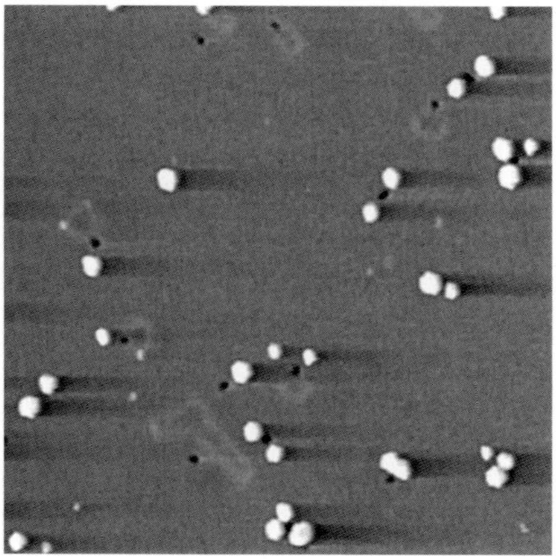

Fig. 5.6. Pt(111) after 5 keV Xe^+ ion bombardment at 300 K with a fluence of 4.1×10^{15} ions/m^2. The fluence corresponds to an average of 14 impacts in the surface area shown (see text). Topograph size is 590 Å × 590 Å

An iteration of well-timed ion pulses is able to change the time evolution of the scattering intensity to the behavior typical for layer–by–layer growth: Continued intensity oscillations are observed (Fig. 5.5d). By using ion pulses also the growth of Cu on Cu(111) [487] and even the heteroepitaxy of Ni on Cu(111) could be turned into layer–by–layer growth [538].

The ion pulses are so efficient in manipulating growth because they give rise to the formation of a huge number of island nuclei. Ion impacts not only cause sputtering, but also induce the formation of adatoms and adatom clusters on the surface [552–558]. As an example, Fig. 5.6 shows a Pt(111) surface after 5 keV Xe^+ ion bombardment at 300 keV, with a dose corresponding to 14 impacts in the area shown. Although the signatures of the impacts differ considerably, they are typically characterized by a small crater surrounded by zero, one or several adatom islands, confined to an area with a diameter of less than 50 Å. Obviously, almost each impact gives rise to an adatom cluster, which acts as a stable nucleus during subsequent growth. Even for relatively low energy bombardment with an average adatom yield around unity [e.g. for 400 eV Ar^+ on Pt(111)] a large density of nuclei is observed [536, 559]. The reason is simply that the distribution of the number of adatoms created per impact is very broad, such that in a significant fraction of the impacts several adatoms are formed simultaneously in a small area. These adatoms are either directly formed as a cluster or nucleate with high probability due to their spatial correlation. As a result, a significant fraction of impacts will lead to stable nuclei.

A real space demonstration for the efficiency of pulsed ion bombardment as a means to manipulate homoepitaxial growth was given for the homoepitaxial growth on Pt(111) [536]. Figure 5.7 compares the morphology after

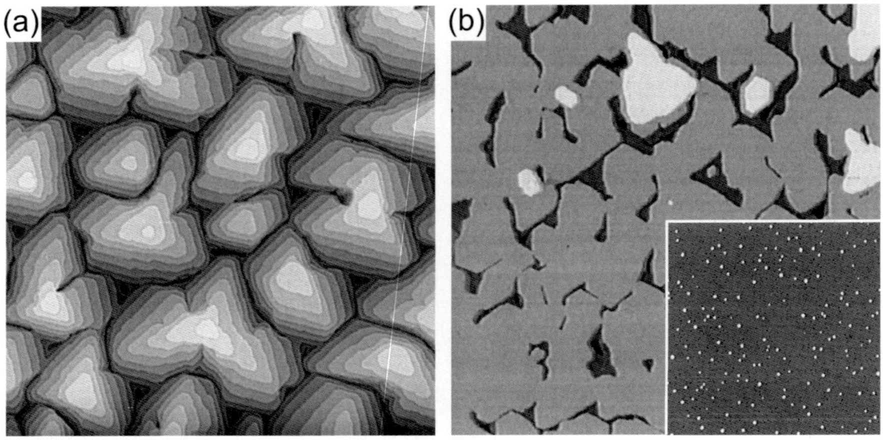

Fig. 5.7. Pt(111) after deposition of 5 ML Pt at 400 K in a CO background pressure of $p_{\mathrm{CO}} \approx 5 \times 10^{-10}$ mbar with $F = 5 \times 10^{-3}$ ML/s^{-1} (**a**) without and (**b**) with 4 keV Ar^+ ion pulses with a fluence of 2×10^{16} ions/m^2 at the beginning of each deposited layer. Topograph size is 2130 Å × 2130 Å. The inset in (**b**) shows the Pt(111) surface after the initial ion pulse on the same scale (Figs. 8a and 8d of [536])

growth on Pt(111) without and with pulses of 4 keV Ar^+ ions at the beginning of the deposition of each monolayer, under otherwise identical growth conditions. For deposition of 5 ML at 400 K in a CO background pressure of about $p_{\mathrm{CO}} \approx 5 \times 10^{-10}$ mbar, the ion pulses reduce the growth number from 0.95 to 0.20. The inset of Fig. 5.7b visualizes the very high density of nuclei resulting from each ion pulse.

Although growth manipulation by ion pulses is simple and efficient, it has the disadvantage of inducing damage into the growing film. Particularly at temperatures below about $0.3\,T_{\mathrm{m}}$, the subsurface vacancies and vacancy clusters created due to the bombardment are not sufficiently mobile to anneal at the surface [554, 555]. They will thus be incorporated into the growing film and lead to a certain film porosity.

5.4 Layer–By–Layer Growth Through Surface Reconstruction

The explanation of the layer–by–layer growth regime in homoepitaxy on Pt(111) which appears in Figs. 4.2b and 4.3 around 600 K was postponed to the present section, because the underlying mechanism fits quite well into the conceptual framework of growth manipulation and surfactant action [332]. The perfection of the layer–by–layer growth on Pt(111) is unique in metal homoepitaxy: After some initial damping within the first few monolayers, the intensity in the TEAS signal recorded during growth exhibits an oscillation

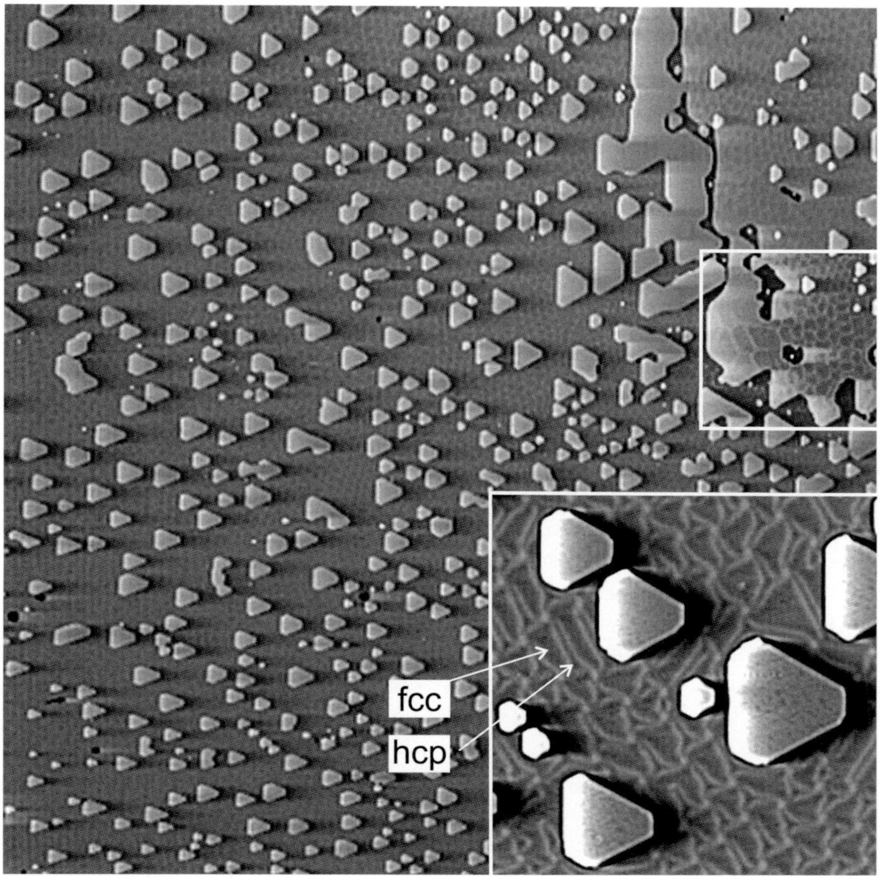

Fig. 5.8. Layer–by–layer morphology on Pt(111) after deposition of 5 ML with $F = 5 \times 10^{-3}\,\mathrm{ML/s}^{-1}$ at 600 K. The inset shows the presence of the reconstruction network on the terrace and its absence on the islands. The reconstruction network is formed by the bright lines separating unfaulted fcc and the narrow, faulted hcp areas. In the contrast enhanced area above the inset the reconstruction is seen to spread onto islands that just coalesced with the preexisting step. Topograph size is 8000 Å × 8000 Å and inset size is 820 Å × 820 Å

amplitude that remains almost constant for more than 150 layers [326]. A large scale STM topograph after the deposition of 5 ML at 600 K shown in Fig. 5.8 exhibits perfect layer–by–layer growth on the length scale defined by the preexisting steps: On this length scale only two uncovered layers are present, the terrace and the islands on the terrace.

A network of lines – the *network reconstruction* – is visible on the terrace level. The lines form a network of Shockley partial dislocations situated between the surface layer and the layer below, which separate areas of unfaulted, regular fcc stacking and areas of faulted hcp stacking for the topmost atomic layer. The property of the network reconstruction which is decisive for its ability to induce layer–by–layer growth is highlighted by the inset of Fig. 5.8: The network reconstruction lines are present only on the terrace, but *not* on the islands. This difference in the reconstruction distribution leads to a natural decoupling of the island nucleation on the terrace from the second layer nucleation on top of islands, in a way that is similar to artificial growth manipulation.

The formation of the reconstruction is a nucleation phenomenon which takes place with a probability much lower than that for the nucleation of islands [330, 331, 560]. The nucleation of the reconstruction does not take place during deposition below 400 K, is still rare between 400 K and 500 K and becomes noticeable only for deposition temperatures above 500 K. A snapshot taken during the initial proliferation of the reconstruction is shown in Fig. 5.9a. After the nucleation of islands on the unreconstructed surface the reconstruction has rushed in from the right, and spreads to the left over the terrace[2]. Once the dense network was formed in the right part of Fig. 5.9a a new phase of nucleation took place, which resulted in an island number density that is higher by a factor of 20 compared to the initial nucleation on the unreconstructed surface.

The new phase of nucleation that occurs after the surface becomes reconstructed is caused by a drastic change in the potential energy landscape for the deposited adatoms. EMT calculations for the potential energy of a Pt adatom on a reconstructed Pt(111) surface are shown in Fig. 5.9b [560]. From left to right the adatom is dragged from an unfaulted area over a reconstruction line into a faulted area. Apparently the adatom potential energy as well as the effective activation barrier for adatom diffusion is increased by about 0.1 eV on the reconstruction line. In other words, the reconstruction partitions the surface into cells, in which the adatom residence time is considerably increased and thus the nucleation probability enhanced. The situation bears some analogy with nucleation on top of an island in the presence of a step edge barrier, as discussed in Sect. 4.3.2, with the important difference, however, that the cell boundary is not a sink for adatoms. A similar partitioning of the surface can be induced by strain relief patterns in heteroepitaxy, which have been used as templates for the growth of ordered island arrays [48]. A characteristic feature of nucleation on such patterned substrates is the occurrence of a regime of growth conditions in which a single island grows within each cell, and the island number density is therefore independent of flux and temperature. This effect was described in Sect. 2.4.6 in the related context of impurity trapping.

After the transient situation during the growth of the first few monolayers, the reconstruction is present on each of the large terraces defined by

[2] The density of the reconstruction lines at a specific location is to some extent a measure for the reconstruction age at that place: Initially the reconstruction lines are only sparsely distributed, but ongoing deposition continues to increase their density to a maximum value [258].

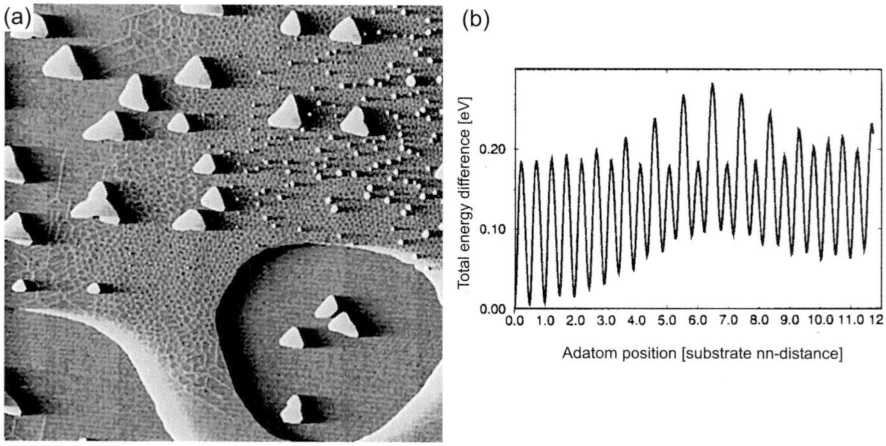

Fig. 5.9. (**a**) Contrast enhanced STM topograph after deposition of 0.15 ML at 640 K. The reconstruction sweeping over the surface from right to left causes a second phase of nucleation, resulting in a drastically increased island density (see text). Topograph size is 8300 Å × 8300 Å(Fig. 9d of [331]). (**b**) Total energy difference calculated by EMT as an adatom is dragged from left to right from an unfaulted, fcc-area over a reconstruction line to a faulted, hcp-area (see text) (Fig. 3 of [560])

the initial concentration of preexisting steps. In the stationary situation the reconstruction does no longer need to be nucleated in order to be present on a large terrace. At the moment when the islands coalesce and come into contact with the neighboring, reconstructed upper terrace, the reconstruction sweeps over the newly formed terrace. The contrast-enhanced box in Fig. 5.8 illustrates the process. Close to the step in the upper right corner (originating from a preexisting step) the reconstruction is seen to spread onto former islands, which just came into contact with the upper, already reconstructed terrace. As the reconstruction spreads with a lateral growth speed of about 5×10^4 Å/ML, during the deposition of a few hundredths of a monolayer after the coalescence of the majority of islands on a terrace, the reconstruction has invaded the entire new terrace.

The information needed to explain the phenomenon of reconstruction induced layer–by–layer growth is now at hand. On each terrace the reconstruction gives rise to an enhancement of the island density by a factor of about 20 compared to the unreconstructed situation. Thus the left hand side of (5.2) is increased by more than three orders of magnitude. As there is no reconstruction on top of the islands, all parameters referring to second layer nucleation on the right hand side of (5.2) remain unchanged. Consequently, the system is strongly driven towards layer–by–layer growth. The effect of the reconstruction is identical to that of the ion pulses given at the beginning of every monolayer as discussed in the preceding section. Compared to the ion

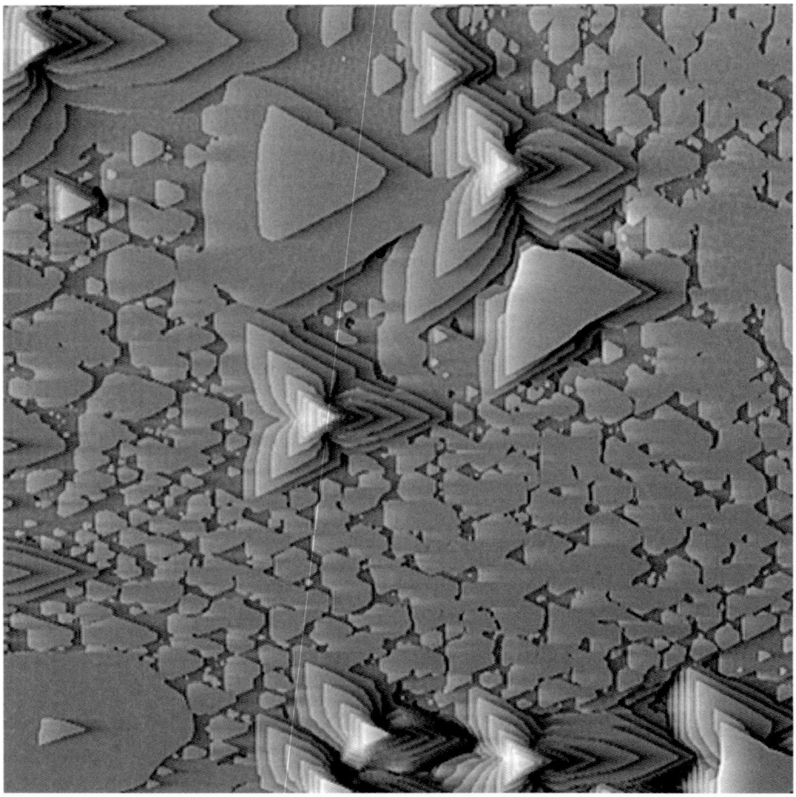

Fig. 5.10. Coexistence of reconstructed and unreconstructed surface areas after deposition of 20 ML at 520 K (see text). Topograph size is 8300 Å × 8300 Å

pulses, the reconstruction has the advantages of being automatically in phase with the layer completion process, and of not inducing any crystal damage.

The one–to–one correspondence between the presence of the reconstruction and the existence of perfect high temperature layer–by–layer growth is underlined by Fig. 5.10. While at 600 K the reconstruction spreads rapidly within a few deposited atomic layers over the entire surface, at lower temperatures coexistence of reconstructed regions with unreconstructed ones may be found even after deposition of several atomic layers. For deposition of 20 ML at 520 K the reconstructed regions are rare, as not more than a few percent of the surface area is reconstructed. Figure 5.10 exhibits such a rare reconstructed region surrounded by unreconstructed areas. The reconstructed part is flat, with the small island spacing characteristic for nucleation on a reconstructed terrace, while the unreconstructed regions display pyramidal mounds on a much larger lengthscale.

Having established the mechanism of reconstruction induced layer–by–layer growth, a number of questions related to the reconstruction itself remain to be addressed: Why is the reconstruction only formed during deposition, and only during deposition above about 400 K? Why is the reconstruction a nucleation phenomenon and why is it growing laterally? Why does the reconstruction not spread across step edges? A full understanding of the details of the kinetics and energetics of the reconstruction has been achieved [115, 330–332, 560, 561], and will be briefly summarized below.

In thermal equilibrium the Pt(111) surface exhibits the network reconstruction above 1330 K, while it is unreconstructed at lower temperatures [562, 563]. The driving force of the reconstruction is the large tensile surface stress in unreconstructed Pt(111) [564,565] (more precisely, the positive strain derivative of the surface free energy), i.e. the fact that surface atoms prefer a shorter in-plane bond length. Therefore, even below 1330 K the reconstruction may be formed under conditions where a high concentration of adatoms is available not only to form islands, but also to incorporate atoms into the surface, thereby releasing tensile stress. Indeed, by adatom incorporation the surface layer in-plane bond length decreases until the surface atomic density has increased by approximately 5%. Although the potential energy of an extra atom in the reconstruction network of the surface layer is about 0.2 eV higher than that of an atom in a kink position (the reconstruction is only metastable), under deposition conditions the incorporation into the surface layer is still a favorable process for an adatom, since its potential energy on the surface is about 1.3 eV higher than that of an atom in the reconstructed surface layer.

In contrast to island nucleation, the reconstruction nucleation probability increases with temperature and becomes appreciable only above 400 K. This is due to the fact that a considerable barrier has to be surmounted for nucleation: The potential energy of the first incorporated atom is about 0.4 eV higher than that of an adatom. Only by the addition of further adatoms the position in the surface layer does it become energetically favorable and a stable reconstruction nucleus is formed. We note that stable reconstruction nuclei may also be formed by single ion impacts, as is illustrated by the faint, ringlike reconstruction lines near impact craters in Fig. 5.6. As reconstruction nucleation requires a higher adatom supersaturation than island nucleation, it is evident that reconstruction nucleation on top of islands cannot take place. This explains the absence of the reconstruction on top of islands.

The reconstruction grows laterally by the addition of adatoms through points of disturbed local symmetry, so called U-turns, which are located at the end of paired reconstruction lines, and densifies by a number of complicated mechanisms [331]. The U-turns must be considered as the repetitive positions for reconstruction growth, similar to the role of kinks in the growth of crystals. As soon as a U-turn runs into a step edge, the energy level of an extra atom in the reconstruction becomes connected to the lower energy level of the kink

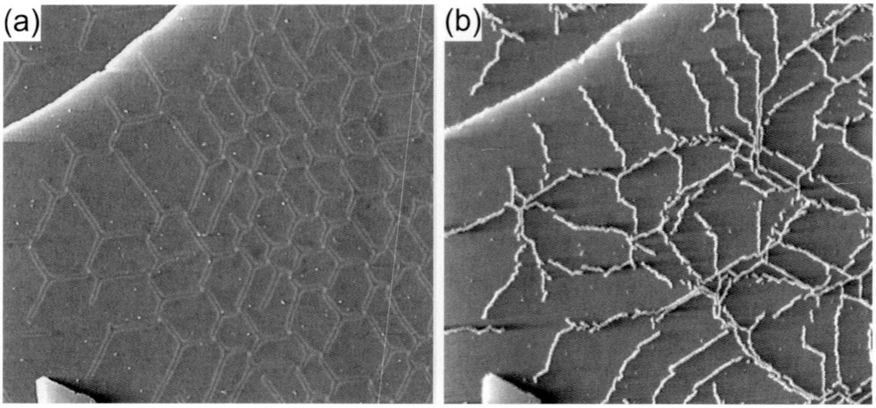

Fig. 5.11. (**a**) STM topograph of a reconstruction network created by a low density of ion impacts for reconstruction nucleation, followed by subsequent Pt deposition for reconstruction growth. (**b**) Same location after exposure to H_2 with $p_{H_2} = 5 \times 10^{-7}$ mbar for 400 s at 30 K. The atoms previously incorporated in the reconstruction were released and form lines of monatomic width. Note that due to the dislocation glide induced by the dynamics of the dereconstruction process, the positions of the lines do not precisely coincide with the initial positions of the reconstruction lines in (**a**). Imaging temperature 30 K, topograph size is 1600 Å × 1600 Å

position *without* a significant barrier in between[3]. Thus atoms are released from the reconstruction, the U-turn retracts and the paired reconstruction lines do *not* grow across the step edge. As a general consequence of this behavior, the reconstruction in the underlying layer is always released by overgrowth, so that no dislocation lines are incorporated into the growing film.

A beautiful illustration of the fact that adatoms are really incorporated into the surface layer is shown in Fig. 5.11. Chemisorption of H at a temperature of 30 K changes the bonding situation in the first layer and makes the reconstructed surface energetically highly unfavorable and dereconstruction takes place. The incorporated atoms are released immediately and form strings of monatomic width on the surface.

***

To conclude this section, we note that a self-regulated decoupling of the nucleation processes on the terrace and on top of islands may take place also on surfaces other that Pt(111), which periodically form specific superstructures or defects during growth. A good example is homoepitaxy on Si(111). Due to the large unit cell of the superstructure, the reconstructed Si(111)-(7 × 7) surface exhibits 49 different domains. In submonolayer deposition ex-

[3] In contrast, as long as the reconstruction is spatially separated from the steps, the two energy levels are only connected through the adatom level which is much higher in energy.

periments preferential nucleation of islands at the boundaries between these domains is observed [541,566]. As the initial reconstruction domains are much larger than the island spacing resulting from homogeneous nucleation within a domain, both homogeneous nucleation within the domains and heterogeneous nucleation at the boundaries occurs initially. After deposition of the first bilayer[4], a new network of domain boundaries results from the coalescence of the islands. This domain boundary network is much denser than on the initial surface, and a considerably higher island number density results from the heterogeneous nucleation at the domain boundaries. In this way the island number density increases, until nucleation takes place exclusively at domain boundaries. Assuming that each island top consists of a single domain of the (7×7) superstructure, it is clear that no nucleation can take place on top of a single island. Irrespective of the existence and magnitude of a step edge barrier, this mechanism thus links nucleation to island coalescence, and minimizes the buildup of surface roughness. It is therefore no surprise that for homoepitaxy on Si(111) long lasting scattering intensity oscillations are observed [567, 568]. Similar arguments may apply to homoepitaxy on many other reconstructed semiconductor surfaces.

5.5 Theory of Pulsed Deposition

We have already seen that the conceptually simplest approach to growth manipulation is through a periodic modulation of the deposition flux: Increasing the flux at the beginning of the growth cycle leads to a large density of first layer nuclei, and drives the system towards layer–by–layer growth. However, the method is rather inefficient due to the weak power law dependence of the island density on the flux.

In this section we explore some consequences of a periodic variation of the deposition flux in a different regime, where the period is small compared to the monolayer deposition time, and hence the synchronization between flux modulation and monolayer completion is not the primary effect. This type of growth is realized e.g. in *pulsed laser deposition* (PLD), where the deposit material is ablated by a pulsed laser, and each pulse typically contains a small fraction of a monolayer. Considerable interest in pulsed deposition was triggered by the observation [569] that PLD is superior to thermal deposition in inducing layer–by–layer growth of heteroepitaxial metal films[5]. Here we discuss pulsed deposition as an illustration of how the introduction of an additional control parameter – the pulse frequency – allows for the separate manipulation of the nucleation processes in the first and second layer.

[4] Due to the diamond structure of Si, a Si island grows on Si(111) in bilayers.

[5] PLD differs from thermal evaporation also in that the deposited atoms have nonthermal energies, and part of the material arrives at the surface in the form of clusters. The effect that this may have on the film morphology will not be addressed here.

The deposition flux is composed of pulses of duration τ_p and pulse repetition rate ν_p. The deposition rate is F_p during a pulse and zero between pulses. The time-averaged deposition flux is then $\bar{F} = F_\mathrm{p}\tau_\mathrm{p}\nu_\mathrm{p}$, and the *pulse intensity*, the number of atoms deposited per unit area during one pulse, is $I_\mathrm{p} = F_\mathrm{p}\tau_\mathrm{p}$. The regime of interest is that of short pulses, $\tau_\mathrm{p}\nu_\mathrm{p} \ll 1$, each of which carries a small fraction of a monolayer, $I_\mathrm{p}\Omega \ll 1$. For example, in the PLD growth of Fe/Cu(111) reported in [569], these parameters take the values $\tau_\mathrm{p}\nu_\mathrm{p} \approx 3 \times 10^{-5}$ and $I_\mathrm{p}\Omega \approx 0.01$. In the following we investigate the consequences of varying the pulse intensity while keeping the fluxes $\bar{F}$ and F_p constant.

To determine the effect of the pulsing on the nucleation in the first layer, the two characteristic time scales τ_p and $1/\nu_\mathrm{p}$ should be compared to the diffusion time τ_D required for an adatom to diffuse to the nearest island, $\tau_D \approx l_D^2/D = 1/DN$. The relative ordering of τ_p, $1/\nu_\mathrm{p}$ and τ_D defines three different regimes [570]. When $\tau_\mathrm{p} \ll 1/\nu_\mathrm{p} \ll \tau_D$, the adatom density changes too slowly to react to the flux modulation, and the nucleation process proceeds as if the surface were exposed to a uniform flux $\bar{F}$. According to the scaling law (2.19), the island number density in the steady state regime (near coalescence) is then given by $N(\bar{F}) \sim (D/\bar{F})^\chi$ independent of I_p. The pulsing becomes relevant when $\tau_D < 1/\nu_\mathrm{p}$, which can be rewritten as

$$I_\mathrm{p} > I_\mathrm{p}^{(1)} \approx \bar{F}/DN(\bar{F}) \sim (\bar{F}/D)^{1-\chi}. \tag{5.3}$$

This condition states that the adatom density generated during one pulse exceeds the typical adatom density that would be present in the steady state nucleation regime under a uniform deposition rate $\bar{F}$ [compare to (2.17)] [571]. Therefore pulsing the deposition rate enhances the island number density, compared to continuous deposition, when (5.3) is fulfilled.

In the opposite limit $\tau_D \ll \tau_\mathrm{p} \ll 1/\nu_\mathrm{p}$, steady state nucleation conditions are established during each pulse, while between the pulses nothing happens, since all free adatoms are absorbed by existing islands on the time scale τ_D. In this case the nucleation process proceeds as under a constant flux $F = F_\mathrm{p}$ (though at a slower rate, since the flux is turned off part of the time) and the island number density is is given by $N(F_\mathrm{p}) \sim (F_\mathrm{p}/D)^\chi$. This regime is reached when

$$I_\mathrm{p} > I_\mathrm{p}^{(2)} \sim (F_\mathrm{p}/D)^{1-\chi}. \tag{5.4}$$

The maximal enhancement factor for the island number density that can be reached by increasing the pulse intensity is given by $N(F_\mathrm{p})/N(\bar{F}) = (F_\mathrm{p}/\bar{F})^\chi = (\tau_\mathrm{p}\nu_\mathrm{p})^{-\chi}$.

In the intermediate regime $I_\mathrm{p}^{(1)} < I_\mathrm{p} < I_\mathrm{p}^{(2)}$ the island number density turns out to be independent of the adatom diffusion coefficient: Since $\tau_\mathrm{p} < \tau_D < 1/\nu_\mathrm{p}$, during the pulses there is not enough time for adatom diffusion to become relevant, while between the pulses diffusion ceases because all free adatoms are captured by the islands. The island number density can then only depend on the pulse intensity. Making a power law ansatz $N \sim I_{\rm p}^a$ and matching it to the known behavior for $I_{\rm p} < I_{\rm p}^{(1)}$ and $I_{\rm p} > I_{\rm p}^{(2)}$ yields

$$N \sim I_{\rm p}^{\chi/(1-\chi)} \quad \text{for } I_{\rm p}^{(1)} < I_{\rm p} < I_{\rm p}^{(2)}. \tag{5.5}$$

Equations (5.3,5.4,5.5) provide a fairly complete description of the dependence of the first layer island density on the pulse parameters[6], which has been confirmed by KMC simulations [570, 571, 573].

Consider next how pulsing the flux affects nucleation on top of the first layer islands [571]. We specialize the discussion to short pulses ($\tau_{\rm p} \to 0$, $F_{\rm p} \to \infty$ at fixed $I_{\rm p}$, and consequently $I_{\rm p}^{(2)} \to \infty$), irreversible nucleation ($i^* = 1$, $\chi = 1/3$) and strong step edge barriers. As was shown in Sect. 4.3.2, the second layer nucleation rate is then determined by the probability for two atoms to be present simultaneously on the island. Second layer nucleation is enhanced by the pulsed flux when the probability of depositing at least two adatoms onto the island in a single pulse becomes sufficiently large. For an island with A adsorption sites the probability of depositing n adatoms in one pulse is given by a Poisson distribution with mean $I_{\rm p} A \Omega$. The probability that $n \geq 2$ is therefore

$$\text{Prob}[n \geq 2] = 1 - \mathrm{e}^{-I_{\rm p} A \Omega}(1 + I_{\rm p} A \Omega) \approx (I_{\rm p} A \Omega)^2 \tag{5.6}$$

for $I_{\rm p} A \Omega \ll 1$. Multiplying this by the number of pulses per unit time yields the second layer nucleation rate for pulsed deposition,

$$\omega_{\rm p} \approx \nu_{\rm p} (I_{\rm p} A \Omega)^2 = \bar{F} \Omega^2 I_{\rm p} A^2. \tag{5.7}$$

This is to be compared to the second layer nucleation rate ω for continuous deposition at flux $\bar{F}$. The theory of Sect. 4.3.2 yields $\omega = (\bar{F} \Omega A)^2 \tau$ where τ denotes the residence time for a single adatom on the island. Pulsing therefore increases the rate of second layer nucleation when

$$I_{\rm p} > I_{\rm p}^{(3)} = \bar{F} \tau, \tag{5.8}$$

which is equivalent to $1/\nu_{\rm p} > \tau$.

Near the onset of the breakdown of layer–by–layer growth, the island spacing l_D and the critical radius R_c for second layer nucleation are, by definition, of similar order of magnitude. In the presence of strong step edge barriers the residence time τ is then much larger than the diffusion time τ_D required to explore a region of size R_c or l_D. This implies that $I_{\rm p}^{(3)} \gg I_{\rm p}^{(1)}$, and hence a range $I_{\rm p}^{(1)} < I_{\rm p} < I_{\rm p}^{(3)}$ of pulse intensities exists where pulsing increases the island number density without affecting the process of second layer nucleation. In this range pulsed deposition thus improves layer–by–layer growth.

---

[6] The coverage dependence of the island density in the intermediate regime $I_{\rm p}^{(1)} < I_{\rm p} < I_{\rm p}^{(2)}$ is rather complex, see [571, 572].

For $I_\mathrm{p} > I_\mathrm{p}^{(3)}$ both l_D and R_c decrease with increasing I_p. To quantify the effect, R_c can be estimated along the lines of Sect. 4.3.3 using the expression (5.7) for the second layer nucleation rate. This yields

$$R_c \approx (\Omega/NI_\mathrm{p})^{1/6}. \tag{5.9}$$

Recalling from (5.5) that (for $i^* = 1$, $\chi = 1/3$) $N \sim I_\mathrm{p}^{1/2}$ in this regime, we see that both R_c and l_D decrease as $I_\mathrm{p}^{-1/4}$, suggesting that pulsing does not further improve layer–by–layer growth when I_p is increased beyond $I_\mathrm{p}^{(3)}$. While these considerations have not yet been tested in detail, KMC simulations of pulsed multilayer deposition do show an improvement of layer–by–layer growth for sufficiently strong step edge barriers [571].

5.6 Kinetic Roughening and Layer–By–Layer Growth

So far the discussion in this chapter has been based on the criterion (4.59), which is a quantitative condition for the onset of mound formation during the growth of the first few layers. Even in the absence of immediate mound formation, however, layer–by–layer growth is almost always a *transient* phenomenon with a limited lifetime. A possible mechanism that causes the breakdown of the layer–by–layer mode after the growth of many layers is mound formation in the weak barrier regime. As was described in Sect. 4.5.2, the maximum number of layers that can be grown before the onset of mound formation is of the order of $(l_D/l_\mathrm{ES})^2$, a large number when $l_D \gg l_\mathrm{ES}$.

In this section the focus will be on the limitations on the duration of layer–by–layer growth which arise from the stochastic nature of the deposition process. The same random fluctuations in the deposition flux that generate self-affine roughness at low temperature (see Sect. 4.8) also limit the ability to grow atomically smooth surfaces at high temperatures, because they cause the layer–by–layer growth cycles in spatially separated regions of the surface to fall out of phase. The quantitative analysis, to be presented below, shows that the effect is quite weak. Nevertheless it is ubiquitously present, and hence any growth manipulation scheme that claims to allow for truly *indefinite* layer–by–layer growth must take it into account [574]. A comparison between the stochastic effects and the mounding instability in the weak barrier regime will be presented at the end of the section (see Fig. 5.12).

A clear manifestation of the transient nature of layer–by–layer growth is the damping of oscillations in the scattering signal that is used to probe the growth mode (see, e.g., Fig. 4.2). Excluding here the damping effects due to the experimental inhomogeneities of the deposition flux (compare Sect. 6.1), the damping is uniquely related to the roughening of the surface: When the surface height distribution becomes much wider than the layer thickness, the relative contributions of odd and even numbered layers to the scattering signal tend to cancel each other and the net signal decreases. A quantitative relation between the surface width W and the anti-phase Bragg intensity I_{anti} introduced in Section 4.1.2 can be derived by evaluating (4.7) under suitable assumptions on the exposed coverages φ_n [575]. It takes the form [576]

$$I_{\text{anti}} \sim \exp(-CW^2), \tag{5.10}$$

where the coefficient $C = 4$ for the Poisson distribution appearing in statistical growth [compare to (4.12) and (4.13)], and $C = \pi^2 \approx 9.87$ for a Gaussian film height distribution [497, 575].

The relation (5.10) implies that the anti-phase scattering intensity decreases indefinitely with increasing surface width. This is generally true if the contribution from any given layer to the scattering intensity vanishes as $\Theta \to \infty$, i.e. if $\varphi_n \to 0$ for all layer indices n, but does not apply to the wedding cake model with top layer nucleation at a critical coverage θ_c described in Sect. 4.2. In that case the top two layers take up a finite fraction of the exposed surface at all times, and hence the intensity oscillation amplitude approaches a nonzero limiting value of order θ_c^2 for $\Theta \to \infty$, although the surface width diverges according to (4.21). A surface covered with wedding cakes growing perfectly in phase would display undamped intensity oscillations of reduced amplitude. In reality the oscillations are damped because of the fluctuations in the deposition flux, and hence in the timing of nucleation events, among different mounds[7]. This is a special case of the general mechanism to be described below, and seems to provide a reasonable description of the damped, low amplitude oscillations characteristic of the mesa mound morphology on Pt(111) (see Sect. 4.1.2).

These considerations suggest a simple strategy for estimating the lifetime of layer–by–layer growth [320]: We determine the contribution W_{stoch} of the fluctuations to the surface width, and posit, based on (5.10), that layer–by–layer growth terminates at a *damping coverage* $\tilde{\Theta}$ defined by the condition

$$W_{\text{stoch}}(\tilde{\Theta}) \approx 1. \tag{5.11}$$

A formal calculation of W_{stoch} can be carried out using the Langevin equations[8] for kinetic roughening introduced in Sect. 4.8.1. However, the main results can be derived from simple arguments.

The key concept needed from the theory of kinetic roughening is the correlation length $\xi(t)$ introduced in (4.104). The correlation length defines the lateral range over which mass transport along the surface is able to average out the fluctuations in the deposition flux; over distances beyond ξ this averaging is no longer effective, and the height fluctuations become uncorrelated. After the growth of a fraction of a monolayer the relevant length scale for

[7] The stochasticity of the nucleation events themselves (*nucleation noise*) is insufficient to break the phase coherence between the mounds, see [339, 577, 578].

[8] To model layer–by–layer growth within the continuum formulation, the discreteness in the vertical direction is accounted for by including terms proportional to $\sin(2\pi h/d)$ on the right hand side of (4.105) [579].

lateral mass transport is obviously the island spacing l_D. The dimensionally correct expression for a correlation length that increases with time according to the power law (4.104) therefore reads

$$\xi \approx l_D \, \Theta^{1/z}. \tag{5.12}$$

The value of the dynamic exponent z will be specified later.

To estimate W_{stoch}, we follow the discussion of noise-induced mound coarsening at the end of Sect. 4.6.2. The surface can be regarded as composed of uncorrelated patches of area ξ^{d_s}, where d_s denotes the dimensionality of the surface[9]. During the growth of Θ layers, the mean number of atoms deposited in one of the patches is $(\xi/a)^{d_s}\Theta$, with a statistical fluctuation of $\pm\sqrt{(\xi/a)^{d_s}\Theta}$. The corresponding height fluctuation is precisely the stochastic contribution W_{stoch} to the surface width. It is obtained by dividing the fluctuation in the number of atoms by the area of the patch, which yields

$$W_{\text{stoch}} \approx (\xi/a)^{-d_s/2}\Theta^{1/2} \approx (l_D/a)^{-d_s/2}\Theta^{(1-d_s/z)/2}. \tag{5.13}$$

If the surface is roughened only by fluctuations, this implies that the surface width $W = W_{\text{stoch}}$ grows as $t^{\alpha/z}$ with $\alpha = (z - d_s)/2$, which proves the relation (4.107).

Combining (5.11) and (5.13) we arrive at the central result of this section, the expression

$$\tilde{\Theta} \approx (l_D/a)^{zd_s/(z-d_s)} \tag{5.14}$$

for the number of layers that can be grown before layer–by–layer growth terminates due to fluctuations. At this time the correlation length (5.12) has reached the value

$$\tilde{l} = \xi(\tilde{\Theta}) \approx l_D (l_D/a)^{d_s/(z-d_s)}, \tag{5.15}$$

which defines the *coherence length* $\tilde{l}$, the lateral length scale over which persistent layer–by–layer growth is possible.

It remains to specify the dynamic exponent z. We are interested primarily in the case of growth without step edge barriers[10], where fluctuations are the only cause of roughening. The analysis of Sect. 4.5.2 then shows that the surface fluctuations can be described by a linear, fourth order Langevin equation[11]. With the addition of fluctuations this becomes the Mullins equa-

---

[9] For $d_s = 1$ the theory can be applied to the growth of island boundaries, see Sect. 3.7.3.

[10] In the other extreme of wedding cake growth without coarsening, the appropriate choice is $1/z = 0$. Equation (5.14) then yields the estimate $\tilde{\Theta} \approx (l_D/a)^{d_s}$ for the damping coverage of the residual intensity oscillations associated with the nucleation of the top terraces.

[11] Strictly speaking nonlinearities, such as the λ_4-term in the CKPZ equation of Table 4.3, should be taken into account. However, in the transient layer–by–layer growth regime the amplitude of the fluctuation-induced roughness is small, and it can be shown that the CKPZ nonlinearity is irrelevant for $\Theta < \tilde{\Theta}$. For this reason the analysis based on the CKPZ equation leads to the same results for $\tilde{\Theta}$ and $\tilde{l}$ as the arguments presented here [320, 579].

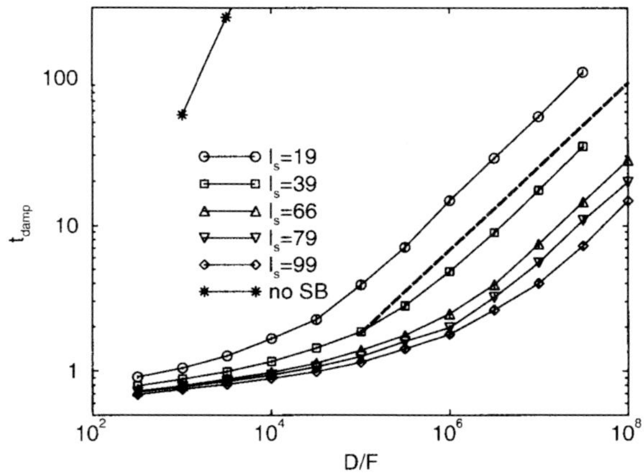

Fig. 5.12. KMC simulation data for the damping coverage $t_{\mathrm{damp}} = \tilde{\Theta}$ as a function of D/F, for different values of the Ehrlich–Schwoebel–length l_{ES}. The figure illustrates the dramatic effect of step edge barriers on the damping of layer-by-layer growth: Without step edge barriers $\tilde{\Theta} \sim (D/F)$ (see the two data points shown as *stars* in the upper left corner), while in the presence of step edge barriers $\tilde{\Theta} \sim (D/F)^{0.57}$ (*bold dashed line*) (Fig. 4.7 of [580])

tion displayed in Table 4.3, which is characterized by $z = 4$. In the physical dimensionality $d_s = 2$, (5.14) and (5.15) then reduce to $\tilde{\Theta} \approx (l_D/a)^4$ and $\tilde{l}/a \approx (l_D/a)^2$. Using the results of nucleation theory derived in Sect. 2.3, the island spacing l_D can be expressed in terms of microscopic parameters. For the case of irreversible nucleation ($i^* = 1$) this yields finally

$$\tilde{\Theta} \sim (D/F)^{2/3}, \quad \tilde{l} \sim (D/F)^{1/3}. \tag{5.16}$$

Extensive tests of the scaling laws (5.14) and (5.15) have been carried out in the framework of KMC simulations. For one-dimensional surfaces the agreement is excellent [317,320]. In two dimensions the available results indicate that (5.16) is not quantitatively correct, but a complete picture is so far lacking. The direct measurement of the damping coverage yields $\tilde{\Theta} \sim D/F$ (see Fig. 5.12), which is consistent with (5.14) if the value $z = 3$ is chosen for the dynamic exponent[12] [580]. This value is also consistent with the behavior of the surface width for coverages beyond $\tilde{\Theta}$, which grows approximately as $W \sim \Theta^{1/6}$. On the other hand, simulations of a closely related model[13]

[12] In [580] the dynamic exponent $z = 3$ is associated with a Langevin equation of the form $\partial h/\partial t = \lambda_3 \nabla^2 |\nabla h| + \delta F$, where the nonlinear term on the right hand side arises from the numerically observed dependence of the adatom density on the surface slope as $n_1(\nabla h) = n_1(0)(1 - C|\nabla h|)$.

[13] The KMC models used in [580] and [581] differ in that [581] allows for fast edge diffusion, while [580] does not. Consequently the submonolayer islands are fractal in [580] and compact in [581]. Through the mechanism described in Sect. 4.9.7, at long times edge diffusion induces a mounding instability in the model of [581].

explicitly confirm the existence of a correlation length given by (5.12) with $z = 4$ in the transient layer–by–layer growth regime.

Irrespective of the precise value of z, the damping coverage predicted by (5.14) is extremely large under realistic conditions. For example, even for a small island spacing of $l_D/a = 20$, typical of Pt(111) homoepitaxy around 150 K (see Fig. 2.8), evaluating (5.14) with $z = 4$ yields $\tilde{\Theta} = 1.6 \times 10^5$. This implies that already a minute step edge barrier is able to drastically reduce the lifetime of layer–by–layer growth: According to the analysis of Sect. 4.5.2, mounds appear at a coverage of order $(l_D/l_{\mathrm{ES}})^2$, which is much smaller than $\tilde{\Theta}$ even if $l_{\mathrm{ES}} \approx a$. The KMC simulation results shown in Fig. 5.12 illustrate this fact[14].

A similar remark applies to the effect of a small miscut. Recall that initial layer–by–layer growth is expected when the spacing l between preexisting steps is much larger than l_D (Sects. 4.1 and 4.7). The scaling picture developed above suggests that a small vicinality will start to affect the growth mode when the correlation length reaches l, at a coverage $\sim (l/l_D)^z$, which is small compared to $\tilde{\Theta}$ if $l \ll \tilde{l}$. For miscuts with $l_D \ll l \ll \tilde{l}$ one therefore expects a regime of transient layer–by–layer growth with a duration determined by the surface vicinality, rather than by the fluctuations in the deposition flux. This is the transition region between layer–by–layer growth and step flow.

[14] Quantitatively, the data shown in Fig. 5.12 suggest that the dependence of the onset time τ^* for mound formation on the growth parameters is described by the relation $\tau^* \sim (l_D^2/l_{\mathrm{ES}})^{1.62}$, rather than by (4.72).

6. Methods

Ein Auge welches sieht, das andre welches fühlt[1]
Paul Klee

The present chapter gives an overview of some of the experimental and theoretical methods which are employed in studies of crystal growth far from equilibrium. A full discussion of these methods is impossible in the framework of this book and for a more complete description the reader may consult the reviews quoted in the text. However, the overview should make the working principles of the methods transparent and allow the reader to understand their meaning in the present context.

6.1 Deposition and Vacuum

The experiments and theories described in this book deal with growth on a surface resulting from the condensation of single atoms (or molecules) out of the gas phase. This method is generally referred to as molecular beam epitaxy (MBE) or physical vapor phase deposition (PVD)[2]. A minimal deposition system for growth studies consists of a sample with a sample holder assembly allowing for temperature control, means for sample cleaning prior to deposition, an evaporation source creating the vapor to be deposited, a surrounding vacuum system, and an instrument for the analysis of growth. In semiconductor industry, MBE growth systems typically comprise of several effusion evaporators, a rotatable sample holder with temperature control and a reflection high energy electron diffraction (RHEED) system monitoring growth, all placed in an efficient vacuum system with a cryo-panel allowing base pressures in the 10^{-11} mbar range [582, 583].

In order to achieve reproducible results and to be able to compare with theoretical predictions, it is necessary to perform the growth experiment under well controlled conditions. In general this means that growth takes place

[1] "One eye that sees, the other that feels"

[2] As distinguished from PVD, chemical vapor phase deposition (CVD) takes place by dissociation of precursor molecules at the hot surface, with one molecular fragment depositing while the other fragments are volatile and do not adhere to the surface.

on a homogeneous, chemically clean surface of well defined crystal orientation and at fixed temperature. In addition, a uniform average rate of chemically well specified atoms is deposited, i.e. the arrival of unwanted atoms or molecules on the surface during growth must be avoided or at least kept at a negligible level. This requires stringent vacuum conditions during all stages of sample cleaning and deposition. According to elementary kinetic gas theory, the flux Φ of particles hitting a surface exposed to a pressure p of a residual gas with molecular mass M is given by

$$\Phi = \frac{p}{\sqrt{2\pi M k_{\mathrm{B}} T}}. \tag{6.1}$$

The coverage Θ of a specific gas resulting from adsorption during a time t is usually measured in monolayers (ML), where one monolayer corresponds to the number of atoms in one atomic substrate layer. For the example of Pt(111) used frequently in this book, 1 ML corresponds to 1.504×10^{19} atoms/m^2. If the average lifetime of the adsorbate species is large compared to t and in the limit of small coverage, the coverage is simply given by

$$\Theta = \Phi S t \Omega, \tag{6.2}$$

where S is the sticking coefficient in the limit of zero coverage of the adsorbate on the surface and Ω is the adsorption site area of an atom in the substrate layer. The relevant time interval t for adsorption prior to deposition is typically the time between the end of sample cleaning and the start of deposition.

A worst case estimate for the tolerable adsorbate coverage can be obtained by considering the following scenario: The adspecies under concern is immobile on the substrate and binds strongly to the deposited atoms. Thus each adsorbate atom or molecule may give rise to the heterogeneous nucleation of an island. In order to be able to measure the island number density N due to homogeneous nucleation – an important quantity in thin film growth as discussed in Chap. 2 – it is thus necessary that the adsorbate concentration is small compared to the expected value of N. As an example, for Pt/Pt(111) the island number density at room temperature is around 10^{-4} islands per adsorption site, and thus the coverage of heterogeneous nucleation centers prior to deposition should certainly be below 10^{-5} ML. Assuming a sticking coefficient[3] $S = 1$, $M = 30$ atomic mass units, and $T = 300\,K$, this coverage is reached in 54 s by exposure to an adspecies partial pressure of 1×10^{-12} mbar. Knowing that under optimal conditions t is of the order of 100 s and the total background pressure is in the low 10^{-11} mbar range, this estimate underlines the necessity to optimize the sample preparation procedure and the vacuum conditions in growth experiments.

[3] Fortunately, for many gases contributing to the residual background pressure $S \ll 1$ at the usual deposition temperatures.

In considering a specific growth experiment it is thus worthwhile to determine the partial pressures of the major components of the background gas prior to and during deposition, and to analyze their behavior with respect to the substrate and thin film materials. A practical procedure for testing the influence of a background gas species is to intentionally double its partial pressure during growth. Significant deviations from the growth behavior under standard conditions then indicate that there is no reason to believe that growth under standard conditions would *not* be influenced by adsorbates.

In some cases (e.g. for Si samples) sample cleaning can be performed by simply heating the sample in a low background pressure. Single crystal metal samples in general require a tedious procedure of sputtering by ion bombardment, heating in specific gases, and annealing. For example, the Pt(111) substrate mentioned frequently in this book is prepared by prolonged sputtering with a mass-separated ion source at 900–1000 K followed by brief annealing at 1300 K. In order to minimize contamination effects, the pressure rise during the last annealing phase prior to deposition must be minimized. This pressure rise is caused by outgassing and desorption of gas species from heated surfaces. Thus it is important to minimize the heated mass and to direct the heating power as efficiently as possible towards the sample.

The purity of a thin film is generally not limited by the purity of the evaporation material, but by the ratio of the rates of arrival of deposited atoms to foreign (impurity) atoms at the growth front. Practically, the magnitude of the deposition rate is limited to values of $10^{-2} - 10^{0}\,\mathrm{ML/s^{-1}}$ by the need to control even the deposition of a fraction of a monolayer with high precision. The purity of the film is therefore ultimately determined by the background pressure during deposition. As evaporation implies heating of the evaporation material and thus also of its surroundings, the power introduced for evaporation and the amount of heated material should be minimized in order to avoid a significant pressure rise during deposition. Commercial evaporators are therefore often not suited for deposition in atomic scale growth experiments. Heating by electron bombardment causes, in addition to temperature-induced desorption and outgassing, also electron-stimulated desorption and therefore induces an often unacceptable pressure increase during deposition [131,132]. Whenever possible, direct current heating of a filament of the evaporation material is the preferred evaporation method, because it dissipates the heating power almost entirely in the evaporation material itself, minimizing the heating of other parts.

A last point to be addressed here is the lateral homogeneity of the depositing particle flux. The evaporated particle flux from a small surface element obeys a cosine distribution with respect to its surface normal [584]. Thus, depending on the geometry and distance of the evaporator, lateral inhomogeneities in the average deposited film thickness on the substrate will arise. While these inhomogeneities are a minor problem for point probes like STM, they may influence the diffraction signal measured during deposition, as a diffracted beam is typically averaging over a sample area with linear dimension of 1 mm [326]. As an example, for layer–by–layer growth the initially simultaneous nucleation on all terraces will be gradually lost due to the deposition flux inhomogeneities. After a deposition time corresponding to a thickness variation of 1 ML over the beam area, the diffraction signal will average over terraces in different stages of evolution between island nucleation and coalescence, and its interpretation is no longer straightforward.

To be quantitative, let the distance from the evaporation source to the sample be R and the size of the observation region (e.g., the spot size in a diffraction experiment) be r. Then elementary geometry shows [585] that the synchronicity between different parts of the spot is lost after deposition of the order of $(R/r)^2$ monolayers if the surface is oriented normal to the beam, while this happens already after deposition of $\sim R/r$ monolayers under conditions of oblique incidence.

6.2 Microscopy

Microscopy is the method *sine qua non* for understanding epitaxial growth. An ideal instrument would allow to follow the motion of atoms and steps (i) on the length scale from Ångström to µm, (ii) at any temperature, (iii) during growth and (iv) with chemical sensitivity. To date such a microscope does not exist.

Scanning probe microscopies and specifically *scanning tunneling microscopy* (STM) have their strength in imaging from the atomic to the mesoscopic length scale, being thus ideally suited to connect atomic scale processes with morphological evolution. Variable temperature scanning tunneling microscopy allowing to study atomic scale processes or growth morphologies in a wide temperature range has become a standard tool by now. Imaging during growth as well as chemical identification, however, can be achieved only under special conditions (see below). Despite its limitations, today the STM is probably the most important instrument for studies of epitaxial growth. Atomic force microscopy (AFM), though generally inferior in resolution compared to STM, extends the range of growth studies to non-conducting samples.

Being restricted to imaging terraces of sizes around 100 Å, *field ion microscopy* (FIM) cannot be used for morphological studies. However, it is an excellent tool for high precision measurements on the atomic scale, allowing the identification and quantitative description of individual atomistic processes.

Electron microscopy is an entire group of microscopies with a wide range of applications. Although all these methods have the great advantage of being suitable for continuous observation during the processing of the sample, surface imaging with electrons, which is required in growth studies, is possible only under specific conditions. One possibility is to use low energy electrons, which are surface sensitive due to their small mean free path [586]. The corresponding instrument is the *low energy electron microscope* (LEEM) [587–590]. It allows the in-situ observation of growth processes with a good lateral resolution of about 50 Å–100 Å. The combination of imaging and simultaneous diffraction makes it specifically suited for heteroepitaxial growth processes and phase transitions. Its limited resolution, however, requires the analysis of growth morphologies which exhibit a large lateral length scale, and thus restricts the method to experiments at elevated temperatures. A second possibility is to use a high energy electron beam impinging at glancing incidence on the sample. Under these conditions the electron beam is surface sensitive, because its mean free path *normal* to the surface is again small. In *reflection electron microscopy* (REM) [591, 592] a foreshortened image of the surface with step contrast and a lateral resolution comparable to LEEM is obtained.

6.2.1 Scanning Tunneling Microscopy

Scanning tunneling microscopy, invented in the 1980's by G. Binnig and H. Rohrer [19, 20], allowed for the first time the imaging of flat surfaces in atomic resolution. Technical aspects, theory and applications of this method are covered in several books [593–595].

In scanning tunneling microscopy a fine metal tip is brought in such close proximity to a conducting sample that the application of a small voltage U_t gives rise to a net tunneling current I_t between the two electrodes (see Fig. 6.1). As shown in Fig. 6.1a, the tunneling current I_t results from the classically forbidden tunneling of sample electrons near the Fermi edge to unoccupied states of the tip (or vice versa, depending on the polarity of U_t) through the potential energy barrier of width Δ (tip–sample separation) and height Φ_w (approximately equal to the average work function of the electrodes). To first approximation the tunneling current depends exponentially

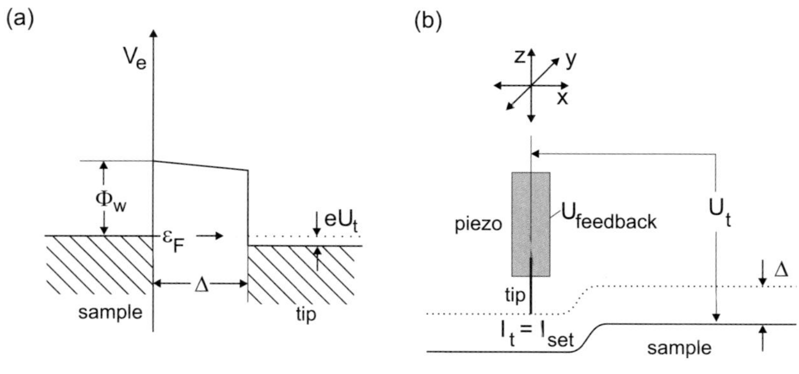

Fig. 6.1. (**a**) Schematic diagram of the one-dimensional potential energy landscape V_e for an electron in a tunneling contact between two metal electrodes (see text). (**b**) Schematic sketch of the principle of STM operation (see text)

on Δ and thus reacts very sensitively to changes in the tip–sample separation. Usually a feedback loop is used to keep I_t at a preset value I_{set} of a few nA (constant current mode) while scanning the tip in the x–y plane across the sample surface. The feedback loop supplies a voltage $U_{feedback}$ to a piezoelectric element carrying the tip, which allows to adjust the tip position during the scanning motion (compare Fig. 6.1b). A map of the feedback voltage $U_{feedback}$, representing the z coordinate of the tip position, versus the x–y position of the tip thus provides a topographic map of the sample surface.

Typically the STM images are represented as grey scale topographs, encoding the z coordinate of the tip in grey values for each x–y pixel. In this representation a bright location corresponds to an elevation and a dark location to a depression in the surface[4]. Due to the extreme sensitivity of the tunneling current to the size of the tunneling gap a resolution in the z-direction of 10^{-2} Å can be achieved straightforwardly, provided that the tunneling gap is sufficiently isolated from the noise and vibrations in the environment. The lateral resolution depends to first approximation on the atomic structure of the tip. If a single atom carries the dominant fraction of the tunneling current, atomic resolution can be achieved.

It is clear that one-dimensional tunneling theory as used implicitly above is unable to provide a satisfactory explanation of STM lateral resolution and the image contrast due to chemically inhomogeneous sample surfaces. Modeling the tunneling tip as an s-wave orbital within a tunneling pertubation theory yields the concise result that in the limit of small U_t the tip follows a contour line of constant surface local density of states near the Fermi energy ϵ_F [596]. This picture suffices to interpret most of the tunneling images presented in this book. Nevertheless, it is far from complete; a satisfactory understanding of atomic resolution STM images of metal surfaces requires a tunneling theory that takes into account the full electronic structure of sample and tip [597].

In atomic scale studies of epitaxial growth, knowledge of the sample temperature in all stages of the experiment, and the ability to perform temperature dependent experiments in a wide temperature range are of key importance. The need for temperature control contradicts the usual scheme of transfering the sample from a preparation stage to an STM stage within the ultra high vacuum system. This scheme would require separate cooling and heating stages with temperature control for sample preparation, transfer and STM imaging, resulting in an extreme apparative effort. Another approach [309] that has been successfully applied in a large part of the experiments de-

---

[4] Alternatively, in this book differentiated topographs allowing a better representation of rough surfaces with many steps are frequently shown. Differentiated topographs appear as if illuminated from the left. Technically they are obtained, e.g., by high-pass filtering the voltage signal $U_{feedback}$ prior to its representation as grey values. The filter transforms a voltage step from 0 to U_0 into a jump from 0 to U_0 followed by an exponential decay with a time constant τ according to $U = U_0 e^{-t/\tau}$.

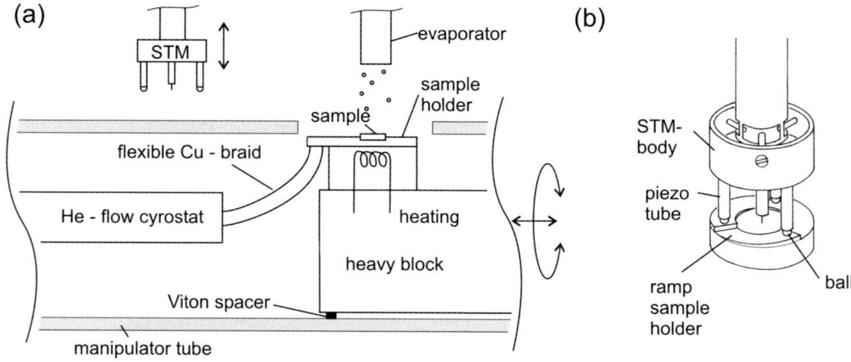

Fig. 6.2. (**a**) Schematic sketch of a sample manipulator for a variable temperature STM system (see text). (**b**) Schematic side view of an inverted beetle type STM lowered onto a triple helix ramp sample holder

scribed in this book relies on the use of a scanning tunneling microscope which can be lowered temporarily onto the sample for investigation (see Fig. 6.2a). This allows the sample to be mounted on a manipulator and to attach a temperature sensor to the sample, so that the temperature can be permanently measured. For temperature control in the range between 20 K and 1500 K the sample can be heated by electron bombardment from the back and cooled by a He flow cryostat. The cryostat is coupled by a flexible Cu braid to the sample, in order to minimize the vibrations transmitted to the sample. The sample with the sample holder is mounted mechanically stiff, but thermally well insulated to a heavy metal block. The key idea of this scheme is to lower the amplitude of the noise transmitted via the Cu braid to the sample assembly, without affecting significantly the lowest achievable sample temperature[5].

The block itself rests on vibration absorbing elastomer spacers in a manipulator tube, which may be translated and rotated in order to move the sample in front of all sample preparation and analysis stages. At a specific upright position the microscope, an inverted beetle STM [599–601], can be lowered onto the sample holder. Coarse approach of tip and sample is accomplished by suitable voltage pulses applied to the three outer piezos, leading via slip-stick motion to a rotation of the microscope down the triple helix ramps of the sample holder, until the tip mounted to the central piezo reaches tunneling contact. Scanning may be performed either by the outer piezo tubes or by the central piezo, while the tip position normal to the surface is adjusted with the central piezo. Using this type of apparatus, STM operation achieving atomic resolution of dense packed metal surfaces is possible in the temperature range from 20 K to 700 K [598].

[5] For an even more effective vibration isolation of the sample from the cryostat, additionally the Cu braid may be clamped thermally well insulated to an intermediate heavy block [598].

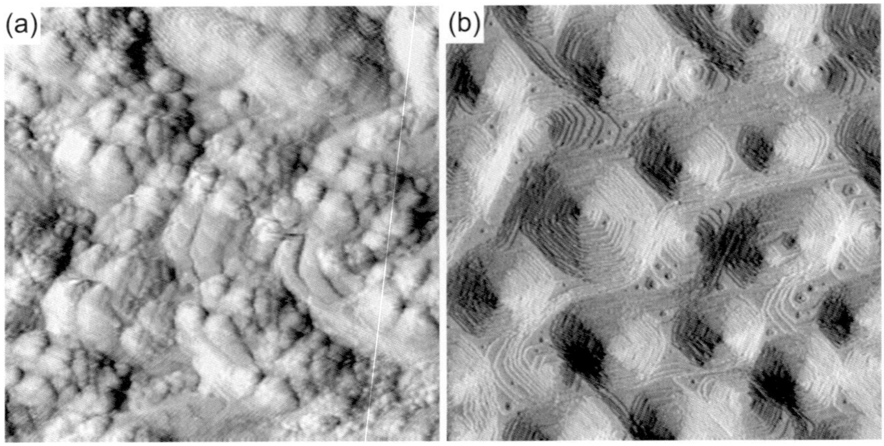

Fig. 6.3. STM topographs of the Pt(111) surface after removal of 50 ML by 600 eV Ar^+ ion bombardment at 625 K (**a**) before and (**b**) after optimization of tip shape. Topograph size is 1650 Å × 1650 Å (Fig. 21 of [309])

Two different types of experiments may be performed with STM systems like the one sketched above. In a *quench experiment* the morphology created by growth at a specific temperature T is quenched to an observation temperature $T' < T$ at which the morphology is frozen. Each quench experiment is only representative of one point in time during the surface morphological evolution, namely the instant at the end of the deposition interval. The principal advantage of this method is that it allows for sufficient time to image many different sample locations and for optimizing the STM imaging conditions. The latter point is of particular importance for rough surfaces which result from the deposition or removal of many atomic layers. Indeed, in order to image such surfaces, adequate optimization of the tip shape is mandatory in order to avoid tip artefacts due to the convolution of tip shape and topography. As an example, Fig. 6.3 compares two images of the same sample after removal of 50 ML by ion erosion before (Fig. 6.3a) and after tip shape optimization[6] (Fig. 6.3b).

In contrast to the quench method, in a *dynamic experiment* the evolution of the surface morphology at a fixed temperature or during a slow temperature variation is followed as a time lapse sequence at a single sample location.

[6] For the investigation of a soft metal (e.g. Pt) the shape of a tip made from a hard material (e.g. tungsten) is simply optimized by gently pushing the tip into the sample. The tip-sample contact usually causes a transfer of sample material to the tip and thus the tip shape is changed. Subsequently the STM is moved to an undamaged area, which is imaged with the newly created tip shape. The procedure is iterated until the imaging quality is satisfactory (no double tips, sharp step edges, etc.). This optimization scheme requires in addition a low mobility of the transferred material at the tip temperature, in order to achieve a stable imaging tip.

Examples for this type of experiment are time lapse sequences during mild temperature variation for the determination of the onset of a diffusion process [602] and time lapse sequences at constant temperature for the observation of step edge fluctuations [231, 255] (see also Sect. 3.2.2) or the motion of individual adatoms or clusters (see Sect. 2.2.3). For the latter purpose an optimized atom tracking scheme may be used, in which the STM does not scan across the surface, but instead a lateral feedback mechanism is used to lock the tip above the object of interest [78]. Therefore atom tracking STM allows diffusion studies with much better time resolution than conventional STM.

Of particular interest for understanding epitaxial growth are time lapse sequences of a sample location *during* deposition, which allow to monitor the entire morphological evolution in a single experiment. This method was successfully applied for homoepitaxial growth studies on Si surfaces [29] and Fe(110) surfaces [491]. Currently, the major drawback of such experiments is their limitation to very low deposition rates, which restricts them to situations where the sensitivity of the growth morphology to contaminations is weak.

Atomic force microscopy (AFM) was invented shortly after the STM [603]. In its simplest version, the method replaces the tunneling current by the force between a cantilever tip and the sample to drive a feedback loop in an arrangement similar to the STM. In this *contact AFM* setup the force is typically measured by light beam deflection from the bending cantilever [604, 605]. In *non-contact AFM* a vibrating cantilever interacts with the surface, resulting in a force gradient measurement that even allows to obtain atomic resolution images [606–608]. The key advantage compared to the STM is that the AFM is applicable to non-conducting samples. By now, the AFM has become the standard materials science tool for surface morphology characterization at ambient conditions also for conducting samples, because under these conditions the tunneling gap is typically unstable due to local variations in conductivity, surface diffusion and inhomogeneous adsorbate layers. For clean, conducting samples in a controlled environment (e.g. ultra high vacuum), however, due to its simplicity the STM is still the tool of ultimate versatility and resolution for surface morphology studies.

6.2.2 Field Ion Microscopy

Field ion microscopy, invented by E.W. Müller in the 1950's, was the first method capable of imaging surfaces with atomic resolution [14, 15]. Several books cover technical issues and the theory and applications of FIM [609–611].

In field ion microscopy an imaging gas – typically He – is supplied to a vacuum system containing a fine metal tip kept at a high positive potential of the order of 10 kV (see Fig. 6.4a). The metal tip to be imaged is etched to a small radius of curvature on the order of 1000 Å. The image gas is dynamically polarized in the electric field of the tip and thus attracted towards it. When a gas atom approaches a critical distance x_c of a few Å above the tip, the gas

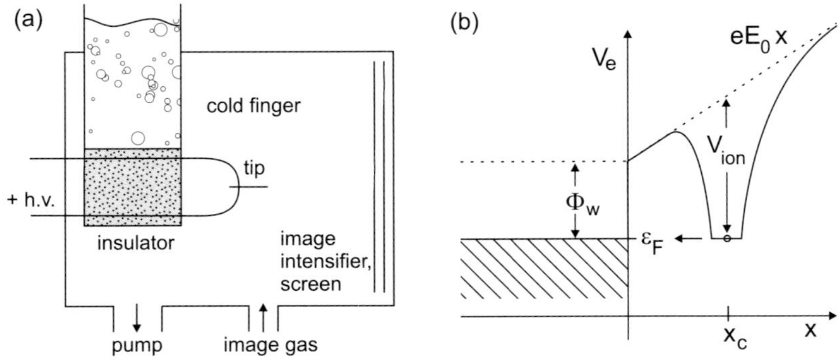

Fig. 6.4. (**a**) Schematic sketch of an elementary FIM apparatus (see text). (**b**) Schematic, one-dimensional diagram of the potential energy V_e of a valence electron of an image gas atom with ionization potential V_{ion} in front of a metal tip (left) with work function Φ_w and in the presence of an electric field of strength E_0 (see text)

atom may be positively ionized with significant probability by the tunneling of an electron into the tip. It follows from the potential energy diagram in Fig. 6.4b that the gas atom can only be ionized around x_c, because at closer distances the electron will not be able to tunnel into unoccupied states of the conduction band of the tip, while at larger distances the increasing width and height of the tunneling barrier suppresses the tunneling probability. The critical distance is defined by the condition $\Phi_w + eE_0x_c = V_{ion}$ which can be read off from the potential energy diagram. Here Φ_w is the tip work function, V_{ion} is the first ionization potential of the gas atom, E_0 is the field strength created by the tip potential and e is the elementary charge. After ionization the gas particles are accelerated away from the tip in the electric field of the tip and hit an image intensifier plate with a phosphor screen, creating a visible, projected image of the tip surface.

Generally the tip is cooled to cryogenic temperatures, as this allows accommodation of the gas atoms to the tip temperature prior to ionization. The thermal velocity component of the gas ions is thereby minimized. Thus the gas ions created at one specific tip location hit the same spot on the phosphor screen with a precision sufficient for atomic resolution of the tip structure. The image contrast itself results from the thickness variation of x_c created by local variations of the field strength E_0. Roughly speaking, E_0 is larger at edges and protrusions, leading thus to a bright appearance of steps and adatoms in the field ion microscope picture (see Fig. 6.5). Raising the tip voltage beyond stable imaging conditions allows to selectively remove protruding atoms from the surface: For these tip atoms the field becomes sufficiently strong to field ionize the atoms themselves. The unique possibility of peeling off tip surface layers by field evaporation is a basic element of atomic

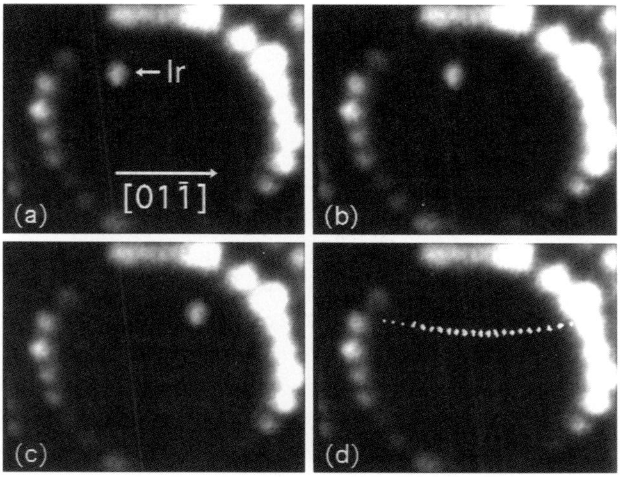

Fig. 6.5. FIM images illustrating site–mapping for an Ir adatom on a Rh(311) apex plane. In (**d**) the coordinates from 300 successive images are superimposed on the substrate image, establishing the one dimensional lattice of adsorption sites for the Ir adatom along a channel of the Rh(311) surface (Fig. 5 of [52])

scale FIM studies, as it allows to create perfect, clean surfaces over and over again.

The methodology used to perform atomic scale studies on the tip apex plane (the outermost, more or less circular atomic layer plane of the tip, bounded by a monatomic step) can be described as follows. After preparation of a clean surface, in the simplest case a single atom is deposited on the apex plane. The tip is imaged at a low temperature T', where the adatom motion is frozen. The position of the bright image spot representing the adatom on the apex plane is recorded. Then, the imaging voltage is turned off completely and the tip is briefly heated for a well defined time t to a temperature $T > T'$. This can be done with high accuracy, because the small mass of the tip allows efficient cooling and heating. After the heating interval, the tip is imaged again at T' and the new position of the adatom is recorded. This procedure is iterated many times and allows to follow the motion of the adatom at the higher temperature T through the low temperature imaging snapshots. Site visitation maps established in this way represent the lattice of possible adsorption sites and its distortion in the FIM image for a given tip, which is a prerequisite for quantitative analysis of the adatom motion. An example of a site visitation map is shown in Fig. 6.5. From the mean square displacement of the adatom per heating interval the adatom diffusion coefficient can be determined (see Sect. 2.2.3). Different or more complex situations investigated by FIM with basically the same methodology concern adatom condensation, the behavior of adatoms at steps, diffusion of clusters, step adatom diffusion as well as adatom–adatom interactions. Examples are mentioned in this book (see Chap. 2) and reviewed in the literature [52, 612, 613].

6.3 Diffraction

The unique capability of microscopy to image real, defective surfaces and to clarify the relation between surface defects and growth is, to some extent, also a disadvantage. Microscopy is always limited to a small field of view and statistically averaged quantities may be obtained only after sufficient, generally time consuming data sampling. In contrast, diffraction methods measuring the global interaction of a beam of particles or radiation with matter automatically yield signals which are averaged over a macroscopic volume or sample surface. Moreover, as the interaction of a particle beam with matter does not require specific instrumental arrangements in the immediate vicinity of the sample (as scanning probe microscopy does), it is well suited for the *in situ* analysis of growth processes, i.e. for measurements during deposition.

In order to be applicable for the analysis of atomic scale structures and processes, the probe particle should possess a wavelength on the order of the interatomic spacing. According to the expression

$$\lambda_{\mathrm{dB}} = \frac{h}{\sqrt{2EM}} \tag{6.3}$$

for the de Broglie wavelength of a particle of mass M, suitable probe particle energies E for atoms and electrons are $E \approx 0.1\,\mathrm{eV}$ and $E \approx 100\,\mathrm{eV}$, respectively. For electromagnetic radiation the relevant wavelengths lie in the x-ray range. In addition, the interaction with the surface must be strong enough to yield a measurable scattering signal; this makes the use of neutrons for surface scattering experiments less suitable.

Thermal, inert, light gas atoms possess not only a suitable wavelength, but are also extremely surface sensitive and destruction free. They are scattered already at a height of about 3Å above the surface due to the Pauli repulsion between the overlapping orbitals of the surface and probe atoms. Sufficiently coherent beams are produced by high pressure expansion of the gas through a nozzle. Usually He atoms are used as probe particles and the method is termed thermal energy atom scattering (TEAS) or Helium atom scattering (HAS) [614]. One benefit specific to thermal energy atom beams interacting with surfaces is their large scattering cross section for point defects on an otherwise flat surface, allowing extremely sensitive studies of the behavior of such defects [615].

Low energy electron diffraction (LEED) is the most suitable method for surface structure determination [616–619]. Low energy electrons are surface sensitive and interact only with the first few layers of the crystal. For surface

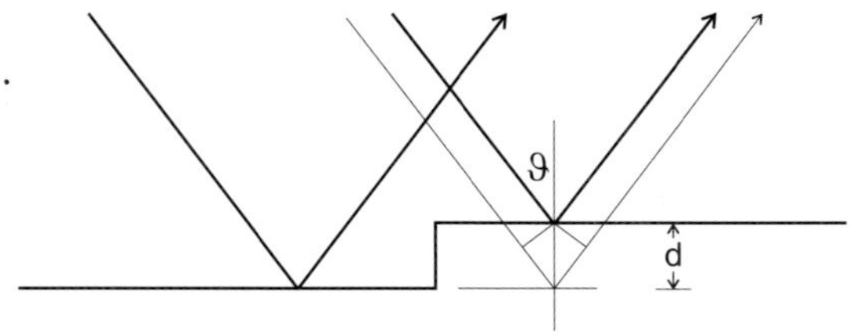

Fig. 6.6. Schematic sketch of the interference of a specularly reflected wave in the presence of a step. Destructive interference will take place if the path difference of the two rays is $2d \cos \vartheta = (n + 1/2)\lambda$ (the path difference is identical to that between the fat and thin right rays), and constructive interference if $2d \cos \vartheta = n\lambda$ (see text)

structure determination in many cases multiple scattering may be neglected (the projection $\boldsymbol{K}_{\|}$ of the scattering vector onto a plane parallel to the surface depends only little on multiple scattering events) and the kinematic scattering approximation may be used for data interpretation. While LEED is traditionally used as a method for surface crystallography, highly coherent beams and precise measurements of the quasi-diffuse intensity surrounding the reciprocal lattice rods which satisfy the two dimensional Bragg condition provide quantitative information on the distribution of surface steps. These measurements are accomplished by spot profile analysis LEED (SPALEED) [620,621]. Besides these low energy methods, also keV electron beams are used for the determination of surface structure and for the monitoring of growth (see below). In order to achieve sufficient surface sensitivity, these beams are directed to the surface at grazing incidence, an arrangement which is particularly compatible with a deposition situation. The method is known as reflection high energy electron diffraction (RHEED) [591, 592].

Although x-rays interact only relatively weakly with matter, the availability of intense x-ray beams from synchrotron radiation facilities has turned them into a practical tool for the investigation of surface morphology [622]. Diffraction at the semi-infinite crystal leads to crystal truncation rods in the diffraction pattern, which contain surface specific structural information. The crystal truncation rods are faint intensity lines between the positions of the Bragg peaks of the three-dimensional lattice in the direction normal to the surface. Similar to LEED, the quasi-diffuse intensity surrounding the crystal truncation rods contains information on the distribution of surface steps. The highly monochromatic synchrotron radiation allows for interference of the backscattered waves to take place over lateral distances as far as μm. A grazing incidence geometry is employed to greatly enhance the surface sensitivity of the x-rays and is again well suited for in situ measurements during growth.

A simple, yet efficient way to investigate growth by diffraction is to monitor the specular scattered Bragg intensity at or near anti-phase scattering conditions. Figure 6.6 visualizes the principle. Neglecting bulk contributions and surface corrugation, for the specularly reflected wave of wavelength λ represented by the two rays reflected at the surface, destructive interference takes place if their path difference is $2d\cos\vartheta = (n+1/2)\lambda$ (anti-phase scattering condition) and constructive interference, if $2d\cos\vartheta = n\lambda$ (in-phase scattering condition), where n is an integer. Using the scattering vector $\boldsymbol{K} = \boldsymbol{k}_f - \boldsymbol{k}_i = (\boldsymbol{K}_{\|}, K_{\perp})$ to define the scattering phase $\Delta\phi = K_{\perp}d/2\pi$, the anti-phase condition reads $\Delta\phi = n + 1/2$ and the in-phase condition $\Delta\phi = n$.

Consider now a two-layer surface with total coverage Θ in the submonolayer regime. The exposed area of the upper level is then Θ and that of the lower level $1-\Theta$. The resulting normalized anti-phase scattering intensity is $I_{\text{anti}} = (1-2\Theta)^2$, a parabola that reaches the value unity at zero and one monolayer and vanishes at half-monolayer coverage. Perfect layer-by-layer growth maintains a two-layer surface indefinitely, with exposed areas $\Theta - [\Theta]$ and $1 - (\Theta - [\Theta])$ for the upper and lower level, respectively. Here $[\Theta]$ denotes the integer part of Θ. Consequently the anti-phase intensity for perfect layer-by-layer (LBL) growth is an oscillatory function with monolayer period, composed of parabolae with sharp peaks at integer values of Θ and zeros at half-integer coverages (see (4.8)).

The actually measured anti-phase intensity oscillations for LBL growth do not show sharp peaks, because the theoretical anti-phase Bragg intensity is smeared out by convolution with the instrument response function, and moreover because perfect LBL does not exist. For a general multi-layer surface the normalized anti-phase Bragg intensity is

$$I_{\text{anti}} = \left| \sum_{n=0}^{\infty} (-1)^n \varphi_n \right|^2 , \tag{6.4}$$

where φ_n denotes the exposed coverage of layer n (see Sect. 4.1.2). A derivation of (6.4) will be given below. As discussed in Sect. 4.2, for rough growth without material transport between different layers, the anti-phase scattering intensity decays exponentially with increasing Θ. Thus measurement of the anti-phase intensity allows to identify the growth mode during the growth process.

From the analysis of the specular reflected scattering intensity, only information on the distribution of the atoms over the different surface layers is obtained. But the analysis of the full scattering intensity also yields information about the lateral distribution of the surface atoms or steps. In order to see how this is accomplished, it is useful to briefly sketch the kinematic scattering theory [620, 623, 624]. Within this approximation the scattering intensity

$$I(\boldsymbol{K}) = N_{\text{s}} \sum_{\boldsymbol{u}} \mathrm{e}^{-\mathrm{i}\boldsymbol{K}\cdot\boldsymbol{u}} C(\boldsymbol{u}) \tag{6.5}$$

is simply the Fourier transform of the two-point correlation function

$$C(\boldsymbol{u}) = \frac{1}{N_{\mathrm{s}}} \sum_{\boldsymbol{x}} f(\boldsymbol{x}) f(\boldsymbol{x}+\boldsymbol{u}), \tag{6.6}$$

where $\boldsymbol{x}$ is a point of the three-dimensional crystal lattice, and $f(\boldsymbol{x})$ takes the value 1 if there is a surface scatterer (i.e., a surface atom[7]) at the position $\boldsymbol{x}$ and 0 if not. This entails the assumption that the scattering factor is the same for all atoms. The sum is over all N_{s} surface atoms. The correlation function $C(\boldsymbol{u})$ is the probability to find two scatterers separated by a displacement vector $\boldsymbol{u}$. Expressing the displacement vector as a sum of an in-plane displacement $\boldsymbol{r}$ and a vertical displacement $\boldsymbol{z} = dn\boldsymbol{e}_z$, where $\boldsymbol{e}_z$ denotes the unit vector in the z-direction, the correlation function can be written as $C(\boldsymbol{u}) = C(\boldsymbol{r}, n)$. In the limit of infinite lateral separation, $|\boldsymbol{r}| \to \infty$, there is no correlation between surface scatterers and thus

$$C(|\boldsymbol{r}| \to \infty, n) \equiv C_\infty(n) = \sum_i \varphi_i \varphi_{i+n} \tag{6.7}$$

which does not depend on $\boldsymbol{r}$. The quantity on the right hand side of (6.7) is the one-dimensional vertical correlation function of the surface, which is equal to the probability of finding two surface atoms in independent columns at a height difference of n atomic layers. Subtracting the limiting value (6.7) from the correlation function, the scattering intensity $I(\boldsymbol{K})$ may be expressed as a sum of two contributions, the Bragg intensity $I_{\mathrm{B}}(\boldsymbol{K})$ and the quasi-diffuse intensity $I_{\mathrm{D}}(\boldsymbol{K})$:

$$I(\boldsymbol{K}) = I_{\mathrm{B}}(\boldsymbol{K}) + I_{\mathrm{D}}(\boldsymbol{K}). \tag{6.8}$$

The Bragg intensity arises from the constant part $C_\infty(n)$ of the correlation function and is given by

$$I_{\mathrm{B}}(\boldsymbol{K}) = \sum_{\boldsymbol{r}} \mathrm{e}^{-\mathrm{i}\boldsymbol{K}_\parallel \cdot \boldsymbol{r}} \left[\sum_{n=-\infty}^{\infty} C_\infty(n) \mathrm{e}^{-\mathrm{i}K_\perp n d} \right]. \tag{6.9}$$

For a given value of $K_\perp$ the expression in square brackets is a constant. Consequently, the Bragg intensity is a lattice of delta functions located at positions where $\boldsymbol{K}_\parallel$ is a reciprocal lattice vector. The magnitude of the intensity depends only on the distribution of the scatterers in the direction normal to the surface, but not on their lateral distribution. The evaluation of (6.9) under anti-phase scattering conditions for $\boldsymbol{K}_\parallel = 0$ yields precisely the anti-phase Bragg intensity I_{anti} as expressed in (6.4).

The quasi-diffuse intensity $I_{\mathrm{D}}(\boldsymbol{K})$ is the Fourier transform of the $\boldsymbol{r}$-dependent part of the correlation function,

$$I_{\mathrm{D}}(\boldsymbol{K}) = \sum_{\boldsymbol{r}} \mathrm{e}^{-\mathrm{i}\boldsymbol{K}_\parallel \cdot \boldsymbol{r}} \sum_{n=-\infty}^{\infty} [C(\boldsymbol{r}, n) - C_\infty(n)] \mathrm{e}^{-\mathrm{i}K_\perp n d}. \tag{6.10}$$

---

[7] If the three-dimensional position vector $\boldsymbol{x}$ is decomposed into a two-dimensional in-plane vector $\boldsymbol{r}$ and a vertical displacement $\boldsymbol{z}$, then $f(\boldsymbol{r}, \boldsymbol{z}) = 1$ iff $h(\boldsymbol{r}) = |\boldsymbol{z}|/d$, where $h(\boldsymbol{r})$ is the height function used elsewhere in this book for the description of surface configurations.

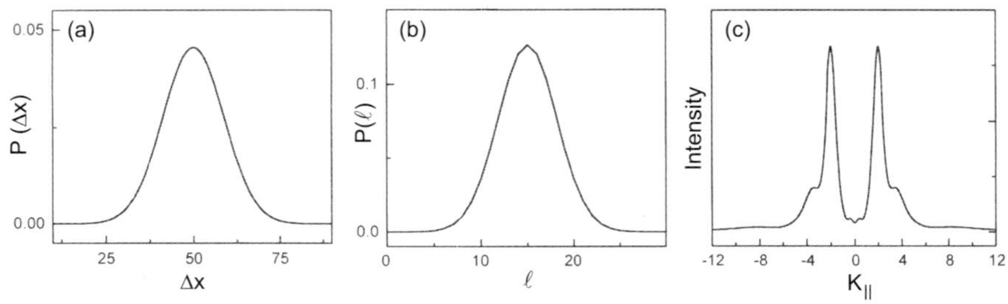

Fig. 6.7. (**a**) One-dimensional Gaussian island separation distribution $P(\Delta x)$, with Δx being the separation between the island centers which has as mean the average island separation $l_D = 50\,a$. (**b**) One-dimensional Gaussian island size distribution $P(\ell)$, with ℓ being the island length with a mean of $15\,a$. (**c**) Calculated quasi-diffuse scattering intensity. The parallel scattering vector $K_\parallel$ is given in % of the surface Brillouin zone ($100\,\%\,\mathrm{BZ} = 2\pi/a$). The diffraction pattern is dominated by the peaks at $K_\parallel = \pm 2\,\%\,\mathrm{BZ}$ due to the mean island separation $l_D = 50\,a$ (Figs. 2.5a, 2.5c and 2.5d of [625])

For a given value of $K_\perp$ the quasi-diffuse intensity is distributed throughout $\boldsymbol{K}_\parallel$-space. The shape of I_D contains information about the lateral distribution of the scatterers, which enters via $C(\boldsymbol{r}, n)$. The quasi-diffuse intensity is maximized for anti-phase scattering conditions.

The strategy for the interpretation of measured quasi-diffuse scattering intensities is to assume a specific surface morphology, to calculate the corresponding correlation function $C(\boldsymbol{r}, n)$, and to convert it into the quasi-diffuse scattering intensity using (6.10). Prior to comparison to any measurement, the calculated intensity has to be convoluted with an appropriate instrument response function, which describes the instrumental broadening of the scattering intensities. Using analytical expressions for $C(\boldsymbol{r}, n)$ derived for specific step distributions, the distribution parameters (such as the mean step separation and the width of the distribution of step separations) may be fitted to yield optimal agreement with the measured intensity. Another approach which is used more and more frequently is to calculate $C(\boldsymbol{r}, n)$ numerically for the hypothetical surface morphology, which may be generated e.g. by a KMC simulation [623, 625].

The interpretation of the quasi-diffuse intensity, though generally not unambiguous, is less arbitrary than the brief sketch presented here may suggest. As one important rule of thumb, a typical distance in the morphology, such as the typical island separation l_D, gives rise to an additional peak in the quasi-diffuse intensity at $K_\parallel = 2\pi/l_D$. An example for the calculated quasi-diffuse intensity corresponding to a one-dimensional island size distribution is shown in Fig. 6.7. Standard methods for the determination of average morphological properties through the analysis of the quasi-diffuse intensity are SPALEED [620] and x-ray diffraction [624, 626, 627], see [473] for an overview.

6.4 Total Energy Calculations

A central theme in this book is the effect that the rates of individual atomistic processes have on the growth of a thin film, and the quantitative extraction of such rates from observations of the growth morphology. It is therefore of great importance to be able to reliably calculate atomistic rates for specific materials, both as an input into realistic growth simulations, and for comparison with experimentally determined numbers. As most processes of interest are thermally activated, the rate Γ of a process can generally be written in the Arrhenius form

$$\Gamma = \Gamma_0 \, \mathrm{e}^{-E_\mathrm{a}/k_\mathrm{B}T}, \tag{6.11}$$

which depends on two parameters, the energy barrier E_a and the attempt frequency Γ_0. Within transition state theory, both quantities can be expressed in terms of the potential energy landscape of the system (see Sect.2.2 for a detailed discussion in the case of surface diffusion). If the total energy of the surface is known for all possible atomic arrangements, the rate of any process connecting two configurations can be obtained by locating the transition path with the lowest saddle point energy[8].

Since the thermal motion of the atoms is slow compared to the dynamics of the electron system, the total energy calculation can be carried out within the *Born-Oppenheimer approximation*, which implies that the electronic ground state energy is computed for a static configuration of nuclei. In principle, the problem is thus reduced to the solution of the stationary Schrödinger equation for the many-electron system; in practice, such a calculation becomes intractable already for a moderate number N_e of electrons. The remarkable recent progress in total energy calculations for large systems has become possible only by the realization of *density functional theory* (DFT) that the ground state energy is determined uniquely by the electron density $n_e(\boldsymbol{x})$. The shift from the ground-state wave function to the electron density implies an enormous advantage, because it transfers the problem from the $3N_e$-dimensional configuration space of the electrons to three-dimensional physical space.

Ideally, total energy calculations should be carried out from first principles (*ab initio*), in the sense that the only input to the calculation is the chemical identity of the atoms involved (i.e., the charges of the nuclei). The basic concepts and procedures involved in ab initio DFT calculations will be described in the next subsection. In many cases it is however desirable to have access to an approximate, computationally less expensive yet reasonably accurate description of the energetics, which can be used in explorative calculations for large numbers of atoms. While the general form of such approximate potentials can be inferred from density functional ideas, they become quantitatively useful only by introducing free parameters which are fitted empirically

[8] The location of transition paths and saddle points in a high-dimensional energy landscape is a demanding numerical task, see [273, 628].

to certain bulk properties of the material of interest. Some aspects of the semi-empirical many-body potentials which have been used in the modeling of metal homoepitaxy will be described in Sect. 6.4.2.

General reviews on total energy calculations in the context of epitaxial growth are given in [273, 275]. The purpose of the present section is merely to familiarize the reader with some of the technical terms which are encountered throughout the book, whenever the results of such calculations are discussed.

6.4.1 Density Functional Theory

The basis of density functional theory[9] are two theorems derived by Hohenberg and Kohn in 1964 [631]. The first theorem states that, for an electron gas in an arbitrary external potential $V(\boldsymbol{x})$ (as given, for example, by the positions of the atomic nuclei), the potential is uniquely determined by the ground state electron density $n_e^0(\boldsymbol{x})$. While the converse of this statement is obvious, the first Hohenberg-Kohn theorem is remarkable, because it implies that the ground state electron density also determines (through the Schrödinger equation) the ground state wavefunction, and hence contains (in principle) all information about the many-body ground state properties. The second theorem is a variational principle which states that the ground state energy can be obtained by minimizing the expression

$$\mathcal{E}_V[n_e(\boldsymbol{x})] = \mathcal{E}_{\mathrm{HK}}[n_e(\boldsymbol{x})] + e \int \mathrm{d}\boldsymbol{x}\, n_e(\boldsymbol{x}) V(\boldsymbol{x}), \tag{6.12}$$

where the Hohenberg-Kohn functional $\mathcal{E}_{\mathrm{HK}}[n_e(\boldsymbol{x})]$ is is *universal*, i.e. independent of the specific system defined by V. The explicit form of $\mathcal{E}_{\mathrm{HK}}$ is not known, but following Kohn and Sham [632] it can be decomposed according to

$$\mathcal{E}_{\mathrm{HK}}[n_e(\boldsymbol{x})] = \mathcal{T}_0[n_e(\boldsymbol{x})] + \frac{1}{2} e^2 \int \mathrm{d}\boldsymbol{x}\, \mathrm{d}\boldsymbol{x}' \,\frac{n_e(\boldsymbol{x}) n_e(\boldsymbol{x}')}{|\boldsymbol{x} - \boldsymbol{x}'|} + \mathcal{E}_{\mathrm{xc}}[n_e(\boldsymbol{x})]. \tag{6.13}$$

Here $\mathcal{T}_0[n_e(\boldsymbol{x})]$ is the kinetic energy of a system of *noninteracting electrons* with the density distribution $n_e(\boldsymbol{x})$, and the second term on the right hand side is the classical Coulomb interaction energy. The remaining contributions to the energy are contained in the *exchange-correlation functional* $\mathcal{E}_{\mathrm{xc}}$. Since the other terms on the right hand side are specified independently, (6.13) effectively *defines* $\mathcal{E}_{\mathrm{xc}}$ in terms of the (unknown) functional $\mathcal{E}_{\mathrm{HK}}$. Of course, such a definition is useful only if accurate approximations to $\mathcal{E}_{\mathrm{xc}}$ can be found. We will return to this point below.

While an explicit formula for the kinetic energy $\mathcal{T}_0$ is not available, its contribution can nevertheless be evaluated exactly by transforming the variational problem $\delta\mathcal{E}_V/\delta n_e(\boldsymbol{x}) = 0$ into a set of equations for noninteracting electrons. These are the Kohn-Sham equations [632]

---

[9] For details see [629, 630].

$$\left[-\frac{\hbar^2}{2m_e}\nabla^2 + V_{\text{eff}}(\boldsymbol{x})\right]\psi_i(\boldsymbol{x}) = \epsilon_i\psi_i(\boldsymbol{x}). \tag{6.14}$$

Despite their superficial similarity to the single particle Schrödinger equation, the meaning of the Kohn-Sham equations (6.14) is fundamentally different, because the effective potential

$$V_{\text{eff}}(\boldsymbol{x}) = e^2\int \mathrm{d}\boldsymbol{x}\,\mathrm{d}\boldsymbol{x}'\,\frac{n_e(\boldsymbol{x}')}{|\boldsymbol{x}-\boldsymbol{x}'|} + \frac{\delta\mathcal{E}_{\text{xc}}}{\delta n_e}(\boldsymbol{x}) + V(\boldsymbol{x}) \tag{6.15}$$

depends on the electron density, which in turn has to be determined from the single particle wavefunctions through

$$n_e(\boldsymbol{x}) = \sum_i f_i|\psi_i(\boldsymbol{x})|^2. \tag{6.16}$$

Here the f_i are Fermi occupation numbers depending on the electron temperature. In the ground state $f_i = 2$ for the $N_e/2$ lowest lying single particle states, each state being filled with two electrons of opposite spin, and $f_i = 0$ for all other states; however, for computational reasons it may be convenient to work at elevated electron temperatures [633]. Together (6.14), (6.15) and (6.16) define a nonlinear problem which has to be solved self-consistently.

To fully specify the problem, an approximation for the exchange-correlation functional $\mathcal{E}_{\text{xc}}$ has to be adopted. A useful starting point for general considerations is the formally exact expression [629]

$$\mathcal{E}_{\text{xc}}[n_e(\boldsymbol{x})] = \frac{1}{2}e^2\int \mathrm{d}\boldsymbol{x}\; n_e(\boldsymbol{x})\int \mathrm{d}\boldsymbol{x}'\;|\boldsymbol{x}-\boldsymbol{x}'|^{-1}n_{\text{xc}}(\boldsymbol{x},\boldsymbol{x}'). \tag{6.17}$$

Here n_{xc} is referred to as the *exchange-correlation hole*, which describes the depletion of the probability of finding another electron at point $\boldsymbol{x}'$ due to the presence of the electron at point $\boldsymbol{x}$. It can be related to the electron pair correlation function of the interacting electron gas, and satisfies the sum rule

$$\int \mathrm{d}\boldsymbol{x}'\; n_{\text{xc}}(\boldsymbol{x},\boldsymbol{x}') = -1. \tag{6.18}$$

According to (6.17), the exchange-correlation energy can be regarded as the Coulomb interaction energy between an electron and its own exchange-correlation hole.

In the *local density approximation* (LDA), the exchange-correlation energy is written as a local average of the exchange-correlation energy $\epsilon_{\text{xc}}(n_e)$ of a uniform electron gas of constant density n_e [632],

$$\mathcal{E}_{\text{xc}}^{\text{LDA}}[n_e(\boldsymbol{x})] = \int \mathrm{d}\boldsymbol{x}\; n_e(\boldsymbol{x})\epsilon_{\text{xc}}(n_e(\boldsymbol{x})). \tag{6.19}$$

Accurate values for $\epsilon_{\text{xc}}(n_e)$ are available from expansions at high and low densities, and from quantum Monte Carlo calculations [634]. The LDA satisfies the sum rule (6.18). Its central assumption is that the electron density varies slowly on the scale $2\pi/q_{\text{F}}$ set by the (local) Fermi wave number

$q_F = (3\pi^2 n_e)^{1/3}$. Corrections to the LDA become appreciable when the dimensionless electron density gradient

$$s = |\nabla n_e|/(2q_F n_e) \tag{6.20}$$

is not small compared to unity. Experience has shown, however, that the LDA yields reliable results far beyond this stringent condition, which is rarely satisfied in applications.

The term *generalized gradient approximation* (GGA) subsumes approximation schemes that seek to improve on the LDA expression (6.19) by including a dependence on the electron density gradient in the integrand. An expansion in leading orders of s is not sufficient for this purpose, because the behavior at small s does not properly reflect the overall s-dependence, and because the resulting functional may violate (6.18) [635]. A class of GGA functionals which are based on first principles only (without any empirical input) and which respect several constraints such as (6.18) was constructed by Perdew and coworkers (the PW91 [635, 636] and PBE [637] functionals). The GGA is usually superior to the LDA in the description of cohesive energies and other bulk properties (see, e.g., [638, 639]); compared to the LDA, which tends towards overbinding, it softens and expands the bonds. For metal surfaces the situation is less clear, because neither the LDA nor the GGA properly account for the long-ranged tail of the exchange-correlation hole, which becomes important in semi-infinite systems.

The actual numerical implementation of a DFT calculation requires several additional technical steps (see e.g. [275, 633, 640]). First, to be able to solve the Kohn-Sham equations (6.14) within a finite plane wave basis set, the system has to be made periodic in all spatial directions by defining a suitable *supercell*. For a typical configuration, such as an adatom on a surface, the supercell consists of a slab of a certain number of atomic layers, which is surrounded by vacuum regions in the vertical direction and continued periodically in the horizontal direction. The lateral extent of the cell has to be sufficiently large so that the interaction of the adatom with its periodic images does not distort the quantities of interest. Similarly the vacuum region has to be chosen sufficiently wide, and the number of layers in the slab should be large enough to adequately represent the behavior of the underlying bulk. Clearly the size of the required supercell increases with the complexity of the atomic configuration of interest. This makes it difficult to treat processes such as the descent of an adatom from a rough step (see Sect. 4.3.4).

Second, rather than dealing with all electrons in the system, it is advantageous to restrict attention to the valence electrons which are actually involved in the bonding. The combined effect of the core electrons and the atomic nuclei on the valence electrons is modeled by suitable *pseudopotentials*. Ideally, these are constructed such as to yield valence electron wave functions which behave as the true wave functions outside of the core region but contain no nodes inside the core. This eliminates the rapid oscillations of the wave functions near the nucleus, and reduces the required size of the plane wave basis.

Pseudopotentials that also reproduce the correct charge density in the core region are called norm-conserving. An important property of a pseudopotential is its transferability, i.e. its ability to mimick the all-electron potential in different atomic environments.

Finally, a variety of numerical algorithms can be employed for the solution of the Kohn-Sham equations [640]. An obvious possibility is to solve the Schrödinger equation (6.14) by matrix diagonalization, using some initial guess for the electron density, and to repeat this procedure iteratively until the self-consistency condition (6.16) is satisfied. Alternative methods aim directly at the minimization of the functional (6.13) with respect to the Kohn-Sham orbitals ψ_i. For this purpose the variational problem is reformulated in terms of some fictitious dynamics for the ψ_i, such as the steepest descent or conjugate gradient approaches.

6.4.2 Semi-Empirical Many Body Potentials

From the experience with the fundamental forces of classical physics we are used to describing the interactions in many body systems through pair potentials. Such an approach is however insufficient to capture many features of metallic bonding even on a qualitative level. For example, within a pair potential model the outermost atomic layer of a semi-infinite crystal is predicted to relax outwards, whereas experimentally inwards relaxation, with a decreased layer spacing compared to the bulk lattice constant, is typically found [641]. This reflects the general failure of pair potentials to account for the increase of the effective bond strength between two atoms with decreasing coordination number [139].

Over the last two decades, much effort has therefore been put into the development of classical interaction potentials for metals which include the most important many-body effects of a full quantum mechanical treatment, while not being significantly more costly to evaluate numerically than a simple pair potential. A variety of different schemes appear in the literature, which have been collectively referred to as *glue models* [642]. Two well-known representatives are the *embedded atom method (EAM)* [643, 644] and the *effective medium theory (EMT)* [641, 645, 646]. The common feature of these models is that each atom in the solid is taken to be embedded in an effective host environment – the effective medium of EMT – which is created by all other atoms, and which can be described by a single parameter. The host, or *reference system*, should be chosen such that it imitates as closely as possible the atomic environment in the real system, and the embedding energy of the atoms in the host should be easily calculable. In most cases the chosen reference system is the uniform electron gas, and the energy is obtained as a function of the embedding electron density. It then turns out that the energy difference between the reference system and the real system can be well approximated by a function of the electron density plus a sum of pair

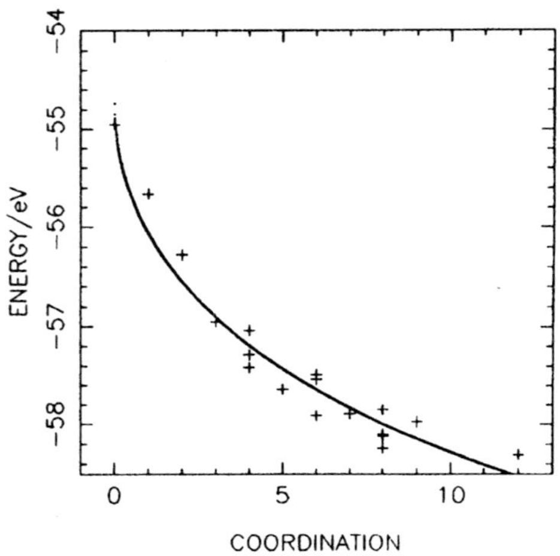

Fig. 6.8. Total energy per atom for eighteen structures of aluminum atoms with fixed interatomic spacing as a function of the coordination number. Symbols show energies computed from DFT using the LDA. Within the glue models, the energies of different structures with the same coordination numbers should coincide. The full line is a fit to (6.23) (Fig. 1 from [642])

interactions. This leads to an expression for the total energy of the solid of the general form

$$E_{\mathrm{tot}} = \sum_i U(\bar{n}_e^i) + \sum_{i,j:i\neq j} \phi(|\boldsymbol{x}_i - \boldsymbol{x}_j|). \tag{6.21}$$

Here the sums run over the atom labels i, j, $\boldsymbol{x}_i$ is the position of nucleus i, $\bar{n}_e^i$ is the embedding electron density for atom i, $U(\bar{n}_e^i)$ is the corresponding embedding energy, and $\phi(r)$ is a pair potential. The first sum contains the energy of the reference system. It is, in a sense, the leading contribution, which is corrected by the sum over the pair potentials. The embedding density $\bar{n}_e^i$ is assumed to be a superposition of contributions from all other atoms, which are functions of the interatomic distances,

$$\bar{n}_e^i = \sum_{j\neq i} \rho(|\boldsymbol{x}_i - \boldsymbol{x}_j|). \tag{6.22}$$

A particular glue model is defined by specifying the three functions U, ϕ and ρ. Different schemes differ in the derivation and physical interpretation of the terms appearing in (6.21), in the functional forms of U, ϕ and ρ, and in the way the parameters entering these functions are determined from experimental data or quantum-mechanical calculations.

Before providing a more detailed description of the EAM and the EMT, it is instructive to see how the many-body energy expression (6.21) accounts for the dependence of the effective bond strength on the coordination number [642]. Consider a structure in which all atoms are equivalent, each atom having a coordination number C, and all interatomic distances take the same value r_0. Then according to (6.21) the energy per atom is $E(C) = U(C\rho(r_0)) + C\phi(r_0)$. For fixed r_0 the pair potential ϕ gives a contribution to the energy that is linear in C, while a nonlinear dependence can arise if the embedding function U is nonlinear in the electron density. Indeed, DFT calculations for atoms embedded in a uniform electron gas show that the embedding energy is nonlinear, with a minimum at a preferred density, for chemically active atoms, while it is linear for the inert gases [647]. This is consistent with the observation that inert gases are well described by pair potentials. For metals, the behavior of $E(C)$ is fairly close to [139, 275]

$$E(C) = E_0 + A\sqrt{C} + BC \tag{6.23}$$

with $A < 0$, $B > 0$ (see Fig. 6.8). The binding energy per atom is a nonlinear, convex function of the coordination number, and hence the energy $E(C-1) - E(C)$ required to break an additional bond decreases with increasing C.

The starting point of the embedded atom method (EAM) is a corollary to the Hohenberg-Kohn theorem, which states that the total energy of a host system with an embedded impurity is given by a universal functional (dependent only on the type of impurity) of the electron density of the unperturbed host [643]. This unknown functional is approximated by an expression of the form (6.21), where the embedding function U is taken from DFT-LDA calculations of atoms embedded into a uniform electron gas [647], and the atomic electron densities entering the embedding density (6.22) are taken from the Hartree-Fock theory for free atoms. The relative contribution of s and d electrons to the atomic densities is treated as a free parameter. The pair potential contribution to (6.21) represents the short-range repulsion between atomic cores, and is assumed to be of the form

$$\phi_{ij}(r) = \frac{Q_i(r)Q_j(r)}{r} \tag{6.24}$$

for the general case of two atoms i, j, possibly of different types, at a distance r. This is simply a screened Coulomb interaction with effective charges Q_i, Q_j. A possible parametrization of the effective charge is $Q(r) = Q_0(1 + c_1 r^{c_2})\exp(-c_3 r)$, where Q_0 is the charge of the outer electrons and c_1, c_2, c_3 are free parameters. A complete set of parameters for a range of fcc metals and their alloys has been determined by fitting to the sublimation energy, equilibrium lattice constant, elastic constants and vacancy-formation energy of the pure metals, and to the heats of solution of the binary alloys [644].

The effective medium theory (EMT) is close in spirit to the EAM, but it places more emphasis on the systematic derivation of the glue model total energy expression from density functional theory [648]. One attempts to construct a reasonable ansatz for the electron density $n_e(\boldsymbol{x})$, and evaluates the Hohenberg-Kohn functional (6.13) for the ansatz density in order to obtain an estimate for the total energy. Since the functional is minimal in the ground state, this estimate is then correct up to second order in the deviations from the true ground state electron density. The ansatz density is of the form [641, 646]

$$n_e(\boldsymbol{x}) = \sum_i \Delta n_e^i(\boldsymbol{x}), \tag{6.25}$$

where $\Delta n_e^i(\boldsymbol{x})$ is the change in the electron density induced at site $\boldsymbol{x}$ by embedding atom i in a uniform electron gas. The embedding density $\bar{n}_e^i$ for atom i is determined self-consistently as the average over the electron density contributions from the other atoms,

$$\bar{n}_e^i = \langle \sum_{j \neq i} \Delta n_e^j(\boldsymbol{x}) \rangle. \tag{6.26}$$

The average is taken over a neutral sphere around atom i, i.e. a sphere for which the total electron charge exactly compensates the charge of the atomic nucleus. The energy of the reference system is the sum of the embedding energies of each of the atoms in a uniform electron gas of density $\bar{n}_e^i$. The leading correction to the energy of the reference system arises from the overlap between the neutral spheres of the different atoms. Using DFT, it can be shown that this correction is well represented by a difference of pair interaction energies [648], leading again to a total energy expression of the general form (6.21).

A simple implementation of EMT that applies to a number of fcc metals can be found in [649]. It is based on the perfect fcc crystal as a reference system, and the environment of each atom i is described by its neutral sphere radius s_i. For the relationship between the neutral sphere radius and the embedding electron density an exponential form is assumed, $\bar{n}_e^i = n_0 \exp[-c_1(s_i - s_0)]$, where s_0 is the lattice constant of the material, and n_0 and c_1 are fit parameters. The functional form for the embedding energy is taken to be

$$U(s_i) = U_0 f(c_2(s_i - s_0)), \quad f(x) = (1 + x)\mathrm{e}^{-x}. \tag{6.27}$$

Here c_2 is another fit parameter, and U_0 is identified with the cohesive energy. Finally, the pair potential describing the overlap of the atomic spheres is also taken to be a pure exponential. The free parameters of the model are determined from the bulk and shear moduli. A large set of energetic and kinetic parameters for a variety of fcc metal surfaces computed within EMT are collected in [92].

We close this subsection with some remarks concerning the accuracy of semi-empirical calculations. A systematic comparison between glue models and LDA density functional theory has been carried out for aluminum [642]. The conclusion of this study is that a rms error of at least 0.1 eV per atom must be expected, *irrespective* of the number of fit parameters contained in the semi-empirical model. Individual energies differ from the DFT values by as much as 0.3 eV. Among different glue schemes, those in which the functions

ρ and ϕ in (6.21) decay exponentially, as in the EMT model described above, show the best performance. Since the energy scale of interest for kinetic processes on metal surfaces lies precisely in the range of a few tenths of electron volts, it is clear that the accuracy of semi-empirical schemes is often insufficient for a meaningful comparison to experiments. The examples of atomic processes discussed in Sects. 2.2.5 and 4.3.4 – surface diffusion and step crossing – illustrate this fact, and underscore the need for first principles DFT calculations.

6.5 Growth Simulations

Depending on the time scales of interest, the atomistic processes involved in crystal growth can be modeled deterministically or stochastically. On the shortest time scale corresponding to the attempt frequency Γ_0 in (6.11), typically on the order of 10^{-12} s, the atom reacts as a classical particle to the forces exerted by its neighbors. In contrast, on the time scale on which transitions between distinct configurations actually occur, the behavior is that of a random process. This is because the separation of time scales between the attempt frequency and the transition rate, $\Gamma \ll \Gamma_0$, which is in fact a precondition for the validity of the expression (6.11), ensures that the system loses its memory between subsequent transitions. The computational methods of choice for simulating these two temporal regimes are Molecular Dynamics (MD) and Kinetic Monte Carlo (KMC), respectively. In the following three subsections the main features of these methods, as well as their limitations, will be described. In the last subsection we mention some recent approaches which attempt to combine the advantages of MD and KMC. For a general overview of simulation methods in crystal growth we refer to [273, 650].

6.5.1 Molecular Dynamics

A molecular dynamics simulation consists of numerically integrating the classical equations of motion for a many-body system [651, 652]. The forces between the atoms can be obtained from a total energy calculation of the kind described in the preceding section. Usually semi-empirical potentials are employed that can be evaluated with modest numerical effort[10]. Once the forces are specified, the equations of motion are discretized in time and solved using some finite difference algorithm. Molecular dynamics simulations can be

[10] In ab initio MD schemes the forces are determined by performing a DFT total energy calculation in each time step [640]. Alternatively the method of Car and Parrinello can be used, in which the minimization of the Hohenberg-Kohn functional is carried out simultaneously with the classical motion of the nuclei by endowing the Kohn-Sham wavefunctions with a fictitious classical dynamics [653].

performed at fixed energy or at fixed temperature, corresponding to the microcanonical and the canonical ensembles, respectively. In the canonical case additional *thermostat* forces which mimick the interaction with a heat bath have to be included.

Achieving a high accuracy in the numerical solution is often not a major concern, since small rounding errors merely take the system from one typical trajectory to another, without affecting the average properties of interest. Nevertheless it is clear that the discretization time step Δt_{dis} has to be small compared to the shortest natural time scale in the system. For atomic processes at solid surfaces, this implies that the time step must be much smaller than the inverse attempt frequency, yielding $\Delta t_{\mathrm{dis}} \leq 10^{-14}$ s. A typical simulation of, say, 1000 atoms can cover on the order of 10^6 time steps. The corresponding real time interval of 10 ns is hardly enough to follow atomic processes during growth, e.g. for a Pt adatom on Pt(111) the time is not even sufficient to perform a single diffusion jump at 300 K.

This *time scale problem* is the main impediment to the application of MD simulations to thermally activated processes such as crystal growth; traditionally the method has been most successful in the modeling of liquids, where this problem does not arise [652]. In MD growth studies the time scale problem has usually been circumvented by using a deposition flux that is many orders of magnitude larger than in reality [153, 654, 655]. Since the relevant parameter controlling the morphology is the *ratio* between the time scales for deposition and for thermal motion, deposition at extremely large fluxes is essentially equivalent to working at extremely low temperatures, where all activated processes are frozen [157]. While such simulations can yield some qualitative insights, it is not possible to extrapolate the results to realistic conditions in a straightforward manner.

Molecular dynamics simulations have nevertheless proved to be highly useful for investigating individual atomistic processes during crystal growth, which proceed on short time scales and involve the motion of several atoms. Examples which are discussed in this book include the transient motion of the deposited atom along the surface (Sect. 2.1) and the processes of funneling and steering which are involved in the condensation at steps and protrusions (Sect. 2.4.1). Molecular dynamics simulations have also been employed to determine the range of validity of the transition state theory for surface diffusion (Sect. 2.2.2), and to identify exotic exchange mechanisms [375, 656].

6.5.2 The Kinetic Monte Carlo Method

A Monte Carlo (MC) simulation is the numerical realization of a stochastic process[11], in which a system evolves over a set of configurations $\{\mathcal{C}\}$ with prescribed transition rates $\Gamma(\mathcal{C} \rightarrow \mathcal{C}')$ for passing from configuration $\mathcal{C}$ to configuration $\mathcal{C}'$. In statistical physics, Monte Carlo simulations have been

[11] Mathematically, such a process is referred to as a *Markov chain* [657].

extensively applied to the evaluation of thermodynamic equilibrium properties of interacting many-body systems [658]. In this context the simulation is used to generate a set of sample configurations with statistical weights corresponding to the correct equilibrium distribution, the canonical Gibbs ensemble. The rates $\Gamma(\mathcal{C} \to \mathcal{C}')$ are therefore chosen mainly according to numerical speed and convenience, and do not need to represent the actual dynamic evolution of the system. The only requirement on the rates is the condition of *detailed balance*,

$$\Gamma(\mathcal{C} \to \mathcal{C}') \exp(-E(\mathcal{C})/k_\mathrm{B}T) = \Gamma(\mathcal{C}' \to \mathcal{C}) \exp(-E(\mathcal{C}')/k_\mathrm{B}T) \qquad (6.28)$$

which ensures that the simulation converges to the desired equilibrium distribution $Z^{-1} \exp[-E(\mathcal{C})/k_\mathrm{B}T]$ [here $E(\mathcal{C})$ denotes the energy of the system in configuration $\mathcal{C}$ and Z is the partition function]. For example, in the Metropolis scheme [659] a possible transition is chosen at random; it is accepted with certainty if it lowers the energy of the system, and accepted with probability $\exp[-\Delta E/k_\mathrm{B}T]$ if it increases the energy by ΔE.

In contrast to this kind of *thermodynamic* MC, *kinetic* Monte Carlo simulations aim at modeling the time evolution of the system. Simulations of nonequilibrium phenomena such as crystal growth are always of the kinetic type. The separation of time scales inherent in the Arrhenius law (6.11) implies that the atomistic processes occurring during crystal growth fit naturally into the MC scheme [660]: The time required for the actual motion across the energy barrier is much shorter than the waiting time between subsequent transitions, and hence it is reasonable to regard the transition itself as instantaneous. In this approximation no two transitions in the system ever occur simultaneously, and the time evolution can be described as a sequence of configurations separated by individual atomistic events. For example, if the configuration $\mathcal{C}'$ emerges from $\mathcal{C}$ by moving an adatom to a neighboring adsorption site, then $\Gamma(\mathcal{C} \to \mathcal{C}')$ is equal to the adatom jump rate; and if the two configurations differ by the presence of an additional adatom at a specified site, then the rate is $\hat{F} = F\Omega$.

Here and in the following a *configuration* refers to a stable arrangement of the surface, in which all adatoms reside at adsorption sites and which can therefore be approximately represented by specifying the occupation or vacancy of the sites of a perfect crystal lattice. This is known as a *lattice gas* representation. In general, there will be a large (but finite) number N_ev of possible events that can occur in a given configuration $\mathcal{C}$, each of which is characterized by its rate $\Gamma_i(\mathcal{C})$, with $i = 1, ..., N_\mathrm{ev}$ labeling the events. The detailed balance condition (6.28) on the rates retains its importance also in the kinetic context, as will be discussed below in Sect. 6.5.3.

Given a set of transition rates for each possible configuration of the surface, the task of the KMC simulation is, first, to generate a sequence of configurations in accordance with the rates, and second, to keep track of the physical time that elapses during the evolution. We first address the second issue [661]. Recall that the precise meaning of the statement that a random process has rate Γ is that the probability for the process to take place in an infinitesimal time interval dt equals Γdt [657]. The probability for *any* event to take place within the time interval dt when the surface is in the configuration $\mathcal{C}$ is therefore $\Gamma_{\text{tot}}(\mathcal{C})dt$, where

$$\Gamma_{\text{tot}}(\mathcal{C}) = \sum_{i=1}^{N_{\text{ev}}} \Gamma_i(\mathcal{C}) \tag{6.29}$$

is the total rate out of configuration $\mathcal{C}$. It is then straightforward to show that the (random) waiting time Δt_W until the next event is exponentially distributed[12] with mean $\Gamma_{\text{tot}}(\mathcal{C})^{-1}$. Thus the waiting time can be obtained from a random number u between 0 and 1 through the relation

$$\Delta t_W = -\Gamma_{\text{tot}}(\mathcal{C})^{-1} \ln u. \tag{6.30}$$

If the fluctuations in the waiting time can be neglected, the mean waiting time $\Gamma_{\text{tot}}(\mathcal{C})^{-1}$ is used instead.

The simplest way to generate a sequence of configurations with weights corresponding to the prescribed transition rates is to pick a possible event i at random, and to accept it with probability $p_i = \Gamma_i/\Gamma_{\text{max}}$, where Γ_{max} is the largest among the Γ_i. If the transition rates are of widely different magnitude, as is usually the case for thermally activated processes, this procedure is very inefficient, because most of the p_i are extremely small, and hence most attempted moves are rejected. This problem can be avoided by using an *event-driven* algorithm, which ensures that a transition is actually performed in each step [662, 663]. In the event-driven approach each event is selected with a probability proportional to its rate, rather than at random. More precisely, a random number u is chosen uniformly between 0 and 1, and that event i_0 is performed for which the condition

$$\sum_{i=1}^{i_0-1} \Gamma_i(\mathcal{C}) \leq u\Gamma_{\text{tot}}(\mathcal{C}) \leq \sum_{i=1}^{i_0} \Gamma_i(\mathcal{C}) \tag{6.31}$$

is satisfied. The disadvantage of this algorithm lies in the time required to search for the event i_0, and in the computational overhead needed to update the list of events after each step. In both cases the efficiency can be improved by subdividing the list of events into subgroups and selecting i_0 in a two-step process; optimal performance is reached when the number of groups is of the order of $\sqrt{N_{\text{ev}}}$ [663]. Further speedup can be achieved by dividing the groups into subgroups and sub-subgroups in a hierarchical fashion. For large systems the most efficient approach is to organize the events on a binary tree, which reduces the search time to $\mathcal{O}(\log N_{\text{ev}})$ [664].

[12] The fact that the waiting time for an event that occurs with constant rate is exponentially distributed is familiar from the physics of radioactive decay. Mathematically, such events form a *Poisson process* [657].

6.5.3 Constructing KMC Models

A kinetic Monte Carlo simulation of crystal growth may serve a range of different scientific purposes [665]. If the goal of the simulation is to describe a particular material, then as many atomistic details as possible should be included in the model. In particular, the correct lattice geometry must be used and the transition rates should be taken from a reliable total energy calculation [660]. If, on the other hand, the goal is to elucidate the influence of certain atomistic processes on qualitative features of the morphology, then the model should be kept as simple as possible to ensure that the effect of varying some model parameter can be clearly identified. By adding further processes as more experimental information becomes available, the model can be successively refined and quantitative agreement with the experiment can be achieved. In the following some examples which illustrate these two possible approaches will be described.

***

The first step in the construction of any KMC model is the selection of the allowed atomistic processes or *moves*. A major difficulty in developing models based on ab initio or semi-empirical total energy calculations is that the number of possible moves – and hence the number of energy barriers that need to be computed – proliferates very rapidly as the number of atoms involved in a process increases[13]. As an example, we consider a class of KMC models for fcc(100) metal surfaces which are based on semi-empirical many-body potentials [375, 667, 668]. The only allowed moves are single atom hops to nearest neighbor sites, and it is assumed that the energy barrier for a hop depends on the occupancy of the seven nearest and next-nearest neighbor sites surrounding the initial and final site of the atom (Fig. 6.9). Since each neighbor site can be either vacant or occupied, the total number of possible environments is $2^7 = 128$, which by symmetry reduces to 72. It is straightforward in principle to compute the energy barriers[14] for these 72 cases and store the corresponding transition rates in a lookup table to be used during the simulation [660]. There is however a possible pitfall that must be avoided [669]: One has to make sure that the set of energy barriers adds up to a consistent energy landscape, so that the detailed balance condition (6.28) is fulfilled. Even if the barriers have been derived from an expression for the total energy of the system, due to the approximations involved in the construction they do not necessarily induce a well-defined configuration energy $E(\mathcal{C})$ on the level of the lattice gas model.

[13] This problem is even more pronounced in the modeling of semiconductor surfaces with complicated reconstructions. In a recent DFT-based KMC study of GaAs growth, 10 different hopping processes for an *isolated* Ga atom had to be taken into account [666].

[14] We assume here that a single attempt frequency is chosen for all processes.

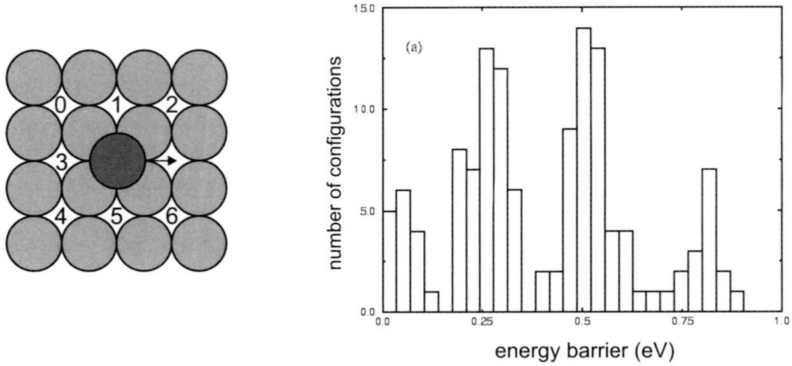

Fig. 6.9. Left: The hopping rate of an atom on a fcc(100) surface is assumed to depend on the occupancy of the seven neighboring sites labeled 0 – 6. Right: Histogram of EAM energy barriers for all possible environments in the case of Cu(100) (Fig. 2a of [670] reprinted with permission from Elsevier Science)

To make the consistency requirement on the energy barriers more explicit, consider first two configurations $\mathcal{C}$ and $\mathcal{C}'$ which are connected by a single move, and denote by $E_\mathrm{a}(\mathcal{C} \to \mathcal{C}')$ the corresponding activation energy. Since the energy barrier is the difference between the energies in the transition state and the initial state, and since the transition state is the same for the process $\mathcal{C} \to \mathcal{C}'$ as for the reverse process $\mathcal{C}' \to \mathcal{C}$, it follows that the difference between the barriers for the two processes equals the energy difference between the two configurations,

$$E_\mathrm{a}(\mathcal{C} \to \mathcal{C}') - E_\mathrm{a}(\mathcal{C}' \to \mathcal{C}) = E(\mathcal{C}') - E(\mathcal{C}). \tag{6.32}$$

Extending this argument to configurations which are connected through a chain of transitions, one finds that the energy difference between the two configurations equals the difference between the sums of the activation energies which are encountered when traversing the chain in the forward and backward directions. The existence of a consistent energy landscape implies that the energy difference between any two configurations must be *independent* of the path in configuration space that is chosen to move from one to the other; in particular, the energy difference along any closed path must vanish. This implies that the sum of activation energies encountered along a closed path must be the same, irrespective of the direction in which the path is traversed[15]. This condition is obviously fulfilled when detailed balance holds in the sense of (6.28). When it is violated, the system can lower its "energy"

[15] If the attempt frequencies are not the same for all processes, then this condition is replaced by the more general requirement that the product of transition rates $\prod_i \Gamma(\mathcal{C}_i \to \mathcal{C}_{i+1})$ along any closed path should be independent of the direction. Within transition state theory, this is equivalent to the existence of a consistent *free* energy landscape.

indefinitely by following loops in configuration space, which introduces an unphysical bias into the simulation [669]. It is particularly important to ensure that detailed balance is preserved when the rates of individual processes are adjusted to fit experimental observations.

A simple way of evading these problems, which moreover reduces the computational effort and allows for a transparent physical interpretation, is to condense the set of barriers into an effective bond-counting model [375, 668]; for such a model the condition of detailed balance is easily checked. For the example of Cu(100), the rationale for this reduction is contained in the histogram in Fig. 6.9, where the set of processes is seen to decompose naturally into four groups each with rather similar activation energies [667, 670]; the same structure is found for other fcc(100) metal surfaces [375,668]. On the simplest level of model construction, identical energy barriers are assigned to all processes within a group. This leads to the expression [668, 670]

$$E_{\mathrm{a}} = E_{\mathrm{D}} + E_{\mathrm{b}}(\eta_3 - \eta_2 - \eta_6) \tag{6.33}$$

for the activation energy of a general process, where η_i is the occupation number of site i (following the labeling of Fig. 6.9) with the possible values zero or one, E_{D} is the energy barrier for diffusion of a free adatom, and E_{b} is an effective bond strength. The parameters E_{D} and E_{b} are fixed by fitting the energy barriers for all processes to the form (6.33). The physical picture behind (6.33) is that a neighboring atom at site 3 increases the barrier because it binds the hopping atom more strongly at its initial site, while atoms at sites 2 and 6 lower the barrier by providing additional coordination in the transition state. The expression (6.33) is as simple as the energy barrier (6.34) used in the standard SOS model (see below), but it contains much more of the correct microscopic physics.

A similar classification of the possible atomistic processes according to the environment of the moving atom can be carried out for fcc(111) surfaces, see Sect. 3.1 and [273, 293]. KMC models for Pt(111) homoepitaxy based on semi-empirical potentials and experimental observations are described in [213,273]. Models based on DFT energetics have been developed for Al growth on Al(111) [266, 275–277]; their performance is discussed in Sect. 3.7.2. Another recent appplication of DFT-based KMC simulations is the modeling of island nucleation in the presence of adatom-adatom interactions [180, 183], see Sect. 2.4.5.

***

The construction of material-specific KMC models based on realistic total energy calculations is a relatively recent development [660]. Much of the earlier KMC work on crystal growth was carried out within the standard solid-on-solid (SOS) model, which represents the crystal by a simple cubic lattice without overhangs. In the most basic version of the model[16], the energy barrier for a diffusion process is assumed to be proportional to the number of bonds in the initial state of the hopping atom [671]. For modeling multilayer growth, an additional parameter controlling the rate of interlayer transport is needed (see Sect. 4.3). This can be included in the model by keeping track of the number of next nearest neighbors of the atom in the layer below (and above) [367]. When the atom approaches a descending step edge this number decreases, which is penalized energetically, causing the atom to deflect. The resulting expression for the energy barrier reads

$$E_{\mathrm{a}} = E_{\mathrm{D}} + E_{\mathrm{b}} n_i + \Delta E_{\mathrm{S}} (m_i - m_f) \Theta_H (m_i - m_f). \tag{6.34}$$

Here n_i is the number of in-plane neighbors of the adatom in the initial position, m_i and m_f is the initial and final number of out-of-plane next nearest neighbors, and the Heaviside function $\Theta_H(x) = 1$ for $x > 0$ and 0 otherwise; it ensures that the second term on the right hand side is present only when $m_i > m_f$. The three energy parameters in (6.34) represent the diffusion barrier of a free adatom, an effective bond strength, and an additional barrier ΔE_{S} for interlayer transport, respectively. Despite its simplicity, the SOS model has been remarkably successful in reaching semi-quantitative agreement with experimental observations, both for semiconductors [672,673] and metals [674]. Because of its lack of realism, however, it is clear that the parameters entering the model cannot be given an unambiguous microscopic interpretation.

There have been several attempts to refine the basic SOS model in order to achieve quantitative agreement with experimental growth morphologies, while maintaining the simplicity of the description in terms of a few model parameters. Examples include the Fe/Fe(001) [391,497] and Ag/Ag(001) [151, 393] systems; we discuss the latter case in somewhat more detail here.

The first step is to use the correct (fcc) crystal structure. Since this makes it necessary to deal with atoms which are not deposited onto stable adsorption sites [fourfold-hollow sites on the fcc(100) surface], a certain amount of downward funneling is naturally included in the model (see Sect. 2.4.1). The barrier for diffusion on the flat terrace can be determined from the island density in the submonolayer regime (see Sect. 2.3), and is fixed at $E_{\mathrm{D}} = 0.4$ eV. Explicit assumptions about the in-plane energetics are avoided by working at sufficiently low temperatures, where the lateral binding of adatoms is irreversible. However, strict irreversibility cannot be maintained at the island edges, since this would result in fractal islands, whereas the experimentally observed islands are compact. Therefore a set of rules for diffusion along step edges are posited: Diffusion to a forward kink is instantanous, while attachment to a backward kink is restricted by an additional kink rounding barrier

[16] We do not discuss here SOS models of the *limited mobility* type, in which the diffusion of an adatom to an incorporation site is an instantanous process occurring immediately after deposition. For reviews on these and related theoretically motivated SOS models see [36, 106, 650].

E_{kr} (compare to Sect. 3.7.1). Finally, interlayer transport is assumed to be facile (without any additional barrier) at kinks, and subject to an additional barrier ΔE_S at straight steps.

This formulation of the model leaves E_{kr} and ΔE_S as the only free parameters, which are determined by comparison with experimental data for films of constant thickness (25 ML) over a range of temperatures (55 K $\leq T \leq$ 300 K) [393]. Setting $E_{kr} = 0.41$ eV and $\Delta E_S = 0.06$–0.07 eV, most features of the experimentally observed morphology are reproduced quite well.

6.5.4 Hybrid Approaches

The KMC method overcomes the time scale problem of molecular dynamics by performing directly the activated moves, without simulating the short time atomic motion between the events. The disadvantage of the KMC approach is, however, that it requires a predetermined set of possible configurations and transitions, which is fixed prior to the simulation. This is problematic in situations where complicated rearrangements involving several atoms play an important role, because such collective moves cannot be foreseen without performing an explicit simulation of the short time dynamics. It would therefore be desirable to devise a simulation scheme which exploits the activated nature of the dynamics, as in KMC, but in which the possible transitions to new configurations are created "on the fly" during the calculation.

In recent years several such schemes have been suggested, which show considerable promise towards more efficient and realistic growth simulations [675]. In the *hyperdynamics* approach to MD [676], the potential energy landscape is deformed by raising the energy within the potential basins while leaving the transition states unchanged, thus lowering the energy barriers. Within transition state theory, the effect of the deformation on the rates can be accounted for and the time scale of the original, undeformed problem can be reconstructed. The *parallel replica MD method* [677] exploits a parallel computer architecture to simulate the dynamics between activated events for several statistically independent replicas of the system; this reduces the waiting time for the next event by a factor which is proportional to the number of replicas.

In *temperature accelerated dynamics* [152] a number of candidate transitions out of a given potential basin are generated by performing the MD simulation at an elevated temperature T_{high} and reflecting the system back into the basin when a transition has occurred. The time t_{low} at which the transition would have occurred at the lower temperature of interest is obtained from the relation

$$t_{\mathrm{low}} = t_{\mathrm{high}} \exp[E_a/k_B(T_{\mathrm{low}}^{-1} - T_{\mathrm{high}}^{-1})], \tag{6.35}$$

which holds within transition state theory. After a sufficiently large sample of possible transitions has been created, the one with the smallest value of

t_{low} is performed and the process starts anew. Using this method the growth of 4 ML Ag on Ag(100) in the temperature range $T = 0 - 70$ K could be simulated with a realistic deposition flux of 0.075 ML/s.

In a related approach, the candidate transitions are generated by launching uphill saddle point searches from the current potential minimum [678]. The searches are performed using the *dimer method* [628], in which two copies of the system form a rigid dimer in configuration space and the dimer is rotated and translated in such a way that it locates saddle points. This method has been applied to the growth of Al/Al(100) at temperatures between 30 and 100 K and a deposition flux of 3.2×10^4 ML/s. Even at these low temperatures and high fluxes, a perfect crystal with a smooth surface was formed due to a multitude of complex, collective restructuring events.

References

1. R.P. Feynman, R.B. Leighton, M. Sands: *The Feynman Lectures on Physics*, Vol. 1 (Addison-Wesley, Palo Alto 1963).
2. S.G. Brush: *The Kind of Motion We Call Heat. A History of the Kinetic Theory of Gases in the 19th Century, Book 2* (North-Holland, Amsterdam 1976).
3. R.W. Cahn: *The Coming of Materials Science* (Pergamon, Oxford 2001).
4. J. Kepler: *Strena seu de Nive sexangula*, (Frankfurt, 1611).
5. F.C. Frank: Contemp. Phys. **23**, 3 (1982).
6. R. Kaischew: J. Cryst. Growth **51**, 643 (1981).
7. M. Volmer: *Kinetik der Phasenbildung*, (Theodor Steinkopff, 1939).
8. W. Burton, N. Carbrera, F. C. Frank: Nature **163**, 398 (1949).
9. A.A. Chernov: Contemp. Phys. **30**, 251 (1989).
10. G.G. Lemmlein, N.V. Gliki: Dokl. Akad. Nauk SSSR **94**, 473 (1954).
11. C.M. Heck: Phys. Rev. **51**, 686 (1937).
12. D.W. Pashley: 'A Historical Review of Epitaxy'. In: *Epitaxial Growth*, ed. by J.W. Matthews (Academic Press, New York 1975) pp. 2–27.
13. B. Lewis, J.C. Anderson: *Nucleation and Growth of Thin Films* (Academic Press, London 1978).
14. E.W. Müller: Z. Physik **131**, 136 (1951).
15. E.W. Müller: J. Appl. Phys. **27**, 474 (1956).
16. G. Ehrlich, F.G. Hudda: J. Chem. Phys. **44**, 1039 (1966).
17. R.L. Schwoebel, E.J. Shipsey: J. Appl. Phys. **37**, 3682 (1966).
18. M.G. Lagally, Z. Zhang: Nature **417**, 907 (2002).
19. G. Binnig, H. Rohrer, Ch. Gerber, E. Weibel: Appl. Phys. Lett. **40**, 178 (1982).
20. G. Binnig, H. Rohrer, C. Gerber, E. Weibel: Phys. Rev. Lett. **49**, 57 (1982).
21. J.-J. Métois, J.-C. Heyraud, A. Pimpinelli: Surf. Sci. **420**, 250 (1999).
22. W.K. Burton, N. Carbrera, F. C. Frank: Phil. Trans. Roy. Soc. A **243**, 299 (1951).
23. W. Kossel: Nachr. Ges. Wiss. Göttingen, math.-phys. Klasse, p. 135 ff. (1927).
24. L. Royer: Bull. Soc. Fr. Min. **51**, 7 (1928).
25. D.J. Eaglesham: J. Appl. Phys. **77**, 3597 (1995).
26. E. Bauer: Z. Kristallogr. **110**, 372 (1958).
27. R. Kern: 'Fundamentals of Epitaxy'. In: *Crystal Growth in Science and Technology*, ed. by H. Arend, J. Hulliger (Plenum Press, New York 1989) pp. 143–165.
28. D.E. Jesson: 'Morphological Evolution of Strained Semiconductor Films'. In: *Morphological Organization in Epitaxial Growth and Removal*, ed. by Z. Zhang, M.G. Lagally (World Scientific, Singapore 1998), pp. 195–212.
29. B. Voigtländer: Surf. Sci. Rep. **43**, 127 (2001).
30. C. Teichert: Phys. Rep. **365**, 335 (2002).
31. I.V. Markov: *Crystal Growth for Beginners: Fundamentals of Nucleation, Crystal Growth and Epitaxy* (World Scientific, Singapore 1995).

32. A.–L. Barabàsi, H.E. Stanley: *Fractal Concepts in Surface Growth*, (Cambridge University Press, 1995).
33. J. Villain, A. Pimpinelli: *Physique de la Croissance Cristalline* (Éditions Eyrolles, Paris 1995).
34. Y. Saito: *Statistical Physics of Crystal Growth* (World Scientific, Singapore 1996).
35. A. Pimpinelli, J. Villain: *Physics of Crystal Growth* (Cambridge University Press, Cambridge 1998).
36. P. Meakin: *Fractals, Scaling and Growth far from Equilibrium* (Cambridge University Press, 1998).
37. J.A. Venables: *Introduction to Surface and Thin Film Processes*, (Cambridge University Press, Cambridge 2000).
38. D.R. Lide (Ed.): *Handbook of Chemistry and Physics*, 78th edition (CRC Press, Boca Ratan 1997) pp. 5–24.
39. R.W. Zwanzig: J. Chem. Phys. **32**, 1173 (1960).
40. B. McCarroll, G. Ehrlich: J. Chem. Phys. **38**, 523 (1963).
41. W.F. Egelhoff, Jr., I. Jacob: Phys. Rev. Lett. **62**, 921 (1989).
42. P.S. Weiss, D.M. Eigler: Phys. Rev. Lett.. **69**, 2240 (1992).
43. D.E. Sanders, A.E. DePristo: Surf. Sci. **254**, 341 (1991).
44. D.E. Sanders, D.M. Halstead, A.E. DePristo: J. Vac. Sci. Technol. A **10**, 1986 (1992).
45. G. de Lorenzi, G. Ehrlich: Surf. Sci. Lett. **293**, L900 (1993).
46. M.J. Abrams, P.S. Weiss: Surf. Sci. **312**, 1 (1994).
47. S.C. Wang, G. Ehrlich: J. Chem. Phys. **94**, 4071(1991).
48. H. Brune: Surf. Sci. Rep. **31**, 121 (1998).
49. A. Hitzke, M. B. Hugenschmidt, R. J. Behm: Surf. Sci. **389**, 8 (1997).
50. A. Gölzhäuser, G. Ehrlich: Phys. Rev. Lett. **77**, 1334 (1996).
51. R. Gomer: Rep. Prog. Phys. **53**, 917 (1990).
52. G. Kellogg: Surf. Sci. Rep. **21**, 1 (1994).
53. T. Ala-Nissila, R. Ferrando, S.C. Ying: Adv. Phys. **51**, 949 (2002).
54. L.J. Lauhon, W. Ho: Phys. Rev. Lett. **85**, 4566 (2000).
55. K.J. Laidler: *Chemical Kinetics*, (McGraw-Hill, London 1965).
56. P. Hänggi, P. Talkner, M. Borkovec: Rev. Mod. Phys. **62**, 251 (1990).
57. H. Eyring: J. Chem. Phys. **3**, 107 (1935).
58. G.H. Vineyard: J. Phys. Chem. Solids **3**, 121 (1957).
59. C. Ratsch, M. Scheffler: Phys. Rev. B **58**, 13163 (1998).
60. C.L. Liu, J.M. Cohen, J.B. Adams, A.F. Voter: Surf. Sci. **253**, 334 (1991).
61. G. Boisvert, N. Mousseau, L.J. Lewis: Phys. Rev. Lett. **80**, 203 (1998).
62. H.A. Kramers: Physica (Utrecht) **7**, 284 (1940).
63. D.E. Sanders, A.E. DePristo: Surf. Sci. Lett. **264**, L169 (1992).
64. G. Boisvert, N. Mousseau, L.J. Lewis: Phys. Rev. B **58**, 12667 (1998).
65. G. Boisvert, L.J. Lewis, A. Yelon: Phys. Rev. Lett. **75**, 469 (1995).
66. S. C. Wang, G. Ehrlich: Surf. Sci. **206**, 451 (1988).
67. K. Kyuno, A. Gölzhäuser, G. Ehrlich: Surf. Sci. **397**, 191 (1998).
68. P.J. Feibelman, J. S. Nelson, G. Kellogg: Phys. Rev. B **49**, 10548 (1994).
69. Y. W. Mo: Phys. Rev. Lett. **71**, 2923 (1993).
70. M. Bott: *Atomare Diffusionsprozesse auf Pt(111) untersucht mit temperaturvariabler Rastertunnelmikroskopie*, Jül-Bericht 3133 (Forschungszentrum Jülich, Jülich 1995).
71. J.A. Stroscio, D. M. Eigler: Science **254**, 1319 (1991).
72. P. Ebert, M.G. Lagally, K. Urban: Phys. Rev. Lett. **70**, 1437 (1993).
73. J. Li, R. Berndt, W.-D. Schneider: Phys. Rev. Lett. **76**, 1888 (1996).

74. K. Morgenstern, G. Rosenfeld, B. Poelsema, G. Comsa: Phys. Rev. Lett. **74**, 2058 (1995).
75. J.M. Carpinelli, B.S. Swartzentruber, Phys. Rev. B **58**, R13423 (1998).
76. L. Bartels, G. Meyer, K.-H. Rieder, D. Velic, E. Knoesel, A. Hotzel, M. Wolf, G. Ertl: Phys. Rev. Lett. **80**, 2004 (1998).
77. T.R. Linderoth, S. Horch, E. Lægsgaard, I. Stensgaard, F. Besenbacher: Phys. Rev. Lett. **78**, 4978 (1997).
78. B.S. Swartzentruber: Phys. Rev. Lett. **76**, 459 (1996).
79. R. van Gastel, E. Somfai, S.B. van Albada, W. van Saarloos, J.W.M. Frenken: Phys. Rev. Lett. **86**, 1562 (2001).
80. M.L. Grant, B.S. Swartzentruber, N.C. Bartelt, J.B. Hannon: Phys. Rev. Lett. **86**, 4588 (2001).
81. H.J.W. Zandvliet, B. Poelsema, B.S. Swartzentruber: Physics Today **54** (7), 40 (2001).
82. M. Bott, M. Hohage, M. Morgenstern, T. Michely, G. Comsa: Phys. Rev. Lett. **76**, 1304 (1996).
83. K. Kyuno, G. Ehrlich: Surf. Sci. **437**, 29 (1999).
84. J. Wollschläger, T. Schmidt, M. Henzler, M. I. Larsson: Surf. Sci. **454-456**, 566 (2000).
85. J.J. Mortensen, B. Hammer, O.H. Nielsen, K.W. Jacobsen, J.K. Nørskov: 'Density Functional Theory Study of Self–Diffusion on the (111) Surfaces of Ni, Pd, Pt, Cu, Ag and Au'. In: *Elementary Processes in Excitations and Reactions on Solid Surfaces, Springer Series in Solid-State Sciences, Vol. 121*, ed. by A. Okiji, H. Kasai, K. Makoshi (Springer, Berlin 1996) pp. 173–182.
86. G. Boisvert, L.J. Lewis, M. Scheffler: Phys. Rev. B **57**, 1881 (1998).
87. P.J. Feibelman: Phys. Rev. Lett. **81**, 168 (1998).
88. D.W. Bassett, P.R. Webber: Surf. Sci. **70**, 520 (1978).
89. R. Wang, K.A. Fichthorn: Surf. Sci. **301**, 253 (1994).
90. M. Villarba, H. Jónsson: Surf. Sci. **317**, 15 (1994).
91. S. Liu, Z. Zhang, J. Norskøv, H. Metiu: Surf. Sci. **321**, 161 (1994).
92. P. Stoltze: J. Condens. Matter **6**, 9495 (1994).
93. Y. Li, A.E. DePristo: Surf. Sci. **351**, 189 (1996).
94. E.P. Münger, V. Chirita, J.E. Greene, J.-E. Sundgren: Surf. Sci. Lett. **355**, L325 (1996).
95. F. Máca, M. Kotrla, O.S. Truishin: Vacuum **54**, 113 (1999).
96. D. Walton: J. Chem. Phys. **37**, 2182 (1962).
97. J. A. Venables: Philos. Mag. **27**, 697 (1973).
98. S. Stoyanov, D. Kashchiev: 'Thin Film Nucleation and Growth Theories: A Confrontation with Experiment'. In: *Current Topics in Materials Science*, Vol. 7, ed. by E. Kaldis (North Holland, Amsterdam 1981) pp. 69–141.
99. J. A. Venables, G. D. T. Spiller, M. Hanbücken: Rep. Prog. Phys. **47**, 399 (1984).
100. H. Brune, G.S. Bales, J. Jacobsen, C. Borragno, K. Kern: Phys. Rev. B **60**, 5991 (1999).
101. K. Kyuno, G. Ehrlich: Phys. Rev. Lett. **84**, 2658 (2000).
102. D.R. Frankl, J.A. Venables: Adv. Phys. **19**, 409 (1970).
103. M.J. Stowell: Phil. Mag. **26**, 349 (1972).
104. G.S. Bales, D.C. Chrzan: Phys. Rev. B. **50**, 6057 (1994).
105. G.S. Bales, A. Zangwill: Phys. Rev. B. **55**, R1973 (1997).
106. J. Krug: Adv. Phys. **46**, 139 (1997).
107. M.C. Bartelt, J.W. Evans: Phys. Rev. B **46**, 12675 (1992).
108. J.G. Amar, F. Family: Phys. Rev. Lett. **74**, 2066 (1995).
109. M.C. Bartelt, J.W. Evans: Phys. Rev. B **54**, R17359 (1996).

110. M.C. Bartelt, J.B. Hannon, A.K. Schmid, C.R. Stoldt, J.W. Evans: Coll. Surf. A **165**, 373 (2000).
111. P.A. Mulheran, J.A. Blackman: Phys. Rev. B **53**, 10261 (1996).
112. P.A. Mulheran, D.A. Robbie: Europhys. Lett. B **49**, 617 (2000).
113. M.N. Popescu, J.G. Amar, F. Family: Phys. Rev. B **64**, 205404 (2001).
114. J.W. Evans, M.C. Bartelt: Phys. Rev. B **66**, 235410 (2002).
115. M. Hohage: *Kinetik atomarer Prozesse auf der Pt(111)-Oberfläche*, Jül-Bericht 3255, (Forschungszentrum Jülich, Jülich, 1996).
116. J.A. Stroscio, D. T. Pierce: Phys. Rev. B **49**, 8522 (1994).
117. P.J. Feibelman: Phys. Rev. B **52**, 12444 (1995).
118. R. Pfandzelter, T. Igel, H. Winter: Phys. Rev. B **62**, R2299 (2000).
119. H. Dürr, J.F. Wendelken, J.-K. Zuo: Surf. Sci. Lett. **328**, L527 (1995).
120. S. Günther, E. Kopatzki, M.C. Bartelt, J.W. Evans, R.J. Behm: Phys. Rev. Lett. **73**, 553 (1994).
121. F. Tsui, J. Wellman, C. Uher, R. Clarke: Phys. Rev. Lett. **76**, 3164 (1996).
122. C.-M. Zhang, M.C. Bartelt, J.-M. Wen, C.J. Jenks, J.W. Evans, P.A. Thiel: Surf. Sci. **406**, 178 (1998).
123. T.R. Linderoth, S. Horch, L. Petersen, S. Helveg, M. Schønning, E. Lægsgaard, I. Stensgaard, F. Besenbacher: Phys. Rev. B **61**, R2448 (2000).
124. P.J. Feibelman: Phys. Rev. B **61**, R2452 (2000).
125. S. Liu, L. Bönig, H. Metiu: Phys. Rev. B **52**, 2907 (1995).
126. M.C. Bartelt, L.S. Perkin, J.W. Evans: Surf. Sci. **344**, L1193 (1995).
127. P.J. Feibelman: Surf. Sci. **423**, 169 (1999).
128. C. Ratsch, A. Zangwill, P. Smilauer, D.D. Vvedensky: Phys. Rev. Lett. **72**, 3194 (1994).
129. C. Ratsch, P. Smilauer, A. Zangwill,D.D. Vvedensky: Surf. Sci. Lett. **329**, L599 (1995).
130. P.J. Feibelman, T. Michely: Surf. Sci. Lett. **492**, L723 (2001).
131. M. Kalff, G. Comsa, T. Michely: Phys. Rev. Lett. **81**, 1255 (1998).
132. M. Kalff: *Morphologische Entwicklung von Pt(111) bei der Pt-Deposition und Ionenstrahlerosion*, Jül-Bericht 3625 (Forschungszentrum Jülich, Jülich, 1999).
133. J. Villain, A. Pimpinelli, L. Tang, D. Wolf: J. Phys. I France **2**, 2107 (1992).
134. G. Boisvert, L.J. Lewis: Phys. Rev. B **59**, 9846 (1999).
135. K. Kyuno, G. Ehrlich: Surf. Sci. **511**, 459 (2002).
136. P.J. Feibelman, T. Michely: Surf. Sci. **501**, 461 (2002).
137. C. Busse, W. Langenkamp, C. Polop, A. Petersen, H. Hansen, U. Linke, P.J. Feibelman, T. Michely: Surf. Sci. Lett. **539**, L560 (2003).
138. L. Pauling: *The Nature of the Chemical Bond* (Cornell University Press, Ithaca, NY, 1960).
139. J.K. Nørskov, K.W. Jacobsen, P. Stoltze, L.B. Hansen: Surf. Sci. **283**, 277 (1993).
140. P.J. Feibelman: Phys. Rev. B. **60**, 4972 (1999).
141. M.C. Fallis, M.S. Daw, C.Y. Fong: Phys. Rev. B **51**, 7817 (1995).
142. S.K. Gupta, B.M. Nappi, K.A. Gingerich: Inorg. Chem **20**, 966 (1981).
143. S.C. Wang, G. Ehrlich: Phys. Rev. Lett. **71**, 4174 (1993).
144. S.C. Wang, G. Ehrlich: Phys. Rev. Lett. **62**, 2297 (1989).
145. S.C. Wang, G. Ehrlich: Phys. Rev. Lett. **70**, 41 (1993).
146. M. Villarba, H. Jónsson: Surf. Sci. **324**, 35 (1995).
147. P. Stoltze, J.K. Nørskov: Phys. Rev. B **48**, 5607 (1993).
148. G. Kellogg: Phys. Rev. Lett. **76**, 98 (1996).
149. J.W. Evans, D.E. Sanders, P.A. Thiel, A.E. DePristo: Phys. Rev. B **41**, 5410 (1990).

150. J.W. Evans: Phys. Rev. B **43**, 3897 (1991).
151. K.J. Caspersen, C.R. Stoldt, A.R. Layson, M.C. Bartelt, P.A. Thiel, J.W. Evans: Phys. Rev. B **63**, 085401 (2001).
152. F. Montalenti, M.R. Sørensen, A.F. Voter: Phys. Rev. Lett. **87**, 126101 (2001).
153. C.L. Kelchner, A.E. DePristo: Surf. Sci. **393**, 72 (1997).
154. J. Yu, J.G. Amar: Phys. Rev. Lett. **89**, 286103 (2002).
155. H. Park, A. Provata, S. Redner: J. Phys. A **24**, L1391 (1991).
156. M. Raible, S.G. Mayr, S.J. Linz, M. Moske, P. Hänggi, K. Samwer: Europhys. Lett. **50**, 61 (2000).
157. F. Montalenti, A.F. Voter: Phys. Rev. B **64**, 081401 (2001).
158. S. van Dijken, L.C. Jorritsma, B. Poelsema: Phys. Rev. Lett. **82**, 4038 (1999).
159. S. van Dijken, L.C. Jorritsma, B. Poelsema: Phys. Rev. B **61**, 14047 (2000).
160. P.J. Feibelman: Phys. Rev. Lett. **65**, 729 (1990).
161. G. Kellogg, P.J. Feibelman: Phys. Rev. Lett. **64**, 3143 (1990).
162. C. Chen, T.T. Tsong: Phys. Rev. Lett. **64**, 3147 (1990).
163. J.D. Wrigley, G. Ehrlich: Phys. Rev. Lett. **44**, 661 (1980).
164. S.C. Wang, G. Ehrlich: Phys. Rev. Lett. **67**, 2509 (1991).
165. C. Itzykson, J.-M. Drouffe: *Statistical Field Theory*, Vol. I (Cambridge University Press, Cambridge 1989).
166. L.H. Tang: J. Phys. I France **3**, 935 (1993).
167. H. Kallabis, P.L. Krapivsky, D.E. Wolf: Eur. Phys. J. B **5**, 801 (1998).
168. J. Krug: Eur. Phys. J. B **18**, 713 (2000).
169. P.L. Krapivsky, J.F.F. Mendes, S. Redner: Phys. Rev. B **59**, 15950 (1999).
170. I. Furman, O. Biham: Phys. Rev. B **55**, 7917 (1997).
171. S.V. Khare, T.L. Einstein: Phys. Rev. B **54**, 11752 (1996).
172. M. Giesen: Prog. Surf. Sci. **68**, 1 (2001).
173. K.H. Lau, W. Kohn: Surf. Sci. **75**, 69 (1978).
174. P. Hyldgaard, M. Persson; J. Phys. Condens. Matter. **12**, L13 (2000).
175. T.T. Tsong: Rep. Progr. Phys. **51**, 759 (1988).
176. F. Watanabe, G. Ehrlich: J. Chem. Phys. **95**, 6075 (1991).
177. J. Repp, F. Moresco, G. Meyer, K.-H. Rieder: Phys. Rev. Lett. **85**, 2981 (2000).
178. N. Knorr, H. Brune, M. Epple, A. Hirstein, M.A. Schneider, K. Kern: Phys. Rev. B **65**, 115420 (2002).
179. A. Bogicevic, S. Ovesson, P. Hyldgaard, B.I. Lundqvist, H. Brune, D.R. Jennison: Phys. Rev. Lett. **85**, 1910 (2000).
180. K.A. Fichthorn, M. Scheffler: Phys. Rev. Lett. **84**, 5371 (2000).
181. S.C. Wang, U. Kürpick, G. Ehrlich: Phys. Rev. Lett. **81**, 4923 (1998).
182. S. Ovesson: Phys. Rev. Lett. **88**, 116102 (2002).
183. K.A. Fichthorn, M.L. Merrick, M. Scheffler: Appl. Phys. A **75**, 17 (2002).
184. J. A. Venables, H. Brune: Phys. Rev. B **66**, 195404 (2002).
185. J.V. Barth, H. Brune, B. Fischer, J. Weckesser, K. Kern: Phys. Rev. Lett. **84**, 1732 (2000).
186. T. Michely, W. Langenkamp, H. Hansen, C. Busse: Phys. Rev. Lett. **86**, 2695 (2001).
187. C. Polop, H. Hansen, C. Busse, T. Michely: Phys. Rev. B **67**, 193405 (2003).
188. D. Kandel: Phys. Rev. Lett. **78**, 499 (1997).
189. M. Kotrla, J. Krug, P. Šmilauer: Surf. Sci. **454-456**, 681 (2000).
190. E. Bauer, H. Poppa : Thin Solid Films **12**, 167 (1972).
191. S. Horch, H.T. Lorensen, S. Helveg, E. Lægsgaard, I. Stensgaard, K.W. Jacobsen, J.K. Norskøv, F. Besenbacher: Nature **398**, 134 (1999).
192. R.T. Tung, W.R. Graham: Surf. Sci. **97**, 73 (1980).
193. R. Stumpf: Phys. Rev. B **53**, R4253 (1996).

194. B. Hammer, O.H. Nielsen, J.K. Nørskov: Catal. Lett. **46**, 31 (1997).
195. G. Kellogg: Phys. Rev. Lett. **72**, 1662 (1994).
196. B. Poelsema, L.K. Verheij, G. Comsa: Phys. Rev. Lett. **49**, 1731 (1982).
197. M. Kotrla, J. Krug, P. Šmilauer: 'Effects of Adsorbates on Submonolayer Growth'. In: *Collective Diffusion on Surfaces: Correlation Effects and Adatom Interactions*, ed. by M.C. Tringides, Z. Chvoj (Kluwer Academic, 2001) pp. 247–257.
198. M. Kotrla, J. Krug, P. Šmilauer: Phys. Rev. B **62**, 2889 (2000).
199. M. Kotrla, J. Krug, P. Šmilauer: Surf. Sci. **482-485**, 840 (2001).
200. J. A. Venables: Physica A **239**, 35 (1997).
201. J. A. Venables, J. H. Harding: J. Cryst. Growth **211**, 27 (2000).
202. C. Polop, H. Hansen, W. Langenkamp, Z. Zhong, C. Busse, M. Kotrla, U. Linke, P.J. Feibelman, T. Michely: submitted.
203. J. Vrijmoeth, H.A. van der Vegt, J. A. Meyer, E. Vlieg, R. J. Behm: Phys. Rev. Lett. **72**, 3843 (1994).
204. S. Oppo, V. Fiorentini, M. Scheffler: Phys. Rev. Lett. **71**, 2437 (1993).
205. G. Büchner: *The Plays of Georg Büchner*, transl. by V. Price (Oxford University Press, 1971).
206. T. Michely, M. Hohage, M. Bott, G. Comsa: Phys. Rev. Lett. **70**, 3943 (1993).
207. I. Sunagawa: *Morphology of Crystals* (Terra Scientific Publishing, Tokyo 1987) p. v.
208. T.A. Witten, L.M. Sander: Phys. Rev. Lett. **47**, 1400 (1981).
209. T.A. Witten, L.M. Sander: Phys. Rev. B **27**, 5686 (1983).
210. J.S. Langer: Rev. Mod. Phys. **52**, 1 (1980).
211. J.C. Heyraud, J.J. Métois: J. Cryst. Growth **50**, 571 (1980).
212. K. Spangenberg: 'Wachstum und Auflösung der Kristalle'. In: *Handwörterbuch der Naturwissenschaften*, Bd. X (Gustav Fischer Verlag, Jena 1934) pp. 362–401.
213. J. Jacobsen, K. W. Jacobsen, P. Stoltze, J.K. Nørskov: Phys. Rev. Lett. **74**, 2295 (1995).
214. G. Kellogg: J. Phys. (Paris), Colloq. C2 **47**, 331 (1986).
215. M. Giesen, G. Schulze Icking-Konert, D. Stapel, H. Ibach: Surf. Sci. **366**, 229 (1996).
216. M. Biebl: Phys. Bl. **52**, 1010 (1996).
217. J.C. Heyraud, J.J. Métois: Surf. Sci. **128**, 334 (1983).
218. A. Pavlovska, D. Dobrev, E. Bauer: Surf. Sci. **314** 341 (1994).
219. C. Herring: 'The Use of Classical Macroscopic Concepts in Surface-Energy Problems'. In: *Structure and Properties of Solid Surfaces*, ed. by R. Gomer, C.S. Smith (University of Chicago Press, Chicago, 1953) pp. 5–81.
220. M. Wortis: 'Equilibrium Crystal Shapes and Interfacial Phase Transitions'. In: *Chemistry and Physics of Solid Surfaces*, Vol. 7, ed. by R. Vanselow, R. Howe (Springer, Berlin 1988) pp. 367–405.
221. G. Wulff: Zeitschr. f. Kristallog. **34**, 449 (1901).
222. P. Nozières: 'Shape and Growth of Crystals'. In: *Solids Far From Equilibrium*, ed. by C. Godrèche (Cambridge University Press, Cambridge 1991) pp. 1–154.
223. K. Arenhold, S. Surnev, H. P. Bonzel, P. Wynblatt: Surf. Sci. **424**, 271 (1999).
224. M.J. Rost, R. van Gastel, J.W.M. Frenken: Phys. Rev. Lett. **84**, 1966 (2000).
225. S. Kodambaka, V. Petrova, S.V. Khare, D.D. Johnson, I. Petrov, J.E. Greene: Phys. Rev. Lett. **88**, 146101 (2002).
226. Y.W. Mo, B.S. Swartzentruber, R. Kariotis, M.B. Webb, M.G. Lagally: Phys. Rev. Lett. **63**, 2393 (1989).
227. T. Michely, G. Comsa: Surf. Sci. **256**, 217 (1991).

228. T. Michely, G. Comsa: 'Two-Dimensional Island Shapes'. In: *Morphological Organization in Epitaxial Growth and Removal* ed. by Z. Zhang, M.G. Lagally (World Scientific, Singapore 1998) pp. 265–306.
229. D.C. Schlößer, L.K. Verheij, G. Rosenfeld, G. Comsa: Phys, Rev. Lett. **82**, 3843 (1999).
230. G. Schulze Icking-Konert, M. Giesen, H. Ibach: Phys. Rev. Lett. **83**, 3880 (1999).
231. M. Giesen, C. Steimer, H. Ibach: Surf. Sci. **471**, 80 (2001).
232. H. Hansen, C. Busse, T. Michely, unpublished.
233. R. Stumpf, M. Scheffler: Phys. Rev. Lett. **72**, 254 (1994).
234. S. Dieluweit, H. Ibach, M. Giesen, T.L. Einstein: Phys. Rev. B **67**, 121410 (2003).
235. D. Chowdhury, D. Stauffer: *Principles of Equilibrium Statistical Mechanics* (Wiley-VCH, Weinheim 2000).
236. P.J. Feibelman: Phys. Rev. B **62**, 17020 (2000).
237. A. Emundts, M. Nowicki, H.P. Bonzel: Surf. Sci. Lett. **496**, L35 (2002).
238. M. Nowicki, C. Bombis, A. Emundts, H.P. Bonzel: Phys. Rev. B **67**, 075405 (2003).
239. K. Morgenstern, G. Rosenfeld, E. Lægsgaard, F. Besenbacher, G. Comsa: Phys. Rev. Lett. **80**, 556 (1998).
240. L. Vitos, H.L. Skriver, J. Kollár: Surf. Sci. **425**, 212 (1999).
241. G. Schulze Icking-Konert, M. Giesen, H. Ibach: Surf. Sci. **398**, 37 (1998).
242. P.J. Feibelman: Phys. Rev. B **60**, 11118 (1999).
243. P.J. Feibelman: Phys. Rev. B **52**, 16845 (1995).
244. P.J. Feibelman: Surf. Sci **463**, L661 (2000).
245. M.J. Rost, T. Michely, G. Comsa: Phys. Rev. B **57**, 1992 (1998).
246. R.K.P. Zia: J. Stat. Phys. **45**, 801 (1986).
247. M. Holzer: Phys. Rev. B **42**, 10570 (1990).
248. A. Kara, S. Durukanoglu, T.S. Rahman: Phys. Rev. B **53**, 15489 (1996).
249. R.C. Nelson, T.L. Einstein, S.V. Khare: Surf. Sci. **295**, 462 (1993).
250. T.–Y. Fu, Y.–R. Tzeng, T.T. Tsong: Phys. Rev. B **54**, 5932 (1996).
251. T.–Y. Fu, Y.–R. Tzeng, T.T. Tsong: Surf. Sci. Lett. **366**, L691 (1996).
252. T.–Y. Fu, T.T. Tsong: Phys. Rev. B **61**, 4511 (2000).
253. N.C. Bartelt, R.M. Tromp, E.D. Williams: Phys. Rev. Lett. **73**, 1656 (1994).
254. H.P. Bonzel, A. Emundts: Phys. Rev. Lett. **84**, 5804 (2000).
255. H.-C. Jeong, E.D. Williams: Surf. Sci. Rep. **34**, 171 (1999).
256. H.P. Bonzel: Phys. Rep., in print.
257. M. Bott, T. Michely, G. Comsa: Surf. Sci. **272**, 161 (1992).
258. M. Hohage, M. Bott, M. Morgenstern, Z. Zhang, T. Michely, G. Comsa: Phys. Rev. Lett. **76**, 2366 (1996).
259. H. Brune, H. Röder, K. Bromann, K. Kern, J. Jacobsen, P. Stoltze, K. Jacobsen, J.K. Nørskov: Surf. Sci. Lett. **349**, 115 (1996).
260. M.J. Rost, T. Michely, G. Comsa: unpublished data.
261. H. Brune, C. Romainczyk, H. Röder, K. Kern: Nature **369**, 469 (1994).
262. R.Q. Hwang, J. Schröder, C. Günther, R.J. Behm: Phys. Rev. Lett. **67**, 3279 (1991).
263. P. Meakin: Phys. Rev. A **33**, 3371 (1986).
264. Z. Zhang, X. Chen, M.G. Lagally: Phys. Rev. Lett. **73**, 1829 (1994).
265. A. Bogicevic, J. Strömquist, B. I. Lundqvist: Phys. Rev. Lett. **81**, 637 (1998).
266. S. Ovesson, A. Bogicevic, B. I. Lundqvist: Phys. Rev. Lett. **83**, 2608 (1999).
267. G.S. Sun: Simulations of Platinum Growth on Pt(111) Using Density Functional Theory and Kinetic Monte Carlo Simulations. PhD thesis, University of Washington, Seattle (2000).

268. A.A. Chernov: *Modern Crystallography III: Crystal Growth* (Springer, Berlin Heidelberg 1984).
269. D.E. Wolf: J. Phys. A **20**, 1251 (1987).
270. J. Krug, H. Spohn: 'Kinetic Roughening of Growing Surfaces'. In: *Solids far from Equilibrium*, ed. by C. Godrèche (Cambridge University Press, Cambridge 1991) pp. 479–582.
271. H. Röder: Microscopic Processes in Heteroepitaxial Growth: Nucleation, Growth, and Alloying of Silver on the (111) Surface of Platinum. PhD Thesis 1288, EPFL, Lausanne (1994).
272. J. Jacobsen, K. W. Jacobsen, J.K. Nørskov: Surf. Sci. **359**, 37 (1996).
273. H. Jónsson: Annu. Rev. Phys. Chem. **51**, 623 (2000).
274. J. Wu, E.G. Wang, K. Varga, B.G. Liu, S.T. Pantelides, Z. Zhang: Phys. Rev. Lett. **89**, 146103 (2002).
275. P. Ruggerone, C. Ratsch, M. Scheffler: 'Density Functional Theory of Epitaxial Growth of Metals'. In: *Growth and Properties of Ultrathin Epitaxial Layers*, Vol. 8, ed. by D.A. King, D.P. Woodruff (Elsevier, Amsterdam 1997) pp. 490–544.
276. P. Ruggerone, A. Kley, M. Scheffler: Progr. Surf. Sci. **54**, 331 (1997).
277. C. Ratsch, P. Ruggerone, M. Scheffler: 'Study of Strain and Temperature Dependence of Metal Epitaxy'. In: *Morphological Organization in Epitaxial Growth and Removal*, ed. by Z. Zhang, M.G. Lagally (World Scientific, Singapore 1998) pp. 3–29.
278. J. Krug: unpublished.
279. E. Kopatzki, S. Günther, W. Nichtl-Pecher, R. J. Behm: Surf. Sci. **284**, 154 (1993).
280. S. Esch, M. Hohage, T. Michely, G. Comsa: Phys. Rev. Lett **72**, 518 (1994).
281. S. Esch: *Modifikation der Wachstumsmorphologie in der Homoepitaxie auf Pt(111)*, Jül-Bericht 3256 (Forschungszentrum Jülich, Jülich 1996).
282. J.S. Luo, R.G. Tobin, D.K. Lambert, G.B. Fisher, C.L. DiMaggio: Surf. Sci. **274**, 53 (1992).
283. M.A. Henderson, A. Szabó, J.T. Yates, Jr.: J. Chem. Phys. **91**, 7245 (1989).
284. H. Steininger, S. Lehwald, H. Ibach: Surf. Sci. **123**, 264 (1982).
285. G. Ertl, M. Neumann, K.M. Streit: Surf. Sci. **64**, 393 (1977).
286. D.M. Collins, W.E. Spicer: Surf. Sci. **69**, 85 (1977).
287. M.R. McClellan, J.L. Gland, F.R. McFeeley: Surf. Sci. **112**, 63 (1981).
288. J.E. Reutt-Robey, D.J. Doren, Y.J. Chabal, S.B. Christmann: Phys. Rev. Lett. **61**, 2778 (1988).
289. H.R. Siddiqui, X. Guo, I. Chorkendorff, J.T. Yates, Jr.: Surf. Sci. Lett. **191**, L813 (1987).
290. H. Hopster, H. Ibach: Surf. Sci. **77**, 109 (1978).
291. S. Liu, Z. Zhang, G. Comsa, H. Metiu: Phys. Rev. Lett. **71**, 2967 (1993).
292. S. Liu, H. Metiu: Surf. Sci. Lett. **405**, L497 (1998).
293. M. Villarba: Atomic scale processes relevant to metal crystal growth. PhD Thesis, University of Washington, Seattle (1995).
294. B. Poelsema, R. L. Palmer, G. Comsa: Surf. Sci. **136**, 1 (1984).
295. O. Lehmann: *Molekularphysik* (Leipzig 1888/1889).
296. W. W. Mullins, R. F. Sekerka, J. Appl. Phys. **34**, 323 (1963).
297. B. Caroli, C. Caroli, B. Roulet. 'Instabilities of Planar Solidification Fronts'. In: *Solids Far From Equilibrium*, ed. by C. Godrèche (Cambridge University Press, Cambridge 1991) pp. 155–296.
298. M. Avignon, B.K. Chakraverty: Proc. Roy. Soc. A **310**, 277 (1969).
299. P. Finnie, Y. Homma: Phys. Rev. Lett. **85**, 3237 (2000).
300. G.S. Bales, D.C. Chrzan: Phys. Rev. Lett. **74**, 4879 (1995).

301. A. Pimpinelli, J. Villain, D. E. Wolf: J. Phys. 1 (Paris) **3**, 447 (1993).
302. A. Pimpinelli, J. Villain, D. E. Wolf, J. J. Métois, J. C. Heyraud, I. Elkinani, G. Uimin: Surf. Sci. **295**, 143 (1993).
303. A. Pimpinelli, R. Ferrando: Phys. Rev. B **60**, 17016 (1999).
304. T. Michely, M. Hohage, G. Comsa: 'Instabilities and Kinetic Anisotropies as Determining Factors of Island Growth-Shapes'. In: *Surface Diffusion: Atomistic and Collective Processes* ed. by M. C. Tringides (Plenum Press, New York 1997) pp. 125–134.
305. E. Brener, H. Müller-Krumbhaar, D. Temkin: Phys. Rev. E **54**, 2714 (1996).
306. J. Tersoff, R. M. Tromp: Phys. Rev. Lett. **70**, 2782 (1993).
307. A. F. Becker, G. Rosenfeld, B. Poelsema, G. Comsa: Phys. Rev. Lett. **70**, 477 (1993).
308. J. Jacobsen: Modeling Metal-on-Metal Growth. PhD Thesis, Technical University of Denmark, Lyngby (1996).
309. T. Michely: *Rastertunnelmikroskopuntersuchung der Morphologie von Pt(111) nach Ionenbeschuss*, Jül-Bericht 2569, (Forschungszentrum Jülich, Jülich, 1991).
310. J. Krug, H. Dobbs: Phys. Rev. Lett. **76**, 4096 (1996).
311. J. Krug, H.T. Dobbs, S. Majaniemi: Z. Phys. B **97**, 281 (1995).
312. S.V. Khare, T.L. Einstein: Phys. Rev. B **57**, 4782 (1998).
313. T. Ihle, C. Misbah, O. Pierre-Louis: Phys. Rev. B **58**, 2289 (1998).
314. M. Giesen-Seibert, F. Schmitz, R. Jentjens, H. Ibach: Surface Science **329**, 47 (1995).
315. G. Kellogg: Surf. Sci. **359**, 237 (1996).
316. J. Kallunki, J. Krug: Surf. Sci. Lett. **523**, L53 (2003).
317. H. Kallabis: *Theoretical Aspects of Crystal Growth*, Jül-Bericht 3484 (Forschungszentrum Jülich, Jülich 1997).
318. R. Stumpf, M. Scheffler: Phys. Rev. B **53**, 4958 (1996).
319. M.C. Bartelt, J.W. Evans: Surf. Sci. **314**, L829 (1994).
320. H. Kallabis, L. Brendel, J. Krug, D.E. Wolf: Int. J. Mod. Phys. B **11**, 3621 (1997).
321. J. Zhong, T. Zhang, Z. Zhang, M.G. Lagally: Phys. Rev. B **63**, 113403 (2001).
322. C.R. Stoldt, K.J. Caspersen, M.C. Bartelt, C.J. Jenks, J.W. Evans, P.A. Thiel: Phys. Rev. Lett. **85**, 800 (2000).
323. B. Poelsema, A.F. Becker, R. Kunkel, G. Rosenfeld, L.K. Verheij, G. Comsa: 'The Role of Kinetic Effects in the Growth of Pt on Pt(111)'. In: *Surface Science: Principles and Applications*, ed. by R.F. Howe, R.N. Lamb, K. Wandelt (Springer, Berlin 1993) pp. 95–104.
324. R. Kunkel, B. Poelsema, L.K. Verheij, G. Comsa: Phys. Rev. Lett. **65**, 733 (1990).
325. J. E. Parmeter, R. Kunkel, B. Poelsema, L. K. Verheij, G. Comsa: Vacuum **41**, 467 (1990).
326. R. Kunkel: *Die Untersuchung der Morphologie einer wachsenden Pt(111) Fläche mittels der Streuung thermischer He Atome*, Jül-Bericht 2526 (Forschungszentrum Jülich, Jülich, 1991).
327. B. Poelsema, R. Kunkel, N. Nagel, A. F. Becker, G. Rosenfeld, L. K. Verheij, G. Comsa: Appl. Phys. A **53**, 369 (1991).
328. B. Poelsema, A. F. Becker, G. Rosenfeld, R. Kunkel, N. Nagel, L. K. Verheij, G. Comsa: Surf. Sci. **272**, 269 (1992).
329. G. Rosenfeld, A. F. Becker, B. Poelsema, L. K. Verheij, G. Comsa: Phys. Rev. Lett. **69**, 917 (1992).
330. M. Bott, M. Hohage, T. Michely, G. Comsa: Phys. Rev. Lett. **70**, 1489 (1993).
331. M. Hohage, T. Michely, G, Comsa: Surf. Sci. **337**, 249 (1995).

332. T. Michely, M. Hohage, S. Esch, G. Comsa: Surf. Sci. Lett. **349**, 89 (1996).
333. M. Kalff, P. Šmilauer, G. Comsa, T. Michely: Surf. Sci. **426**, L447 (1999).
334. J. Krug, P. Politi, T. Michely: Phys. Rev. B **61**, 14037 (2000).
335. P.I. Cohen, G.S. Petrich, P.R. Pukite, G.J. Whaley, A.S. Arrott: Surf. Sci. **216**, 222 (1989).
336. M.P. Seah: Surf. Sci. **32**, 703 (1972).
337. J.D. Weeks, G.H. Gilmer: 'Dynamics of Crystal Growth'. In: *Advances in Chemical Physics*, Vol. XL, ed. by I. Prigogine, S.A. Rice (John Wiley, New York 1979) pp. 157–228.
338. J. Krug: J. Stat. Phys. **87**, 505 (1997).
339. I. Elkinani, J. Villain: J. Phys. I **4**, 949 (1994).
340. K. Meinel, M. Klaua, H. Bethge: J. Cryst. Growth **89**, 447 (1988).
341. E.Z. Luo, J. Wollschläger, F. Wegner, M. Henzler: Appl. Phys. A **60**, 19 (1995).
342. W.C. Elliott, P.F. Miceli, T. Tse, P.W. Stephens: Phys. Rev. B **54**, 17938 (1996).
343. M. Albrecht, H. Fritzsche, U. Gradmann: Surf. Sci. **294**, 1 (1993).
344. J. Krug, P. Kuhn: 'Second Layer Nucleation and the Shape of Wedding Cakes'. In: *Atomistic Aspects of Epitaxial Growth*, ed. by M. Kotrla, N.I. Papanicolaou, D.D. Vvedensky, L.T. Wille (Kluwer, Dordrecht 2002) pp. 145–163.
345. D. Cherns: Phil. Mag. **36**, 1429 (1977).
346. S.C. Wang, T.T. Tsong: Surf. Sci. **121**, 85 (1982).
347. G. Leonardelli, E. Lundgren, M. Schmid: Surf. Sci. **490**, 29 (2001).
348. P. Politi, G. Grenet, A. Marty, A. Ponchet, J. Villain: Phys. Rep. **324**, 271 (2000).
349. R. Ghez, S.S. Iyer: IBM J. Res. Dev. **32**, 804 (1988).
350. R.L. Schwoebel: J. Appl. Phys. **40**, 614 (1969).
351. R.E. Caflisch, W. E, M.F. Gyure, B. Merriman, C. Ratsch: Phys. Rev. E **59**, 6879 (1999).
352. O. Pierre-Louis: Phys. Rev. Lett. **87**, 106104 (2001).
353. P. Politi, J. Villain: Phys. Rev. B **54**, 5114 (1996).
354. J. Tersoff, A.W. Denier van der Gon, R.M. Tromp: Phys. Rev. Lett. **72**, 266 (1994).
355. S. Harris: Phys. Rev. B **52**, 16793 (1995).
356. J. Rottler, P. Maass: Phys. Rev. Lett. **83**, 3490 (1999).
357. S. Heinrichs, J. Rottler, P. Maass: Phys. Rev. B **62**, 8338 (2000).
358. O. Biham, I. Furman, V. Pirronello, G. Vidali: Astrophys. J. **553**, 595 (2001).
359. J. Krug: Phys. Rev. E **67**, 065102(R) (2003).
360. C. Castellano, P. Politi: Phys. Rev. Lett. **87**, 056102 (2001).
361. P. Politi, C. Castellano: Phys. Rev. E **66**, 031605 (2002).
362. P. Politi, C. Castellano: Phys. Rev. E **66**, 031606 (2002).
363. P.Politi, C. Castellano: Phys. Rev. B **67**, 075408 (2003).
364. K. Kyuno, G. Ehrlich: Phys. Rev. Lett. **81**, 5592 (1998).
365. S. Liu, H. Metiu: Surf. Sci. **359**, 245 (1996).
366. J.A. Meyer, J. Vrijmoeth, H.A. van der Vegt, E. Vlieg, R.J. Behm: Phys. Rev. B **51**, 14790 (1995).
367. P. Šmilauer, S. Harris: Phys. Rev. B **51**, 14798 (1995).
368. I. Markov: Phys. Rev. B **54**, 17930 (1996).
369. K. Bromann, H. Brune, H. Röder, K. Kern: Phys. Rev. Lett. **75**, 677 (1995).
370. J. Krug: Phys. Rev. Lett. **87**, 149601 (2001).
371. R. Ferrando, G. Tréglia: Surf. Sci. **377-379**, 843 (1997).
372. P. Politi: J. Phys. I France **7**, 797 (1997).
373. L. Hansen, P. Stoltze, K.W. Jacobsen, J.K. Nørskov: Phys. Rev. B **44**, 6523 (1991).

374. B.D. Yu, M. Scheffler: Phys. Rev. Lett. **77**, 1095 (1996).
375. J. Merikoski, I. Vattulainen, J. Heinonen, T. Ala-Nissila: Surf. Sci. **387**, 167 (1997).
376. R. Ferrando, G. Tréglia: Phys. Rev. Lett. **76**, 2109 (1996).
377. J. Krug, M. Schimschak: J. Phys. I **5**, 1065 (1995).
378. J. Krug, M. Plischke, M. Siegert: Phys. Rev. Lett. **70**, 3271 (1993).
379. J. Villain: J. Phys. I France **1**, 19 (1991).
380. G.S. Bales: Surf. Sci. **356**, L439 (1996).
381. W.W. Mullins: J. Appl. Phys. **30**, 77 (1959).
382. J. A. Stroscio, D. T. Pierce, M.D. Stiles, A. Zangwill, L.M. Sander: Phys. Rev. Lett. **75**, 4246 (1995).
383. J. Krug: Physica A **263**, 170 (1999).
384. J.-K. Zuo, J.F. Wendelken: Phys. Rev. Lett. **78**, 2791 (1997).
385. J.E. Van Nostrand, S.J. Chey, D.G. Cahill: Phys. Rev. B **57**, 12536 (1998).
386. M.F. Gyure, J.J. Zinck, C. Ratsch, D.D. Vvedensky: Phys. Rev. Lett. **81**, 4931 (1998).
387. J. Krug, M. Rost: Phys. Rev. B **60**, R 16334 (1999).
388. C. Castellano, J. Krug: Phys. Rev. B **62**, 2879 (2000).
389. Z. Zhang, J. Detch, H. Metiu: Phys. Rev. B **48**, 4972 (1993).
390. J. Krug: 'Continuum Equations for Step Flow Growth'. In: *Dynamics of Fluctuating Interfaces and Related Phenomena*, ed. by D. Kim, H. Park, B. Kahng (World Scientific, Singapore 1997) pp. 95–113.
391. J. G. Amar, F. Family: Phys. Rev. B **54**, 14071 (1996).
392. M.C. Bartelt, J.W. Evans: Surf. Sci. **423**, 189 (1999).
393. K.J. Caspersen, A.R. Layson, C.R. Stoldt, V. Fournee, P.A. Thiel, J.W. Evans: Phys. Rev. B **65**, 193407 (2002).
394. A.J. Bray: Adv. Phys. **43**, 357 (1994).
395. H.-J. Ernst, F. Fabre, R. Folkerts, J. Lapujoulade: Phys. Rev. Lett. **72**, 112 (1994).
396. M.V. Ramana Murty, T. Curcic, A. Judy, B.H. Cooper, A.R. Woll, J.D. Brock, S. Kycia, R.L. Headrick: Phys. Rev. Lett. **80**, 4713 (1998).
397. K. Thürmer, R. Koch, M. Weber, K.H. Rieder: Phys. Rev. Lett. **75**, 1767 (1995).
398. T. Michely, M. Kalff, G. Comsa, M. Strobel, K.-H. Heinig: Phys. Rev. Lett. **86**, 2589 (2001).
399. T. Michely, M. Kalff, G. Comsa, M. Strobel, K.-H. Heinig: 'Coarsening Mechanisms in Surface Morphological Evolution'. In: *Atomistic Aspects of Epitaxial Growth*, ed. by M. Kotrla, N.I. Papanicolaou, D.D. Vvedensky, L.T. Wille (Kluwer, Dordrecht 2002) pp. 185–196.
400. M. Kalff, G. Comsa, T. Michely: Surf. Sci. **486**, 103 (2001).
401. L.-H. Tang, P. Šmilauer, D.D. Vvedensky: Eur. Phys. J. B **2**, 409 (1998).
402. C. Herring: J. Appl. Phys. **21**, 301 (1950).
403. J.G. Amar: Phys. Rev. B **60**, R11317 (1999).
404. S. Schinzer, M. Kinne, M. Biehl, W. Kinzel: Surf. Sci. **439**, 191 (1999).
405. M. Ahr, M. Biehl, M. Kinne, W. Kinzel: Surf. Sci. **465**, 339 (2000).
406. P. Šmilauer, D.D. Vvedensky: Phys. Rev. B **52**, 14263 (1995).
407. D. Moldovan, L. Golubovic: Phys. Rev. E **61**, 6190 (2000).
408. J. Krug: 'Nonequilibrium Surface Dynamics with Volume Conservation'. In: *Nonequilibrium Statistical Mechanics in One Dimension*, ed. by V. Privman (Cambridge University Press, Cambridge 1997) pp. 305–327.
409. G. S. Bales, A. Zangwill: Phys. Rev. B **41**, 5500 (1990).
410. M. Rost, P. Šmilauer, J. Krug: Surf. Sci. **369**, 393 (1996).
411. L. Schwenger, R.L. Folkerts, H.-J. Ernst: Phys. Rev. B **55**, R7406 (1997).

412. T. Maroutian, L. Douillard, H.-J. Ernst: Phys. Rev. Lett. **83**, 4353 (1999).
413. T. Maroutian, L. Douillard, H.-J. Ernst: Phys. Rev. B **64**, 165401 (2001).
414. C. Schelling, G. Springholz, F. Schäffler: Phys. Rev. Lett. **83**, 995 (1999).
415. P. Tejedor, P. Šmilauer, C. Roberts, B.A. Joyce: Phys. Rev. B **59**, 2341 (1999).
416. J. Mysliveček, C. Schelling, F. Schäffler, G. Springholz, P. Šmilauer, J. Krug, B. Voigtländer: Surf. Sci. **520**, 193 (2002).
417. A. Pimpinelli, A. Videcoq: Surf. Sci. **445**, L23 (2000).
418. N. Cabrera, D.A. Vermilyea: 'The Growth of Crystals from Solution'. In: *Growth and Perfection of Crystals*, ed. by R. Doremus, B. Roberts, D. Turnbull (Wiley, New York 1958) pp. 393–408.
419. J.P. v.d. Eerden, H. Müller-Krumbhaar: Phys. Rev. Lett. **57**, 2431 (1986).
420. D. Kandel, J. Weeks: Phys. Rev. B **49**, 5554 (1994).
421. J. Krug: Europhys. Lett. **60**, 788 (2002).
422. M. Vladimirova, A. De Vita, A. Pimpinelli: Phys. Rev. B **64**, 245420 (2001).
423. N. Néel, T. Maroutian, L. Douillard, H.-J. Ernst: submitted.
424. P. Politi, J. Krug: Surf. Sci. **446**, 89 (2000).
425. O. Pierre-Louis, C. Misbah: Phys. Rev. B **58**, 2259 (1998).
426. A. Pimpinelli, I. Elkinani, A. Karma, C. Misbah, J. Villain: J. Phys. Condens. Matter **6**, 2661 (1994).
427. F. Gillet, O. Pierre-Louis, C. Misbah: Eur. Phys. J. B **18**, 519 (2000).
428. M. Giesen-Seibert, H. Ibach: Surface Science **316**, 205 (1994).
429. J. Kallunki, J. Krug: Phys. Rev. E **62**, 6229 (2000).
430. O. Pierre-Louis, M. R. D'Orsogna, T. Einstein: Phys. Rev. Lett. **82**, 3661 (1999).
431. M.V. Ramana Murty, B.H. Cooper: Phys. Rev. Lett. **83**, 352 (1999).
432. J. Kallunki, J. Krug, M. Kotrla: Phys. Rev. B **65**, 205411 (2002).
433. M. Rusanen, I. T. Koponen, J. Heinonen, T. Ala-Nissila: Phys. Rev. Lett. **86**, 5317 (2001).
434. J. Kallunki: Growth instabilities of vicinal crystal surfaces during Molecular Beam Epitaxy. PhD Thesis, University of Duisburg-Essen (2003).
435. B.B. Mandelbrot: *The Fractal Geometry of Nature* (W.H. Freeman, San Francisco 1982).
436. O. Malcai, D.A. Lidar, O. Biham, D. Avnir: Phys. Rev. E **56**, 2817 (1997).
437. D.Avnir, O. Biham, D. Lidar, O. Malcai: Science **279**, 39 (1998).
438. H.-N. Yang, Y.-P. Zhao, G.-C. Wang, T.-M. Lu: Phys. Rev. Lett. **76**, 3774 (1996).
439. M. Lütt, J.P. Schlomka, M. Tolan, J. Stettner, O.H. Seeck, W. Press: Phys. Rev. B **56**, 4085 (1997).
440. M. Kardar, G. Parisi, Y.C. Zhang: Phys. Rev. Lett. **56**, 889 (1986).
441. S.F. Edwards, D.R. Wilkinson: Proc. R. Soc. London A **381**, 17 (1982).
442. D.E. Wolf, J. Villain: Europhys. Lett. **13**, 389 (1990).
443. S. Das Sarma, P. Tamborenea: Phys. Rev. Lett. **66**, 325 (1991).
444. L. Golubović, R. Bruinsma: Phys. Rev. Lett. **66**, 321 (1991).
445. H.K. Janssen: Phys. Rev. Lett. **78**, 1082 (1997).
446. C.-S. Chin, M. den Nijs: Phys. Rev. E **59**, 2633 (1999).
447. M. Lässig: Phys. Rev. Lett. **80**, 2366 (1998).
448. M. den Nijs: 'Is There a Z-Theorem for Dynamic Critical Exponents?'. In: *Dynamics of Fluctuating Interfaces and Related Phenomena*, ed. by D. Kim, H. Park, B. Kahng (World Scientific, Singapore 1997) pp. 272–292.
449. Z.-W. Lai, S. Das Sarma: Phys. Rev. Lett. **66**, 2348 (1991).
450. L.H. Tang, T. Nattermann: Phys. Rev. Lett. **66**, 2899 (1991).
451. M. Moseler, O. Rattunde, J. Nordiek, H. Haberland: Nucl. Instr. and Meth. B **164-165**, 522 (2000).

452. T. Halpin-Healy, Y.C. Zhang: Phys. Rep. **254**, 215 (1995).
453. J. Krug: Phys. Rev. A **36**, 5465 (1987).
454. A. Ballestad, B.J. Ruck, M. Adamcyk, T. Pinnington, T. Tiedje: Phys. Rev. Lett. **86**, 2377 (2001).
455. J. Krug: J. Phys. A **22**, L769 (1989).
456. N.G. Nakhodkin, A.I. Shaldervan: Thin Solid Films **10**, 109 (1972).
457. M. Schimschak, J. Krug: Phys. Rev. B **52**, 8550 (1995).
458. S. Das Sarma, C.J. Lanczycki, S.V. Ghaisas, J.M. Kim: Phys. Rev. B **49**, 10693 (1994).
459. L.H. Tang, T. Nattermann: Phys. Rev. A **45**, 7156 (1992).
460. C.E. Botez, K. Li, E.D. Lu, W.C. Elliott, P.F. Miceli, E.H. Conrad, P.W. Stephens: Appl. Phys. Lett. **81**, 4718 (2002).
461. C.E. Botez, P.F. Miceli, P.W. Stephens: Phys. Rev. B **64**, 125427 (2001).
462. C.E. Botez, W.C. Elliott, P.F. Miceli, P.W. Stephens: Phys. Rev. B **66**, 075418 (2002).
463. T. Salditt, T.H. Metzger, J. Peisl: Phys. Rev. Lett. **73**, 2228 (1994).
464. J. Krim, G. Palasantzas: Int. J. mod. Phys. B **9**, 599 (1995).
465. G. Palasantzas, J. Krim: Phys. Rev. Lett. **73**, 3564 (1994).
466. A.C. Dürr, F. Schreiber, K.A. Ritley, V. Kruppa, J. Krug, H. Dosch, B. Struth: Phys. Rev. Lett. **90**, 016104 (2003).
467. M. Siegert, M. Plischke: Phys. Rev. E **53**, 307 (1996).
468. G. Lengel, R.J. Phaneuf, E.D. Williams, S. Das Sarma, W. Beard, F.G. Johnson: Phys. Rev. B **60**, R8469 (1999).
469. S. Majaniemi, T. Ala-Nissila, J. Krug: Phys. Rev. B **53**, 8071 (1996).
470. J. G. Amar, F. Family: Surf. Sci. **365**, 177 (1996).
471. Y.-P. Zhao, H.-N. Yang, G.-C. Wang, T.-M. Lu: Phys. Rev. B **57**, 1922 (1998).
472. J. Wollschläger, E.Z. Luo, M. Henzler: Phys. Rev. B **57**, 15541 (1998).
473. Y.-P. Zhao, G.-C. Wang, T.-M. Lu: *Characterization of Amorphous and Crystalline Rough Surfaces – Principles and Applications* (Academic Press, New York 2001).
474. G. Rosenfeld, B. Poelsema, G. Comsa: 'Epitaxial growth modes far from equilibrium'. In: *Growth and Properties of Ultrathin Epitaxial Layers*, Vol. 8, ed. by D.A. King, D.P. Woodruff (Elsevier, Amsterdam 1997) pp. 490–544.
475. Y. Suzuki, H. Kikuchi, N. Koshizuka: Jap. J. Appl. Phys. **27**, L1175 (1988).
476. P. Bedrossian, B. Poelsema, G. Rosenfeld, L.C. Jorritsma, N.N. Lipkin, G. Comsa: Surf. Sci. **334**, 1 (1995).
477. L.J. Gómez, S. Bourgeal, J. Ibáñez, M. Salmerón: Phys. Rev. B **31**, 2551 (1985).
478. D.A. Steigerwald, W.F. Egelhoff: Surf. Sci. Lett. **192**, L887 (1987).
479. J.J. DeMiguel, A. Sánchez, A. Cebollada, J.M. Gellego, J. Ferròn, S. Ferrer: Surf. Sci. **189/190**, 1062 (1987).
480. H.-J. Ernst, F. Fabre, J. Lapujoulade: Surf. Sci. Lett. **275**, L682 (1992).
481. S.T. Purcell, B. Heinrich, A.S. Arrott: Phys. Rev. B **35**, 6458 (1987).
482. D.K. Flynn-Sanders, J.W. Evans, P.A. Thiel: Surf. Sci. **289**, 75 (1993).
483. J.A. Stroscio, D. T. Pierce, R. A. Dragoset: Phys. Rev. Lett. **70**, 3615 (1993).
484. H.A. van der Vegt, H.M. van Pinxteren, M. Lohmeier, E. Vlieg, J.M.C. Thornton: Phys. Rev. Lett. **68**, 3335 (1992).
485. M. Stanley, C. Papageorgopoulos, K.R. Roos, M.C. Tringides: Surf. Sci. **355**, L264 (1996).
486. M. Henzler: Surf. Sci. **298**, 369 (1993).
487. W. Wulfhekel, N.N. Lipkin, J. Kliewer, G. Rosenfeld, L.C. Jorritsma, B. Poelsema, G. Comsa: Surf. Sci. **348**, 227 (1996).

488. P.C. Dastoor, J. Ellis, A. Reichmuth, H. Bullman, B. Holst, W. Allison: Surf. Rev. Lett. **1**, 509 (1994).
489. H.A. van der Vegt, M. Breeman, S. Ferrer, V.H. Etgens, X. Torelles, P. Fajardo, E. Vlieg: Phys. Rev. B **51**, 14806 (1995).
490. A. Steltenpohl, N. Memmel: Surf. Sci. **402-404**, 277 (1998).
491. U. Köhler, C. Jensen, C. Wolf, A.C. Schindler, L. Brendel, D.E. Wolf: Surf. Sci. **454-456**, 676 (2000).
492. F.Buatier de Mongeot, G. Costantini, C. Boragno, U. Valbusa: Phys. Rev. Lett. **84**, 2445 (2000).
493. C. DeGiorgi, P. Aihemaiti, F.Buatier de Mongeot, C. Boragno, R. Ferrando, U. Valbusa: Surf. Sci. **487**, 49 (2001).
494. A. Videcoq, F. Hontinfinde, R. Ferrando: Surf. Sci. **515**, 575 (2002).
495. L. Golubović, A. Levandovsky, D. Moldovan: Phys. Rev. Lett. **89**, 266104 (2002).
496. E. Lundgren, B. Stanka, G. Leonardelli, M. Schmid, P. Varga: Phys. Rev. Lett. **82**, 5068 (1999).
497. M.C. Bartelt, J.W. Evans: Phys. Rev. Lett. **75**, 4250 (1995).
498. J. G. Amar, F. Family: Phys. Rev. B **52**, 13801 (1995).
499. C.-M. Zhang, M.C. Bartelt, J.-M. Wen, C.J. Jenks, J.W. Evans, P.A. Thiel: J. Cryst. Growth **174**, 851 (1997).
500. R. Gerlach, T. Maroutian, L. Douillard, D. Martinotti, H.-J. Ernst: Surf. Sci. **480**, 97 (2001).
501. W.F. Chung, M.S. Altman: Phys. Rev. B **66**, 075338 (2002).
502. H. Bethge: 'Molecular Kinetics on Steps'. In: *Kinetics of Ordering and Growth at Surfaces*, ed. by M.G. Lagally (Plenum Press, New York 1990) pp. 125–144.
503. K.R. Roos, M.C. Tringides: Phys. Rev. Lett. **85**, 1480 (2000).
504. K. Morgenstern, G. Rosenfeld, G. Comsa: Phys. Rev. Lett. **76**, 2113 (1996).
505. M. Giesen, G. Schulze Icking-Konert, H. Ibach: Phys. Rev. Lett. **80**, 552 (1998).
506. M. Giesen, H. Ibach: Surf. Sci. **464**, L697 (2000).
507. K. Morgenstern, G. Rosenfeld, G. Comsa, M.R. Sørensen, B. Hammer, E. Lægsgaard, F. Besenbacher: Phys. Rev. B **63**, 045412 (2001).
508. P.J. Feibelman: Surf. Sci. Lett. **478**, L349 (2001).
509. M. Giesen, H. Ibach: Surf. Sci. **431**, 109 (1999).
510. J. G. Amar, F. Family: Phys. Rev. Lett. **77**, 4584 (1996).
511. P. Punyindu Chatraphorn, Z. Toroczkai, S. Das Sarma: Phys. Rev. B **64**, 205407 (2001).
512. C. Ratsch, M.C. Wheeler, M.F. Gyure: Phys. Rev. B **62**, 12636 (2000).
513. A. Pimpinelli, V. Tonchev, A. Videcoq, M. Vladimirova: Phys. Rev. Lett. **88**, 206103 (2002).
514. M. Siegert, M. Plischke: Phys. Rev. Lett. **73**, 1517 (1994).
515. M. Siegert: Phys. Rev. Lett. **81**, 5481 (1998).
516. M.D. Johnson, C. Orme, A.W. Hunt, D. Graff, J. Sudijono, L.M. Sander, B.G. Orr: Phys. Rev. Lett. **72**, 116 (1994).
517. P. Manneville: *Dissipative Structures and Weak Turbulence* (Academic Press, San Diego 1990).
518. M. Siegert: Physica A **239**, 420 (1997).
519. M. Rost, J. Krug: Phys. Rev. E **55**, 3952 (1997).
520. P. Politi: Phys. Rev. E **58**, 281 (1998).
521. P. Politi, A. Torcini: J. Phys. A **33**, L77 (2000).
522. R.V. Kohn, X. Yan: Comm. Pure Appl. Math. (2003).
523. L. Golubović: Phys. Rev. Lett. **78**, 90 (1997).
524. P. Šmilauer, M. Rost, J. Krug: Phys. Rev. E **59**, R6263 (1999).

525. J.R. Arthur: Surf. Sci. **500**, 189 (2002).
526. B. Heinrich, J.A.C. Bland (Eds.): *Ultrathin Magnetic Structures, Vol. I and Vol. II* (Springer, Berlin 1994).
527. J.S. Moodera, L.R. Kinder, T.M. Wong, R. Meservey: Phys. Rev. Lett. **74**, 3273 (1995).
528. W.H. Butler, X.-G. Zhang, T.C. Schulthess, J.M. MacLaren: Phys. Rev. B **63**, 054416 (2001).
529. E.Y. Tsymbal, D.G. Pettifor: Phys. Rev. B **58**, 432 (1998).
530. W.J. Gallagher, S.S.P. Parkin, Y. Lu, X.P. Bian, A. Marley, K.P. Roche, R.A. Altman, S.A. Rishton, C. Jahnes, T.M. Shaw, G. Xiao: J. Appl. Phys, **81**, 3741 (1997).
531. C. Teichert, J. Barthel, H.P. Oepen, J. Kirschner: Appl. Phys. Lett. **74**, 588 (1999).
532. D.J. Eaglesham, F.C. Unterwald, D.C. Jacobsen: Phys. Rev. Lett. **70**, 966 (1993).
533. W.F. Egelhoff, Jr., D.A. Steigerwald: J. Vac. Sci. Technol. A **7**, 2167 (1989).
534. M. Copel, M.C. Reuter, E. Kaxiras, R.M. Tromp: Phys. Rev. Lett. **63**, 632 (1989).
535. G. Rosenfeld, R. Servaty, C. Teichert, B. Poelsema, G. Comsa: Phys. Rev. Lett. **71**, 895 (1993).
536. S. Esch, M. Breeman, M. Morgenstern, T. Michely, G. Comsa: Surf. Sci. **365**, 187 (1996).
537. G. Rosenfeld, N.N. Lipkin, W. Wulfhekel, J. Kliewer, K. Morgenstern, B. Poelsema, G. Comsa: Appl. Phys. A **61**, 455 (1995).
538. W. Wulfhekel, I. Beckmann, N.N. Lipkin, G. Rosenfeld, B. Poelsema, G. Comsa: Appl. Phys. Lett. **69**, 3492 (1996).
539. V.A. Markov, O.P. Pchelyakov, L.V. Sokolov, S.I. Stenin, S. Stoyanov: Surf. Sci. **250**, 229 (1991).
540. M. Horn-von Hoegen, H. Pietsch: Surf. Sci. Lett. **321**, L129 (1994).
541. B. Voigtländer, A. Zinner: Surf. Sci. Lett. **292**, 775 (1993).
542. M. Breeman: Aspects of Stimulated Layer-By-Layer Growth on Metals. PhD Thesis, Rijksuniversiteit Groningen (1993).
543. M. Breeman, G.T. Barkema, M.H. Langelaar, D.O. Boerma: Thin Solid Films **272**, 195 (1996).
544. H.A. van der Vegt, J. Alvarez, X. Torelles, S. Ferrer, E. Vlieg: Phys. Rev. B **52**, 17443 (1995).
545. J. Camarero, J. Ferrón, V. Cros, L. Gómez, A.L. Vázquez de Parga, J.M. Gallego, J.E. Prieto, J.J. de Miguel, R. Miranda: Phys. Rev. Lett. **81**, 850 (1998).
546. M. Horn-von Hoegen, J. Falta, M. Copel, R.M. Tromp: Appl. Phys. Lett. **66**, 487 (1995).
547. D. Kandel, E. Kaxiras: Solid State Physics **54**, 219 (2000).
548. J. Camarero, T. Graf, J.J. de Miguel, R. Miranda, W. Kuch, M. Zharnikov, A. Dittschar, C.M. Schneider, J. Kirschner: Phys. Rev. Lett. **76**, 4428 (1996).
549. V. Scheuch, K. Potthast, B. Voigtländer, H.P. Bonzel: Surf. Sci. **318**, 115 (1994).
550. H. Wolter, M. Schmidt, K. Wandelt: Surf. Sci. **298**, 173 (1993).
551. P.J. Feibelman, S. Esch, T. Michely: Phys. Rev. Lett. **77**, 2257 (1996).
552. T. Michely, G. Comsa: Phys. Rev. B **44**, 8411 (1991).
553. C. Teichert, M. Hohage, T. Michely, G. Comsa: Phys. Rev. Lett. **72**, 1682 (1994).
554. T. Michely, C. Teichert: Phys. Rev. B **50**, 11156 (1994).
555. M. Morgenstern, T. Michely, G. Comsa: Phil. Mag. A **79**, 775 (1999).
556. H. Gades, H.M. Urbassek: Phys. Rev. B **50**, 11167 (1994).

557. M. Ghaly, R.S. Averback: Phys. Rev. Lett. **72**, 364 (1994).
558. M. Ghaly, K. Nordlund, R.S. Averback: Phil. Mag. A **79**, 795 (1999).
559. S. Esch, M. Bott, T. Michely, G. Comsa: Appl. Phys. Lett. **67**, 3209 (1995).
560. J. Jacobsen, K.W. Jacobsen, P. Stoltze: Surf. Sci. **317**, 8 (1994).
561. M. Hohage, M. Morgenstern, G. Comsa, U. Linke, T. Michely, unpublished.
562. A.R. Sandy, S.G.J. Mochrie, D.M. Zehner, G. Grübel, K.G. Huang, D.Gibbs: Phys. Rev. Lett. **68**, 2192 (1992).
563. G. Grübel, K. Huang, D. Gibbs, D.M. Zehner, A.R. Sandy, S.J.G. Mochrie: Phys. Rev. B **48**, 18119 (1993).
564. R.J. Needs, M. Mansfield: J. Phys. Cond. Mat. **1**,7555 (1989).
565. P.J. Feibelman: Phys. Rev. B **51**, 17867 (1995).
566. U.K. Köhler, J.E. Demuth, R.J. Hamers: J. Vac. Sci. Technol. A **7**, 2860 (1989).
567. T. Sakamoto, N.J. Kawai, T. Nakagawa, K. Ohta, T. Kojima: Appl. Phys. Lett. **47**, 617 (1985).
568. T. Sakamoto, N.J. Kawai, T. Nakagawa, K. Ohta, T. Kojima, G. Hashiguchi: Surf. Sci. **174**, 651 (1986).
569. H. Jenniches, M. Klaua, H. Höche, J. Kirschner: Appl. Phys. Lett. **69**, 3339 (1996).
570. P. Jensen, B. Niemeyer: Surf. Sci. **384**, L823 (1997).
571. B. Hinnemann, H. Hinrichsen, D.E. Wolf: Phys. Rev. E **67**, 011602 (2003).
572. B. Hinnemann, H. Hinrichsen, D.E. Wolf: Phys. Rev. Lett. **87**, 135701 (2001).
573. N. Combe, P. Jensen: Phys. Rev. B **57**, 15553 (1998).
574. J. Jacobsen, J.P. Sethna: Surf. Sci. **411**, L858 (1998).
575. J.W. Evans, M.C. Bartelt: Langmuir **12**, 217 (1996).
576. H.C. Kang, J.W. Evans: Surf. Sci. **271**, 321 (1992).
577. D.E. Wolf: 'Computer Simulations of Molecular Beam Epitaxy'. In: *Scale Invariance, Interfaces and Nonequilibrium Dynamics*, ed. by M. Droz, A. McKane, J. Vannimenus, D.E. Wolf (Plenum Press, New York 1995) pp. 215–248.
578. E. Somfai, D.E. Wolf, J. Kertész: J. Phys. I France **6**, 393 (1996).
579. M. Rost, J. Krug: J. Phys. I France **7**, 1627 (1997).
580. L. Brendel: Lebensdauer lagenweisen Kristallwachstums. PhD Thesis, University of Duisburg (2001).
581. R.S. Ross, M.F. Gyure: Phys. Rev. B **61**, 8602 (2000).
582. E.H.C. Parker (Ed.): *The Technology and Physics of Molecular Beam Epitaxy* (Plenum Press, New York 1985).
583. R.F.C. Farrow (Ed.): *Molecular Beam Epitaxy* (Noyes Publications, Park Ridge 1995).
584. R. Glang: 'Vacuum Evaporation'. In: *Handbook of thin film technology*, ed. by L.I. Maissel and R. Glang (McGraw-Hill, New York 1970), pp. 1.4–1.130.
585. D.E. Wolf: 'Adatom Diffusion and Epitaxial Growth'. In: *Dynamics of Fluctuating Interfaces and Related Phenomena*, ed. by D. Kim, H. Park, B. Kahng (World Scientific, Singapore 1997) pp. 173–205.
586. M.P. Seah, W.A. Dench: Surf. Interface Anal. **1**, 2 (1979).
587. W. Telieps, E. Bauer: Ultramicros. **17**, 57 (1985).
588. R.M. Tromp, M.C. Reuter: Ultramicros. **50**, 171 (1993).
589. L.H Veneklasen: Rev. Sci. Instrumen. **63**, 5513 (1992).
590. R.M. Tromp, M. Mankos, M.C. Reuter, A.C. Ellis, M. Copel: Surf. Rev. Lett. **5**, 1189 (1998).
591. P.K. Larsen, P.J. Dobson (Eds.): *Reflection High-Energy Electron Diffraction and Reflection Electron Imaging of Surfaces* (Plenum Press, New York 1987).
592. T. Hsu, Ed.: L. Electron Microsc. Tech. **20**, part 4 (1992).

593. C.J. Chen: *Introduction to Scanning Tunneling Microscopy* (Oxford University Press, Oxford 1993).
594. J.A. Stroscio, W.J. Kaiser: *Scanning Tunneling Microscopy* (Academic Press, San Diego 1993).
595. H.-J. Güntherodt, R. Wiesendanger (Eds.): *Scanning Tunneling Microscopy I-III* (Springer, Berlin 1994, 1995, 1996).
596. J. Tersoff, D.R. Hamann: Phys. Rev. B **31**, 805 (1985).
597. C.J. Chen: 'Unified Perturbation Theory for STM and SFM'. In: *Scanning Tunneling Microscopy III, Springer Series in Surface Science Vol. 29*, ed. by R. Wiesendanger, H.–J. Güntherodt (Springer, Berlin 1993).
598. M. Bott, T. Michely, G. Comsa: Rev. Sci. Instrumen. **66**, 4135 (1995).
599. K. Besocke: Surf. Sci. **181**, 145 (1987).
600. T. Michely, K.-H. Besocke, M. Teske: J. Microsc. **152**, 77 (1988).
601. J. Frohn, J.F. Wolf, K. Besocke, M. Teske: Rev. Sci. Instrumen. **60**, 1200 (1989).
602. M. Morgenstern, T. Michely, G. Comsa: Phys. Rev. Lett. **79**, 1305 (1997).
603. G. Binnig, C.F. Quate, Ch. Gerber: Phys. Rev. Lett. **56**, 930 (1986).
604. G. Meyer, N.M. Amer: Appl. Phys. Lett. **53**, 1045 (1988).
605. B. Drake, C.B. Prater, A.L. Weisenhorn, S.A.C. Gould, T.R. Albrecht, C.F. Quate, D.S. Cannell, H.G. Hansma, P.K. Hansma: Science **243**, 1586 (1989).
606. T.R. Albrecht, P. Grütter, D. Horne, D. Rugar: J. Appl. Phys. **69**, 668 (1991).
607. F. Giessibl: Science **267**, 68 (1995).
608. S. Morita, R. Wiesendanger, E. Meyer (Eds.): *Noncontact Atomic Force Micrsocopy* (Springer, Berlin 2002).
609. E.W. Müller, T.T. Tsong: *Field Ion Microscopy: Principles and Applications* (Elsevier, New York 1969).
610. E.W. Müller, T.T. Tsong: Prog. Surf. Sci. **4**, 1 (1973).
611. R. Wagner: *Field Ion Microscopy* (Springer, Berlin 1982).
612. G. Ehrlich: Surf. Sci. **299/300**, 628 (1994).
613. G. Ehrlich: Surf. Sci. **333**, 865 (1995).
614. G. Comsa: Surf. Sci. **299/300**, 77 (1994).
615. B. Poelsema, G. Comsa: *Scattering of Thermal Energy Atoms* (Springer, Berlin 1989).
616. J.B. Pendry: Surf. Sci. **299/300**, 375 (1994).
617. K. Heinz: Surf. Sci. **299/300**, 433 (1994).
618. P.M. Marcus: Surf. Sci. **299/300**, 447 (1994).
619. M.A. Van Hove, G.A. Somorjai: Surf. Sci. **299/300**, 487 (1994).
620. M. Horn-von Hoegen: Z. Kristallogr. **214**, 591 (1999).
621. M. Henzler: Surf. Sci. **168**, 744 (1986).
622. S. Ferrer, Y. Petroff: Surf. Sci. **500**, 605 (2002).
623. A.K. Myers-Beaghton: Surf. Sci. **241**, 439 (1991).
624. S. Dietrich, A. Haase: Phys. Rep. **260**, 1 (1995).
625. L. Jorritsma: Growth Anisotropies in Cu(001) Homoepitaxy. PhD Thesis, University of Twente, Enschede (1997).
626. V. Holý, U. Pietsch, T. Baumbach: *High-Resolution X-Ray Scattering from Thin Films and Multilayers* (Springer, Berlin 1999).
627. S.K. Sinha: J. Phys. III France **4**, 1543 (1994).
628. G. Henkelman, H. Jónsson: J. Chem. Phys. **111**, 7010 (1999).
629. R.O. Jones, O. Gunnarsson: Rev. Mod. Phys. **61**, 689 (1989).
630. R.M. Dreizler, E.K.U. Gross: *Density Functional Theory* (Springer, Berlin 1990).
631. P. Hohenberg, W. Kohn: Phys. Rev. **136**, 864B (1964).

632. W. Kohn, L.J. Sham: Phys. Rev. **140**, 1133A (1965).
633. M. Bockstedte, A. Kley, J. Neugebauer, M. Scheffler: Comp. Phys. Comm. **107**, 187 (1997).
634. D.M. Ceperley, B.J. Alder: Phys. Rev. Lett. **45**, 566 (1980).
635. J.P. Perdew, J.A. Chevary, S.H. Vosko, K.A. Jackson, M.R. Pederson, D.J. Singh, C. Fiolhais: Phys. Rev. B **46**, 6671 (1992).
636. J.P. Perdew, K. Burke, Y. Wang: Phys. Rev. B **54**, 16533 (1996).
637. J.P. Perdew, K. Burke, M. Ernzerhof: Phys. Rev. Lett. **77**, 3865 (1996).
638. M. Fuchs, M. Bockstedte, E. Pehlke, M. Scheffler: Phys. Rev. B **57**, 2134 (1998).
639. P.H.T. Philipsen, E.J. Baerends: Phys. Rev. B **61**, 1773 (2000).
640. M.C. Payne, M.P. Teter, D.C. Allan, T.A. Arias, J.D. Joannopoulos: Rev. Mod. Phys. **64** 1045 (1992).
641. K.W. Jacobsen: Comments Cond. Mat. Phys. **14**, 129 (1988).
642. I.J. Robertson, V. Heine, M.C. Payne: Phys. Rev. Lett. **70**, 1944 (1993).
643. M.S. Daw, M.I. Baskes: Phys. Rev. B **29**, 6443 (1984).
644. S.M. Foiles, M.I. Baskes, M.S. Daw: Phys. Rev. B. **33**, 7983 (1986).
645. J.K. Nørskov, N.D. Lang: Phys. Rev. B **21**, 2131 (1980).
646. K.W. Jacobsen, J.K. Nørskov, M.J. Puska: Phys. Rev. B **35**, 7423 (1987).
647. M.J. Puska, R.M. Nieminen, M. Manninen: Phys. Rev. B **24**, 3037 (1981).
648. N. Chetty, K. Stokbro, K.W. Jacobsen, J.K. Nørskov: Phys. Rev. B **46**, 3798 (1992).
649. K.W. Jacobsen, P. Stoltze, J.K. Nørskov: Surf. Sci. **366**, 394 (1996).
650. M. Kotrla: Comp. Phys. Comm. **97**, 82 (1996).
651. J.M. Haile: *Molecular Dynamics Simulations. Elementary methods* (Wiley, New York 1992).
652. D.C. Rapaport: *The Art of Molecular Dynamics Simulation* (Cambridge University Press, 1995).
653. M. Parrinello: Solid State Comm. **102**, 107 (1997).
654. G.H. Gilmer, M.H. Grabow, A.F. Bakker: Mater. Sci. Eng. B **6**, 101 (1990).
655. C.M. Gilmore, J.A. Sprague: Phys. Rev. B **44**, 8950 (1991).
656. G.A. Evangelakis, N.I. Papanicolaou: Surf. Sci. **347**, 376 (1996).
657. N.G. van Kampen: *Stochastic Processes in Physics and Chemistry* (North-Holland, Amsterdam 1981).
658. D.P. Landau, K. Binder: *A Guide to Monte Carlo Simulations in Statistical Physics* (Cambridge University Press, Cambridge 2000).
659. N.C. Metropolis, A.W. Rosenbluth, M.N. Rosenbluth, A.M. Teller, E. Teller: J. Chem. Phys. **21**, 1087 (1953).
660. A.F. Voter: Phys. Rev. B **34**, 6819 (1986).
661. K.A. Fichthorn, W.H. Weinberg: J. Chem. Phys. **95**, 1090 (1991).
662. A.B. Bortz, M.H. Kalos, J.L. Lebowitz: J. Comp. Phys. **17**, 10 (1975).
663. P.A. Maksym: Semicond. Sci. Technol. **3**, 594 (1988).
664. J.L. Blue, I. Beichl, F. Sullivan: Phys. Rev. E **51**, R867 (1995).
665. H. Metiu, Y.-T. Lu, Z. Zhang: Science **255**, 1088 (1992).
666. P. Kratzer, M. Scheffler: Phys. Rev. Lett. **88**, 036102 (2002).
667. M. Breeman, G.T. Barkema, D.O. Boerma: Surf. Sci. **303**, 25 (1994).
668. H. Mehl, O. Biham, I. Furman, M. Karimi: Phys. Rev. B **60**, 2106 (1999).
669. H. Mehl, O. Biham: *Self-consistency of energy barriers in Monte Carlo simulations*, unpublished draft.
670. O. Biham, I. Furman, M. Karimi, G. Vidali, R. Kennett, H. Zeng: Surf. Sci. **400**, 29 (1998).
671. G.H. Gilmer, P. Bennema: J. Appl. Phys. **43**, 1347 (1972).
672. S. Clarke, M. Wilby, D.D. Vvedensky: Surf. Sci. **255**, 91 (1991).

673. P. Šmilauer, D.D. Vvedensky: Phys. Rev. B **48**, 17603 (1993).
674. P. Šmilauer, M.R. Wilby, D.D. Vvedensky: Phys. Rev. B **47**, 4119 (1993).
675. A.F. Voter, F. Montalenti, T.C. Germann: Annu. Rev. Mater. Res. **32**, 321 (2002).
676. A.F. Voter: Phys. Rev. Lett. **78**, 3908 (1997).
677. A.F. Voter: Phys. Rev. B **57**, R13985 (1998).
678. G. Henkelman, H. Jónsson: 'Long–Time–Scale Simulations of Al(100) Crystal Growth'. In: *Atomistic Aspects of Epitaxial Growth*, ed. by M. Kotrla, N.I. Papanicolaou, D.D. Vvedensky, L.T. Wille (Kluwer, Dordrecht 2002) pp. 63–74.

Index

SPRINGER SERIES IN SURFACE SCIENCES

Editors: G. Ertl, H. Lüth and D.L. Mills

Founding Editor: H.K.V. Lotsch

1 **Physisorption Kinetics**
By H. J. Kreuzer, Z. W. Gortel

2 **The Structure of Surfaces**
Editors: M. A. Van Hove, S. Y. Tong

3 **Dynamical Phenomena at Surfaces, Interfaces and Superlattices**
Editors: F. Nizzoli, K.-H. Rieder, R. F. Willis

4 **Desorption Induced by Electronic Transitions, DIET II**
Editors: W. Brenig, D. Menzel

5 **Chemistry and Physics of Solid Surfaces VI**
Editors: R. Vanselow, R. Howe

6 **Low-Energy Electron Diffraction**
Experiment, Theory
and Surface Structure Determination
By M. A. Van Hove, W. H. Weinberg, C.-M. Chan

7 **Electronic Phenomena in Adsorption and Catalysis**
By V. F. Kiselev, O. V. Krylov

8 **Kinetics of Interface Reactions**
Editors: M. Grunze, H. J. Kreuzer

9 **Adsorption and Catalysis on Transition Metals and Their Oxides**
By V. F. Kiselev, O. V. Krylov

10 **Chemistry and Physics of Solid Surfaces VII**
Editors: R. Vanselow, R. Howe

11 **The Structure of Surfaces II**
Editors: J. F. van der Veen, M. A. Van Hove

12 **Diffusion at Interfaces: Microscopic Concepts**
Editors: M. Grunze, H. J. Kreuzer, J. J. Weimer

13 **Desorption Induced by Electronic Transitions, DIET III**
Editors: R. H. Stulen, M. L. Knotek

14 **Solvay Conference on Surface Science**
Editor: F. W. de Wette

15 **Surfaces and Interfaces of Solids**
By H. Lüth*)

16 **Atomic and Electronic Structure of Surfaces**
Theoretical Foundations
By M. Lannoo, P. Friedel

17 **Adhesion and Friction**
Editors: M. Grunze, H. J. Kreuzer

18 **Auger Spectroscopy and Electronic Structure**
Editors: G. Cubiotti, G. Mondio, K. Wandelt

19 **Desorption Induced by Electronic Transitions, DIET IV**
Editors: G. Betz, P. Varga

20 **Scanning Tunneling Microscopy I**
General Principles and Applications to Clean and Adsorbate-Covered Surfaces
Editors: H.-J. Güntherodt, R. Wiesendanger
2nd Edition

21 **Surface Phonons**
Editors: W. Kress, F. W. de Wette

22 **Chemistry and Physics of Solid Surfaces VIII**
Editors: R. Vanselow, R. Howe

23 **Surface Analysis Methods in Materials Science**
Editors: D. J. O'Connor, B. A. Sexton, R. St. C. Smart
2nd Edition

24 **The Structure of Surfaces III**
Editors: S. Y. Tong, M. A. Van Hove,
K. Takayanagi, X. D. Xie

25 **NEXAFS Spectroscopy**
By J. Stöhr

26 **Semiconductor Surfaces and Interfaces**
By W. Mönch
3rd Edition

27 **Helium Atom Scattering from Surfaces**
Editor: E. Hulpke

28 **Scanning Tunneling Microscopy II**
Further Applications
and Related Scanning Techniques
Editors: R. Wiesendanger, H.-J. Güntherodt
2nd Edition

29 **Scanning Tunneling Microscopy III**
Theory of STM
and Related Scanning Probe Methods
Editors: R. Wiesendanger, H.-J. Güntherodt
2nd Edition

30 **Concepts in Surface Physics**
By M. C. Desjonquères, D. Spanjaard*)

31 **Desorption Induced by Electronic Transitions, DIET V**
Editors: A. R. Burns, E. B. Stechel, D. R. Jennison

32 **Scanning Tunneling Microscopy and Its Applications**
By C. Bai
2nd Edition

33 **Adsorption on Ordered Surfaces of Ionic Solids and Thin Films**
Editors: H.-J. Freund, E. Umbach

34 **Surface Reactions**
Editor: R. J. Madix

35 **Applications of Synchrotron Radiation**
High-Resolution Studies of Molecules
and Molecular Adsorbates on Surfaces
Editor: W. Eberhardt

36 **Kinetics of Metal-Gas Interactions at Low Temperatures: Hydriding, Oxidation, Poisoning**
By E. Fromm

37 **Magnetic Multilayers and Giant Magnetoresistance**
Fundamentals and Industrial Applications
Editor: U. Hartmann*)

*) Available as a textbook

Printing: Saladruck Berlin
Binding: Stürtz AG, Würzburg